Vectors, Matrices and Geometry

Vectors, Matrices and Geometry

K.T. Leung and S.N. Suen

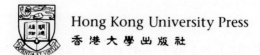

Hong Kong University Press

香港大學出版社

ISBN 962 209 360 4

Published by Hong Kong University Press,
University of Hong Kong,
139 Pokfulam Road,
Hong Kong

Printed in Hong Kong by ColorPrint Production Ltd.

CONTENT

Content

PREFACE

With the present volume the 3-book series on elementary mathematics is now complete. Like its two predecessors, *Fundamental Concepts of Mathematics* and *Polynomials and Equations*, the present book is addressed to a large readership comprising Sixth Form students, first-year undergraduates and students of Institutes of Education.

In Chapter One vectors in the plane are simply defined as ordered pairs of their components. Their addition and scalar multiplication are accordingly given in terms of their components. As soon as the fundamental properties of these operations are established our treatment of the linear algebra of R^2 becomes increasingly independent of the use of components, and the majority of the theorems are seen as logical consequences of the said fundamental properties. Important ideas of linear algbera are then introduced in the context of the familiar plane geometry. In turn the newly obtained methods of linear algebra are used to explore some less familiar areas of plane geometry.

The same approach is used in Chapter Two to study the linear algbera of R^3. In this way the mathematical ideas are reinforced while concepts particular to R^3 are introduced in the context of solid geometry. It has been our experience that the student will gain a deeper understanding and a firmer grasp by moving frequently to and fro between algebra and geometry and by proceeding from a lower dimension to a higher dimension.

Two chapters on analytic geometry follow. Chapter Three starts off with a definition of the conic as the locus of a point given in relation to a fixed line and a fixed point. The equations of conics in standard position are derived and discussed. In the second part of the chapter it is shown in two different ways, algebraic and geometric respectively, that the conics are precisely the curves cut out by a plane on a cone. The third part introduces translations and rotations of the plane. Their associated coordinate transformations are then used to show that all quadratic curves are conic sections. A similar study of the quadric surfaces, quadratic equations and coordinate transformations in space is carried out in Chapter Four. However the results of these two chapters will not be used again in the rest of the book. In fact after the preparation of Chapters One and Two, the student would have no difficulty in tackling higher dimensional linear spaces.

In Chapter Five the notions of linear combination and linear independence are, once again, carefully extended while new ideas such as subspace, base and dimension are introduced. Here the elementary row transformations on matrices provide an effective mean for the evaluation of dimension while a matrix is only used to write down the components of a finite number of vectors in an orderly manner. The algebra of matrices is treated separately in Chapter Six together with determinants of orders 2 and 3. Different ways of obtaining determinants of higher orders are suggested. The book concludes with a fairly complete treatment of linear equations in Chapter Seven.

I am indebted to Mr. S.N. Suen who has once again provided an excellent set of exercises. I also take this opportunity to offer my sincere appreciation and thanks to Ms. Mandy Shing for setting the whole text on AMS-Tex, Mr. Edward Lau for the line drawings and Dr. N.K. Tsing for the computer graphic.

K.T. Leung

University of Hong Kong
June 1994

CHAPTER ONE

VECTORS AND GEOMETRY IN THE PLANE

In school geometry, points on the plane are represented by pairs of real numbers which are called coordinates, and algebraic operations are carried out on the individual coordinates to discover properties of geometric configurations. For example, given two points P and Q represented by (a, b) and (c, d) respectively, the straight line passing through P and Q consists of points X whose coordinates (x, y) satisfy the polynomial equation $(y - b)(c - a) = (x - a)(d - b)$ and the slope of the line is given by the algebraic expression $(d - b)/(c - a)$ in the coordinates of P and Q. We note that here algebraic operations are not carried out on the points P, Q and X but on their coordinates a, b, c, d, x, y.

There are other ways in which algebra can be used in the study of plane geometry. For example, in an earlier book of the present series (see Section 7.14 of *Fundamental Concepts of Mathematics*) the correspondence between points of the plane and complex numbers is used. This correspondence enables us to use a single symbol for a complex number to refer a pair of coordinates. Moreover, it also allows us to operate directly on the complex numbers which now stand for points on the plane, so that the manifold properties of the complex number system are at our disposal in the study of plane geometry. Unfortunately this method cannot be extended to the study of geometry in space because we do not have a similar system of 3-dimensional numbers.

In this chapter we shall explore yet another approach to plane geometry by using vectors in the plane. Vectors in the plane are another kind of algebraic entities and, like complex numbers, they have many properties that are useful for the study of plane geometry. With vectors we also have the same advantage of using a single symbol to refer a pair of components. Furthermore the notion of

1

vectors in the plane lends itself for easy generalization to the notion of vectors in spaces of higher dimensions.

1.1 Vectors in the plane

A vector is usually described as a quantity that has a magnitude and a direction. For example in secondary school physics a displacement, a velocity and a force acting at a point are vectors while a mass and a temperature are not. Though it is possible to give an accurate description of magnitude and direction and use it to define vectors, it becomes a cumbersome task to generalize the notion of direction in higher dimensional spaces. Instead we shall lay down a simple definition of vector in terms of numbers and explain later the meaning of magnitude and direction of a vector.

1.1.1 DEFINITION *A vector in the plane is an ordered pair of real numbers. The numbers of the ordered pair are called the components of the vector. Vectors shall be denoted by bold-faced letters and square brackets shall be used to enclose the two components. Accordingly the real numbers a_1, a_2 are the components of the vector* $\mathbf{a} = [a_1, a_2]$.

1.1.2 REMARKS   While bold-faced types are used for vectors in this book, they will be difficult in handwritten work. Students may find it easier to indicate a vector by placing an arrow over or a bar below the letter. Any other consistent usage is just as satisfactory.

Being ordered pairs of real numbers, two vectors are equal if and only if they have identical components: *for vectors* $\mathbf{a} = [a_1, a_2]$ *and* $\mathbf{b} = [b_1, b_2]$, $\mathbf{a} = \mathbf{b}$ *if and only if* $a_1 = b_1$ *and* $a_2 = b_2$.

Vectors in the plane can be visualized as arrows in the cartesian plane. To present the vector $\mathbf{a} = [a_1, a_2]$ graphically, we draw a directed segment or an arrow from the origin O to the point A with coordinates (a_1, a_2). In this way every vector is represented by an arrow with initial point at the origin O. Conversely every arrow with initial point at O represents an vector in the plane. In Figure 1-1 the arrows representing the vectors $\mathbf{b} = [1, 3]$, $\mathbf{c} = [-2, 4]$, $\mathbf{d} = [4, 2]$ are drawn.

2

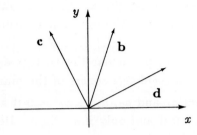

Fig 1-1

1.1.3 REMARKS   The reader will have no doubt noticed that an ordered pair of real numbers a_1 and a_2 can be a point A in the cartesian plane, in this case $A = (a_1, a_2)$, or a vector a in the plane, in this case $\mathbf{a} = [a_1, a_2]$. Seen from this point of view there is no essential difference between a point and a vector. The reason that they are still viewed as distinct entities lies in the different ways in which they shall be treated. The difference will become more apparent as the course progresses. For the time being it is sufficient to keep in mind that points are geometric entities while vectors are algebraic quantities. Consequently we shall adopt the following convention of notation to distinguish these two ways of using one and the same ordered pair.

1.1.4 CONVENTION   *An ordered pair of real numbers a_1 and a_2 may either represent a point or a vector. As a point it will be denoted by a captial letter and the coordinates are enclosed in parentheses:*

$$A = (a_1, a_2) \ .$$

As a vector it will be denoted by the lower case bold-faced type of the same letter and the components are enclosed in brackets:

$$\mathbf{a} = [a_1, a_2] \ .$$

Given a vector $\mathbf{a} = [a_1, a_2]$, the directed segment OA representing a has length $\sqrt{a_1{}^2 + a_2{}^2}$ by Pythagoras' theorem. This non-negative quantity is called the *magnitude* of the vector a and will be denoted by $|\mathbf{a}|$:

$$|\mathbf{a}| = \sqrt{a_1{}^2 + a_2{}^2} \ .$$

3

For the vectors in Figure 1-1, we get $|\mathbf{b}| = \sqrt{[1,3]} = \sqrt{10}$, $|\mathbf{c}| = \sqrt{[-2,4]} = \sqrt{20}$, $|\mathbf{d}| = \sqrt{[4,2]} = \sqrt{20}$. Vectors of magnitude 1 are called *unit vectors*. For example $[1/\sqrt{2}, -1\sqrt{2}]$, $[1/2, \sqrt{3}/2]$, $[0,-1]$ and $[\cos\theta, \sin\theta]$ are all unit vectors. In particular the vectors $\mathbf{e}_1 = [1,0]$ and $\mathbf{e}_2 = [0,1]$ are called the *unit coordinate vectors* of the plane.

Now for any real numbers a_1 and a_2, $a_1{}^2 + a_2{}^2 = 0$ if and only if $a_1 = a_2 = 0$. Therefore $|\mathbf{a}| = 0$ if and only if $\mathbf{a} = [0,0]$. Hence the only vector with zero magnitude is the vector with zero components. This vectors is called the *zero vector* and is denoted by $\mathbf{0}$:

$$\mathbf{0} = [0,0] .$$

1.1.5 THEOREM *Let* $\mathbf{a}$ *be vector. Then* $|\mathbf{a}| = 0$ *if and only if* $\mathbf{a} = \mathbf{0}$.

The zero vector $\mathbf{0}$ is represented by a degenerated segment with both initial and terminal points at the origin O. All other vectors in the plane are called *non-zero vectors*. Each non-zero vector $\mathbf{a}$ determines a non-degenerate directed segment OA. The length $|OA|$ of the directed segment is the magnitude of the vector $\mathbf{a}$ and the direction of the directed segment is the direction of the vector $\mathbf{a}$. Therefore a vector defined by 1.1.1 has a magnitude and a direction in accordance with the physical description of vectors in elementary physics.

1.2 The vector space $\mathbf{R}^2$

We mentioned earlier that the main advantage of vector geometry over coordinate geometry is that the vectors in the plane constitute an algebraic system similar to the system of complex numbers so that the rich algebraic properties of vectors would be at our disposal for the study of plane geometry. We shall now elaborate on the algebraic properties of vectors.

Let $\mathbf{a} = [a_1, a_2]$ and $\mathbf{b} = [b_1, b_2]$ be vectors in the plane. The *sum* of $\mathbf{a}$ and $\mathbf{b}$ is the vector

$$\mathbf{a} + \mathbf{b} = [a_1 + b_1, a_2 + b_2] .$$

The algebraic operation of forming the sum is called the *addition* of vectors. A second algebraic operation on vectors is the *scalar multiplication* by which a vector $\mathbf{a} = [a_1, a_2]$ is multiplied by a real number r to form a vector

$$r\mathbf{a} = [ra_1, ra_2]$$

called the *scalar multiple* of the vector $\mathbf{a}$ by the *scalar r*.

For example if $\mathbf{a} = [1, 3]$ and $\mathbf{b} = [-2, 4]$, then

$$\mathbf{a} + \mathbf{b} = [1, 3] + [-2, 4] = [1 - 2, 3 + 4] = [-1, 7]$$
$$3\mathbf{a} = 3[1, 3] = [3, 9]$$
$$(-1)(\mathbf{a} + \mathbf{b}) = [1, -7] \ .$$

In particular the vector $(-1)\mathbf{a} = [-a_1, -a_2]$ will be denoted by $-\mathbf{a}$:

$$-\mathbf{a} = (-1)\mathbf{a} \ .$$

Moreover subtraction of vectors is defined as

$$\mathbf{b} - \mathbf{a} = \mathbf{b} + (-\mathbf{a}) \ .$$

The addition of vectors can be illustrated by the so-called parallelogram law of mechanics. Let the vectors $\mathbf{a}$ and $\mathbf{b}$ be represented by the directed segments OA and OB. If the sum $\mathbf{a} + \mathbf{b} = \mathbf{c}$ is represented by the directed segment OC, then $OACB$ is a parallelogram in the plane. (Figure 1-2)

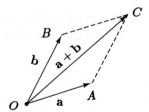

Fig 1-2

The scalar multiple $r\mathbf{a} = \mathbf{d}$ of a non-zero vector $\mathbf{a}$ is represented by a directed segment OD with D lying on the straight line that passes through O and A. The position of D on this straight line depends on the scalar r. Indeed if $1 < r$, then A lies between O and D;

if $1 = r$, then $D = A$;

if $0 < r < 1$, then D lies between O and A;

if $0 = r$, then $D = O$;

if $r < 0$, then O lies between D and A.

In all cases the length $|OD|$ is $|r|$ times the length $|OA|$. Thus the effect of the multiplication of a by r is a scaling of the vector a; hence the factor r is called the *scalar*. We note that ra and a have the same direction if $r > 0$ and they have opposite directions if $r < 0$. Figure 1-3 illustrates several scalar multiples.

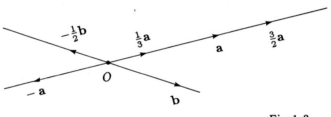

Fig 1-3

Let us now combine the three main components of vector algebra, namely the vectors, the addition and the scalar multiplication, in the following definition.

1.2.1 DEFINITION *The set* $\mathbf{R}^2$ *of all vectors in the plane together with the addition of vectors and the scalar multiplication will be called the vector space of the plane.*

According to the definition, elements of the vector space $\mathbf{R}^2$ are vectors $\mathbf{a} = [a_1, a_2]$. In relation to the vector space $\mathbf{R}^2$, all real numbers are called *scalars*. The scalars themselves are not elements of the vector space $\mathbf{R}^2$ but they appear as components of vectors and in the formation of scalar multiples.

1.2.2 REMARKS Readers may recall that a complex number $z = x + iy$ is also written as an ordered pair (x, y) of its real and imaginary parts. Hence there is a natural correspondence between complex numbers and vectors,

6

matching $a = a_1 + ia_2$ with $\mathbf{a} = [a_1, a_2]$. Under this correspondence the sum of two complex numbers is matched with the sum of the corresponding vectors, and the product of a complex number by a *real* number is matched with the corresponding scalar multiple. In this sense, when restricted to the addition and the multiplication by real numbers the system $\mathbf{C}$ of complex numbers is essentially the same as the vector space $\mathbf{R}^2$. Therefore certain properties of complex numbers can be translated into algebraic properties of vectors. Some of these are listed in following theorem.

1.2.3 THEOREM *Let* $\mathbf{a}$, $\mathbf{b}$ *and* $\mathbf{c}$ *be vectors of the vector space* $\mathbf{R}^2$, *and let* r *and* s *be scalars. Then the following statements hold.*

(1) $\mathbf{a} + \mathbf{b} = \mathbf{b} + \mathbf{a}$.

(2) $(\mathbf{a} + \mathbf{b}) + \mathbf{c} = \mathbf{a} + (\mathbf{b} + \mathbf{c})$.

(3) *There exists a unique vector* $\mathbf{0}$ *such that* $\mathbf{a} + \mathbf{0} = \mathbf{a}$.

(4) *For* $\mathbf{a}$ *there is a unique vector* $-\mathbf{a}$ *such that* $\mathbf{a} + (-\mathbf{a}) = \mathbf{0}$.

(5) $(rs)\mathbf{a} = r(s\mathbf{a})$.

(6) $(r + s)\mathbf{a} = r\mathbf{a} + s\mathbf{a}$.

(7) $r(\mathbf{a} + \mathbf{b}) = r\mathbf{a} + r\mathbf{b}$.

(8) $1\mathbf{a} = \mathbf{a}$.

The validity of these statements is self-evident and its proof is left to the readers as an exercise. Each statement is a translation of a corresponding property of the number system $\mathbf{C}$ and can be referred by the same name. For example (1) is the *commutative law* of addition of vectors and (8) is the *unitary property* of scalar multiplication. So far the restricted number system $\mathbf{C}$ and the vector space $\mathbf{R}^2$ are completely similar, but taken as a whole the number system $\mathbf{C}$ includes the multiplication of complex numbers of which the vector space $\mathbf{R}^2$ does not have an exact counterpart. Therefore further attempts at identifying the two systems should not be made.

To illustrate the usefulness of Theorem 1.2.3 let us give two proofs of a theorem, one referring to components and the other making full use of vector notations.

1.2.4 THEOREM *Let* $\mathbf{a}$ *be a vector of the vector space* $\mathbf{R}^2$ *and* r *be a scalar. Then* $r\mathbf{a} = \mathbf{0}$ *if and only if* $r = 0$ *or* $\mathbf{a} = \mathbf{0}$.

FIRST PROOF In this first proof we shall work with vectors in the plane explicitly as ordered pairs of real numbers. Let $\mathbf{a} = [a_1, a_2]$. Then

$$0\mathbf{a} = [0 \cdot a_1, 0 \cdot a_2] = [0, 0] = \mathbf{0} \ ;$$
$$r\mathbf{0} = [r \cdot 0, r \cdot 0] = [0, 0] = \mathbf{0} \ .$$

Therefore if $r = 0$ or $\mathbf{a} = \mathbf{0}$, then $r\mathbf{a} = \mathbf{0}$. Conversely let $r\mathbf{a} = \mathbf{0}$. The given scalar r is either zero or non-zero. In the former case we have nothing more to prove. Assume $r \neq 0$. Then it follows from $\mathbf{0} = r\mathbf{a} = [ra_1, ra_2]$ that $ra_1 = ra_2 = 0$. Hence $a_1 = a_2 = 0$, proving that $\mathbf{a} = \mathbf{0}$. The proof is complete.

SECOND PROOF In this alternate proof we shall only use some of the eight properties of the vector space $\mathbf{R}^2$ listed in Theorem 1.2.3 without referring to components of vectors.

If $r = 0$, then $r + r = r$. Therefore by (6) of 1.2.3

$$r\mathbf{a} = (r + r)\mathbf{a} = r\mathbf{a} + r\mathbf{a} \ .$$

If $\mathbf{a} = \mathbf{0}$, then $\mathbf{a} + \mathbf{a} = \mathbf{a}$ by (3) of 1.2.3. Furthermore by (7) of 1.2.3

$$r\mathbf{a} = r(\mathbf{a} + \mathbf{a}) = r\mathbf{a} + r\mathbf{a} \ .$$

Thus in either case $r\mathbf{a} = r\mathbf{a} + r\mathbf{a}$. Adding $-r\mathbf{a}$ to both side, using (2), (3) and (4), we obtain

$$\mathbf{0} = r\mathbf{a} + (-r\mathbf{a}) = (r\mathbf{a} + r\mathbf{a}) + (-r\mathbf{a}) = r\mathbf{a} + (r\mathbf{a} + (-r\mathbf{a}))$$
$$= r\mathbf{a} + \mathbf{0} = r\mathbf{a} \ .$$

Therefore if $r = 0$ or $\mathbf{a} = \mathbf{0}$, then $r\mathbf{a} = \mathbf{0}$.

Conversely let $r\mathbf{a} = \mathbf{0}$. It is sufficient to show if $r \neq 0$, then $\mathbf{a} = \mathbf{0}$. Indeed if $r \neq 0$, then $1/r$ is a scalar and by (5), (8) and the first part of the present theorem

$$\mathbf{a} = 1\mathbf{a} = (\frac{1}{r} \cdot r)\mathbf{a} = \frac{1}{r}(r\mathbf{a}) = \frac{1}{r}\mathbf{0} = \mathbf{0} \ .$$

The proof is complete.

Similarly we can give two proofs of the following theorem.

1.2.5 THEOREM *Let $\mathbf{a}$ be a vector and r a scalar. Then*

$$(-r)\mathbf{a} = r(-\mathbf{a}) = -(r\mathbf{a}) \ .$$

First Proof Let $\mathbf{a} = [a_1, a_2]$. Then

$$(-r)\mathbf{a} = (-r)[a_1, a_2] = [-ra_1, -ra_2]$$
$$r(-\mathbf{a}) = r[-a_1, -a_2] = [-ra_1, -ra_2]$$
$$-(r\mathbf{a}) = -[ra_1, ra_2] = [-ra_1, -ra_2] \ .$$

Therefore all three vectors are identical.

Second Proof We note that $-(r\mathbf{a})$ is the unique vector such that $(r\mathbf{a}) + (-(r\mathbf{a})) = \mathbf{0}$ as stipulated in property (4) of 1.2.3. Therefore it suffices to show that

$$r\mathbf{a} + ((-r)\mathbf{a}) = \mathbf{0} \qquad \text{and} \qquad r\mathbf{a} + (r(-\mathbf{a})) = \mathbf{0} \ .$$

For the former, by (6) of 1.2.3 and 1.2.4,

$$r\mathbf{a} + ((-r)\mathbf{a}) = (r - r)\mathbf{a} = 0\mathbf{a} = \mathbf{0} \ .$$

For the latter,
$$r\mathbf{a} + (r(-\mathbf{a})) = r(\mathbf{a} - \mathbf{a}) = r\mathbf{0} = \mathbf{0} \ .$$

Therefore $(-r)\mathbf{a} = r(-\mathbf{a}) = -(r\mathbf{a})$. The proof is complete.

We shall have more comments on these proofs later in 1.3.6. For the present, we just note that the properties of 1.2.4 and 1.2.5 are consequences of properties (1) – (8) of 1.2.3.

EXERCISES

1. Let $\mathbf{a} = [2, 3]$, $\mathbf{b} = [-1, 4]$, and $\mathbf{c} = [0, 6]$. Compute
   (a) $\mathbf{a} - 2\mathbf{b}$,
   (b) $3\mathbf{b} + 5\mathbf{c}$,
   (c) $-4\mathbf{c} - 3\mathbf{a}$,
   (d) $3(\mathbf{b} - 7\mathbf{c})$,
   (e) $2\mathbf{b} - (\mathbf{a} + \mathbf{c})$.
2. Find $\mathbf{a}$ if $\mathbf{a}$ satisfies

$$2[1, 2] - [2, -1] + \mathbf{a} = 5\mathbf{a} + [4, 1] \ .$$

3. Find real numbers c_1 and c_2 such that

$$c_1[1, -4] + [-3, -8] = c_2[-3, 2] \ .$$

4. Is it true that if **a** is a scalar multiple of **b**, then **b** is a scalar multiple of **a** for any vectors **a** and **b** in **R**? Prove or give a counter example.

5. For real numbers m, n and vectors **a**, **b**, prove that
   (a) if $m\mathbf{a} = n\mathbf{a}$ and $\mathbf{a} \neq \mathbf{0}$, then $m = n$;
   (b) if $m\mathbf{a} = m\mathbf{b}$ and $m \neq 0$, then $\mathbf{a} = \mathbf{b}$.

6. Express vectors **a** and **b** in terms of **u** and **v** where

$$3\mathbf{a} - 5\mathbf{b} = \mathbf{u}$$
$$-2\mathbf{a} + 4\mathbf{b} = \mathbf{v} \ .$$

7. By expressing **a**, **b**, and **c** in their components, prove Theorem 1.2.3.

1.3 Linear combinations

Let $\mathbf{a} = [a_1, a_2]$ and $\mathbf{b} = [b_1, b_2]$ be two vectors of the vector space **R**^{2}. We multiply the vectors **a** and **b** by the scalars r and s respectively to obtain scalar multiples $r\mathbf{a}$ and $s\mathbf{b}$; adding them, we get the vector

$$r\mathbf{a} + s\mathbf{b} = [ra_1, ra_2] + [sb_1, sb_2] = [ra_1 + sb_1, ra_2 + sb_2]$$

which is called a *linear combination* of the vectors **a** and **b** (see Figure 1-4).

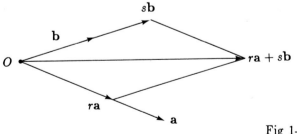

Fig 1-4

Clearly using other scalars r and s, we obtain other linear combinations of the same vectors **a** and **b**. For example, if $\mathbf{a} = [1, 3]$ and $\mathbf{b} = [-2, 4]$, then

10

$$2\mathbf{a} + 3\mathbf{b} = [2, 6] + [-6, 12] = [-4, 18]$$
$$-3\mathbf{a} + 2\mathbf{b} = [-3, -9] + [-4, 8] = [-7, -1]$$
$$1\mathbf{a} - 3\mathbf{b} = 1\mathbf{a} + (-3)\mathbf{b} = [1, 3] + [6, -12] = [7, -9]$$

are all linear combinations of the vectors $\mathbf{a} = [1, 3]$ and $\mathbf{b} = [-2, 4]$.

Observing that

$$\mathbf{a} + \mathbf{b} = 1\mathbf{a} + 1\mathbf{b}, \quad r\mathbf{a} = r\mathbf{a} + 0\mathbf{b} \quad \text{and} \quad s\mathbf{b} = 0\mathbf{a} + s\mathbf{b} ,$$

we conclude that the sum $\mathbf{a} + \mathbf{b}$ and the scalar multiples $r\mathbf{a}$ and $s\mathbf{b}$ are all linear combinations of the vectors $\mathbf{a}$ and $\mathbf{b}$. In other words the formation of linear combinations includes the addition and the scalar multiplication as special cases. Consequently many fundamental properties of vector algebra can be formulated in term of linear combinations. Linear combinations of three or more vectors $\mathbf{a}$, $\mathbf{b}$, $\mathbf{c}$, ... can be defined similarly, for example if r, s and t are scalars then the vector $r\mathbf{a} + s\mathbf{b} + t\mathbf{c}$ is a linear combination of the vectors $\mathbf{a}$, $\mathbf{b}$ and $\mathbf{c}$.

The unit coordinate vectors $\mathbf{e}_1 = [1, 0]$ and $\mathbf{e}_2 = [0, 1]$ play a rather distinguished role with respect to the formation of linear combinations. Firstly it follows from

$$\mathbf{a} = [a_1, a_2] = a_1[1, 0] + a_2[0, 1] = a_1\mathbf{e}_1 + a_2\mathbf{e}_2$$

that every vector $\mathbf{a}$ of the vector space $\mathbf{R}^2$ is a linear combination of the vectors $\mathbf{e}_1$ and $\mathbf{e}_2$ (see Figure 1-5). For this property that every vector of $\mathbf{R}^2$ is a linear combination of $\mathbf{e}_1$ and $\mathbf{e}_2$, we say that the vectors $\mathbf{e}_1$ and $\mathbf{e}_2$ *generate* the vector space $\mathbf{R}^2$.

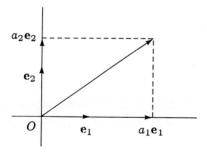

Fig 1-5

Another important property of the vectors e_1 and e_2 is that none of them is a scalar multiple of the other. Indeed $e_1 = re_2$ would mean that $1 = r0$ and $0 = r1$ for some real number r which is impossible; therefore e_1 is not a scalar multiple of e_2. Similarly e_2 is not a scalar multiple of e_1. This property of e_1 and e_2 is referred to as the *linear independence* of the vectors e_1 and e_2. In general, we have the following definition.

1.3.1 DEFINITION *Two vectors* **a** *and* **b** *of the vector space* $\mathbf{R}^2$ *are said to be linearly independent if* **a** *is not a scalar multiple of* **b** *and* **b** *is not a scalar multiple of* **a**. *The vectors* **a** *and* **b** *are said to be linearly dependent if they are not linearly independent. In other words,* **a** *and* **b** *are linearly dependent if and only if* **a** *is a scalar multiple of* **b** *or* **b** *is a scalar multiple of* **a**.

If one of the vectors is the zero vector then **a** and **b** are linearly dependent because in this case $\mathbf{a} = 0\mathbf{b}$ or $\mathbf{b} = 0\mathbf{a}$. Also if $\mathbf{a} = \mathbf{b}$, then **a** and **b** are linearly dependent. Therefore linear independence is a property of certain pairs of distinct non-zero vectors. Geometrically, all scalar multiples of a fixed non-zero vector are represented by directed segments lying on one straight line. Therefore two non-zero vectors **a** and **b** are linearly dependent if and only if their corresponding segments OA and OB lie on one straight line, or equivalently the point O, A and B are collinear. In terms of components the condition for linear dependence is given in the following theorem.

1.3.2 THEOREM *For two vectors* $\mathbf{a} = [a_1, a_2]$ *and* $\mathbf{b} = [b_1, b_2]$ *to be linearly dependent it is necessary and sufficient that* $a_1b_2 - a_2b_1 = 0$.

PROOF Suppose that $a_1b_2 - a_2b_1 = 0$. If $a_1 = a_2 = b_1 = b_2 = 0$, then $\mathbf{a} = \mathbf{0}$ and $\mathbf{b} = \mathbf{0}$. Therefore **a** and **b** are linearly dependent. If it is not true that $a_1 = a_2 = b_1 = b_2 = 0$, then at least one among the four real numbers a_1, a_2, b_1, b_2 must be non-zero. In other words, a_1, a_2, b_1 or b_2 is non-zero. If $a_1 \neq 0$, then it follows from $a_1b_2 - a_2b_1 = 0$ that $b_2 = a_2b_1/a_1$ and

$$\mathbf{b} = [b_1, b_2] = [a_1b_1/a_1, a_2b_1/a_1] = (b_1/a_1)\mathbf{a} \ .$$

Similarly $\mathbf{b} = (b_2/a_2)\mathbf{a}$ if $a_2 \neq 0$. In exactly the same way we can show that **a** is a scalar multiple of **b** if b_1 or b_2 is non-zero. Therefore in all cases **a** and **b** are linearly dependent.

Conversely suppose that **a** and **b** are linearly dependent. If $\mathbf{a} = r\mathbf{b}$, then $a_1 = rb_1$ and $a_2 = rb_2$. Hence $a_1b_2 - a_2b_1 = rb_1b_2 - rb_2b_1 = 0$. Similarly if $\mathbf{b} = s\mathbf{a}$ then $a_1b_2 - a_2b_1 = 0$. The proof of the theorem is now complete.

The theorem picks out the expression $a_1b_2 - a_2b_1$ for the determination of linear dependence of **a** and **b**. Here this expression plays a role similiar to that of the discriminant $b^2 - 4ac$ by which we know if the quadratic equation $ax^2 + bx + c = 0$ has real roots. Indeed $a_1b_2 - a_2b_1$ is called a *determinant* and is denoted by $\begin{vmatrix} a_1 & b_1 \\ a_2 & b_2 \end{vmatrix}$. A more detailed study of determinants is given in Chapter Six. Meanwhile we look at another condition for linear dependence which is formulated in terms of linear combinations.

1.3.3 THEOREM *For two vectors* **a** *and* **b** *of the linear space* $\mathbf{R}^2$ *to be linearly dependent it is necessary and sufficient that* $r\mathbf{a} + s\mathbf{b} = \mathbf{0}$ *for some scalars r and s which are not both zero.*

PROOF Suppose that $r\mathbf{a} + s\mathbf{b} = \mathbf{0}$ where r and s are not both zero. Then $\mathbf{a} = (-s/r)\mathbf{b}$ if $r \neq 0$ or $\mathbf{b} = (-r/s)\mathbf{a}$ if $s \neq 0$. Therefore **a** and **b** are linearly dependent.

Conversely suppose that **a** and **b** are linearly dependent. Then either $\mathbf{b} = r\mathbf{a}$ or $\mathbf{a} = s\mathbf{b}$. In the former case $r\mathbf{a} + (-1)\mathbf{b} = \mathbf{0}$. In the latter case $(-1)\mathbf{a} + s\mathbf{b} = \mathbf{0}$. In either case the condition of the theorem is satisfied because -1 is a non-zero scalar.

Let us summarize the discussion into the following theorems.

1.3.4 THEOREM *Let* $\mathbf{a} = [a_1, a_2]$ *and* $\mathbf{b} = [b_1, b_2]$ *be vectors of the vector space* $\mathbf{R}^2$. *Then the following statements are equivalent.*

(1) **a** *and* **b** *are linearly dependent.*
(2) *The points* $O = (0,0)$, $A = (a_1, a_2)$ *and* $B = (b_1, b_2)$ *are collinear.*
(3) $a_1b_2 - a_2b_1 = 0$.
(4) *There exist scalars r and s, not both being zero, such that* $r\mathbf{a} + s\mathbf{b} = \mathbf{0}$.

1.3.5 THEOREM *Let* $\mathbf{a} = [a_1, a_2]$ *and* $\mathbf{b} = [b_1, b_2]$ *be vectors of the vector space* $\mathbf{R}^2$. *Then the following statements are equivalent.*

(1) **a** *and* **b** *are linearly independent.*

13

(2) Both **a** and **b** are non-zero, and the points $O = (0,0)$, $A = (a_1, a_2)$ and $B = (b_1, b_2)$ are distinct and not collinear.

(3) $a_1 b_2 - a_2 b_1 \neq 0$.

(4) If $r\mathbf{a} + s\mathbf{b} = 0$ for scalars r and s, then $r = s = 0$.

1.3.6 REMARK   The observant reader will have noticed that there are two kinds of proofs. In a proof of the first kind, we rely on Definition 1.1.1 and make heavy use of components of vectors. For example the proof of Theorem 1.3.2 is of the first kind. In a proof of the second kind we do not explicitly rely on Definition 1.1.1, but only use the fundamental properties (1) - (8) of the vector space $\mathbf{R}^2$ listed in Theorem 1.2.3 without referring to components. Therefore a theorem, for example Theorem 1.3.3, which is substantiated by a proof of the second kind is in fact a logical consequence of the said properties (1) - (8) alone, and lends itself to convenient generalization for higher dimensional vector spaces.

EXERCISES

1. Express the following vectors as linear combination of $[1, 0]$ and $[1, 1]$.
   (a) $[3, 2]$. (b) $[0, 0]$. (c) $[-1, 1]$. (d) $[0, 1]$.

2. Show that $[3, 7]$ can be expressed as a linear combination of $[1, 0]$ and $[2, 1]$, but cannot be so for $[1, 3]$ and $[-2, -6]$. Interpret the result geometrically.

3. Show that each of the following pairs of vectors are linearly independent.
   (a) $[1, 2]$, $[-1, 1]$. (b) $[2, 0]$, $[3, 5]$.

4. Find the values of real number x so that the following pairs of vectors are linearly dependent.
   (a) $[1, x]$, $[2, 4]$. (b) $[x, x^2]$, $[-2, 4]$.
   (c) $[1, x]$, $[x, 8]$. (d) $[1, x]$, $[x, x^2]$.

5. Express $[x_1, x_2]$ as a linear combination of $[3, 2]$ and $[-1, 5]$.

6. Let $\mathbf{a} = [a_1, a_2]$, $\mathbf{b} = [b_1, b_2]$ be two linearly independent vectors in $\mathbf{R}^2$. Show that for any $\mathbf{x}$ in $\mathbf{R}^2$, $\mathbf{x}$ can be expressed as a unique linear combination of **a** and **b**; in other words there are unique scalars r and s such that $\mathbf{x} = r\mathbf{a} + s\mathbf{b}$.

7. Given that vectors **a** and **b** are linearly independent, and that $\mathbf{c} = \mathbf{a} + 2\mathbf{b}$, $\mathbf{d} = 2\mathbf{a} - 3\mathbf{b}$. Are **c** and **d** linearly independent?

8. Let **a** and **b** be two linearly independent vectors. Prove the following statements.

   (a) Both **a** and **b** are non-zero vectors.

   (b) If $r\mathbf{a} + s\mathbf{b} = t\mathbf{a} + u\mathbf{b}$, then $r = t$ and $s = u$.

   (c) The vectors **a** and $r\mathbf{a} + \mathbf{b}$ are linearly independent for all scalars r.

   (d) The vectors **a** and $r\mathbf{a} + s\mathbf{b}$ are linearly dependent if and only if $s = 0$.

9. Is it true that **a** and **b** are linearly dependent if for non-zero scalars r and s, $r\mathbf{a} + s\mathbf{b} = \mathbf{0}$? Prove or give a counter example.

10. Is it true that **a** and **b** are linearly dependent if and only if for non-zero scalars r and s, $r\mathbf{a} + s\mathbf{b} = \mathbf{0}$? Prove or give a counter example.

1.4 Plane geometry

By now we have developed just about enough material to carry out a study of plane geometry through the language of vectors. It is not our intention to reformulate a large part of secondary school plane geometry because this would be time-consuming and unproductive. Here we shall establish a link between the geometry of the plane and the algebra of the vector space $\mathbf{R}^2$. By this link, we hope to demonstrate, on the one hand, the usefulness of vector algebra and to provide, on the other hand, algebraic notions with geometric meaning.

Let $A = (a_1, a_2)$ be a point in the cartesian plane. We follow Convention 1.1.4 to denote by $\mathbf{a} = [a_1, a_2]$ the vector represented by the directed segment OA where $O = (0,0)$ and $A = (a_1, a_2)$. The vector **a** shall be referred to as the *position vector* of the point A. Thus points in the plane are associated with their position vectors, which are in turn represented by directed segments with a common initial point at the origin O. We shall find it useful to consider also directed segments in arbitrary position.

1.4.1 CONVENTION *Let $A = (a_1, a_2)$ and $B = (b_1, b_2)$ be two given points in the plane. The directed segment AB with initial point at A and terminal point at B shall be represented by the vector $\mathbf{b} - \mathbf{a} = [b_1 - a_1, b_2 - a_2]$. This vector is called the displacement vector from A to B and is denoted by $\overrightarrow{AB}$.*

In this way every directed segment in the plane is associated with a unique vector (i.e. its displacement vector), although many different directed segments may be associated with a single vector. For example $\overrightarrow{AB} = \overrightarrow{OC}$ where $C = (b_1 - a_1, b_2 - a_2)$. Readers may wish to note that the notions of position vectors and displacement vectors correspond roughly to what physicists call bound vectors and free vectors respectively.

In the following examples some well-known geometric properties of points and lines are expressed in terms of vectors to illustrate the usefulness of the conventions 1.1.4 and 1.4.1.

1.4.2 EXAMPLE   Two distinct points A and B and the origin O are collinear if and only if the position vectors **a** and **b** are linearly dependent.

1.4.3 EXAMPLE   Three distinct points A, B and C are collinear if and only if the displacement vectors $\overrightarrow{AB}$ and $\overrightarrow{AC}$ are linearly dependent.

1.4.4 EXAMPLE $ABCD$ is a parallelogram if and only if $\overrightarrow{AB} = \overrightarrow{DC}$. In this case $\overrightarrow{AD} = \overrightarrow{BC}$.

1.4.5 EXAMPLE   The length of the line segment AB is equal to the magnitude $|\overrightarrow{AB}|$ of the displacement vector $\overrightarrow{AB}$: $|AB| = |\overrightarrow{AB}|$.

1.4.6 EXAMPLE   The midpoint M of the segment AB has the position vector $\frac{1}{2}(\mathbf{a} + \mathbf{b})$ (see Figure 1-6).

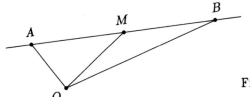

Fig 1-6

1.4.7 EXAMPLE X is a point of the segment AB if and only if its position vector has the form $(1-t)\mathbf{a} + t\mathbf{b}$ where $0 \le t \le 1$. In this case $|AX| = t|AB|$ (see Figure 1-7).

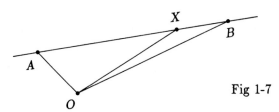

Fig 1-7

16

We shall leave the proofs of the above examples to the readers as exercises and proceed to make use of them to prove a few well-known theorems of plane geometry.

1.4.8 THEOREM *The diagonals of a parallelogram bisect each other.*

PROOF Let $ABCD$ be a parallelogram. If M_1 and M_2 are the mid-points of the diagonals AC and BD respectively, then their position vectors are $\mathbf{m}_1 = \frac{1}{2}(\mathbf{a}+\mathbf{c})$ and $\mathbf{m}_2 = \frac{1}{2}(\mathbf{b}+\mathbf{d})$. Now by 1.4.4. $\mathbf{b}-\mathbf{a} = \mathbf{c}-\mathbf{d}$. Therefore $\mathbf{a}+\mathbf{c} = \mathbf{b}+\mathbf{d}$ and hence $\mathbf{m}_1 = \mathbf{m}_2$. This means that the midpoints M_1 and M_2 of the diagonals are identical because they have identical position vector. Therefore the diagonals bisect each other.

We have therefore proved that if a quadrilateral is a parallelogram, then its diagonals bisect each other. Readers will have no difficulty in proving the converse using a similar argument.

1.4.9 THEOREM *If the diagonals of a quadrilateral bisect each other, then it is a parallelogram.*

Let us consider the straight line L that passes through two given distinct points A and B. In Example 1.4.7 we find that points X of L between A and B are given by

$$\mathbf{x} = (1-t)\mathbf{a} + t\mathbf{b}, \quad 0 \leq t \leq 1$$

in terms of their position vectors. Here the real number t serves as a parameter which determines the position of the point X. For example $X = A$ when $t = 0$; $X = B$ when $t = 1$; X is the midpoint between A and B when $t = 1/2$; X is two-thirds from A to B when $t = 2/3$.

If the parameter t is allowed to assume arbitrary value where is the point X with position vector $\mathbf{x} = (1-t)\mathbf{a} + t\mathbf{b}$? We may use Example 1.4.3 to find an answer. Indeed $\overrightarrow{AX} = \mathbf{x} - \mathbf{a} = t(\mathbf{b}-\mathbf{a})$ and $\overrightarrow{AB} = \mathbf{b}-\mathbf{a}$ are linearly dependent; therefore X lies on L. Conversely if X is any point on L then $\overrightarrow{AX}$ and $\overrightarrow{AB}$ are linearly dependent. Since $\overrightarrow{AB} \neq \mathbf{0}$, we have $\overrightarrow{AX} = t\overrightarrow{AB}$ for some scalar t. It now follows from $\mathbf{x} - \mathbf{a} = t(\mathbf{b}-\mathbf{a})$ that $\mathbf{x} = (1-t)\mathbf{a} + t\mathbf{b}$. We have proved the following theorem.

1.4.10 THEOREM *Let A and B be two distinct points. The straight line passing through A and B consists of all points X whose position vectors have the form*

$$\mathbf{x} = (1 - t)\mathbf{a} + t\mathbf{b}$$

We shall call the above expression of $\mathbf{x}$ a *parametric representation* of (the point X on) the line L; here the scalar t is called the *parameter* of the representation. The position of the point X on L will depend on the value of the parameter t. For example

> if $t < 0$, then A lies between X and B;
>
> if $0 = t$, then $A = X$;
>
> if $0 < t < 1$, then X lies between A and B;
>
> if $t = 1$, then $X = B$;
>
> if $1 < t$, then B lies between A and X.

Clearly Theorem 1.4.10 may be used to some advantage in proving theorems on the concurrence of three or more straight lines. Take for example the following theorem in Euclid's *Elements*.

1.4.11 THEOREM *The medians of a triangle are concurrent.*

PROOF Let ABC be a triangle. Denote the midpoints of the sides BC, CA and AB by D, E, and F respectively. Then

$$\mathbf{d} = \frac{1}{2}(\mathbf{b} + \mathbf{c}), \quad \mathbf{e} = \frac{1}{2}(\mathbf{c} + \mathbf{a}) \quad \text{and} \quad \mathbf{f} = \frac{1}{2}(\mathbf{a} + \mathbf{b}).$$

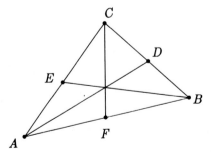

Fig 1-8

By 1.4.10 points of the medians AD, BE and CF are respectively represented by position vectors

$$\mathbf{x} = (1-t)\mathbf{a} + \frac{t}{2}(\mathbf{b}+\mathbf{c})$$
$$\mathbf{x} = (1-u)\mathbf{b} + \frac{u}{2}(\mathbf{c}+\mathbf{a})$$
$$\mathbf{x} = (1-v)\mathbf{c} + \frac{v}{2}(\mathbf{a}+\mathbf{b}).$$

in the parameters t, u and v. The three medians are concurrent if and only if they have a point in common. The point Y called the *centroid* of the triangle ABC with position vector

$$\mathbf{y} = \frac{1}{3}\mathbf{a} + \frac{1}{3}\mathbf{b} + \frac{1}{3}\mathbf{c}$$

is a common point of the medians corresponding to the values of the parameters $t = u = v = \frac{2}{3}$. Therefore the proof is complete.

Similiar method may be used to prove the following slightly more complicated theorem.

1.4.12 THEOREM *Let* $B_1, B_2, \cdots, B_6$ *be the midpoints of the consecutive sides of a hexagon* $A_1 A_2 \cdots A_6$. *The medians of the triangle* $B_1 B_3 B_5$ *and the medians of the triangle* $B_2 B_4 B_6$ *are concurrent.*

We may find it convenient to use the following theorem in a proof of collinearity of three points.

1.4.13 THEOREM *Three points* X, Y *and* Z *are collinear if and only if there are scalars* u, v, *and* w, *not all equal to zero such that*

$$u\mathbf{x} + v\mathbf{y} + w\mathbf{z} = 0 \quad \text{and} \quad u + v + w = 0 \ .$$

PROOF For the trivial case where the points are not distinct, we may, for example, take $u = 1$, $v = -1$ and $w = 0$ if $X = Y$. Let us assume that the three points are distinct. If the points are collinear, then by 1.4.10, $\mathbf{z} = (1-t)\mathbf{x}+t\mathbf{y}$. Therefore $(1-t)\mathbf{x}+t\mathbf{y}-\mathbf{z} = 0$ and $(1-t)+t-1 = 0$. Conversely let u, v and w be scalars, not all equal to zero, such that $u\mathbf{x} + v\mathbf{y} + w\mathbf{z} = 0$ and $u+v+w = 0$. Suppose that $u \neq 0$, then $\mathbf{x} = -\frac{1}{u}(v\mathbf{y}+w\mathbf{z}) = -\frac{v}{u}\mathbf{y}-\frac{w}{u}\mathbf{z}$. It follows from $u + v + w = 0$ that $-\frac{v}{u} - \frac{w}{u} = 1$. Putting $t = -\frac{w}{u}$, we can write $\mathbf{x} = (1-t)\mathbf{y}+t\mathbf{z}$. Therefore X lies on the line passing through Y and Z; hence the three points are collinear. Similar argument applies for $v \neq 0$ or $w \neq 0$. The proof is now complete.

Let us use this theorem to prove the classical Menelaus Theorem. The theorem is not included in Enclid's *Elements* and is attributed to the Greek geometer Menelaus of Alexandria of the first century AD. In order to formulate the theorem properly we need the notion of directed ratio of three points.

1.4.14 DEFINITION *Let A and B be two distinct points and L the straight line passing through A and B. If Z is a point on L and different from B, then $\overrightarrow{ZB} \neq 0$, and the vectors $\overrightarrow{AZ}$ and $\overrightarrow{ZB}$ are linearly dependent. The unique scalar r such that $\overrightarrow{AZ} = r\overrightarrow{ZB}$ is called the directed ratio of Z relative to A and B, and is denoted by $dr(A, B; Z)$.*

Fig 1-9

The directed ratio $dr(A, B; Z)$ determines the position of the point Z on L. For example if $0 < dr(A, B; Z)$ then Z lies between A and B and if $dr(A, B; Z) = 1$ then Z is the midpoint of A and B. We notice that $dr(A, B; B)$ is undefined and $dr(A, B, Z) \neq -1$, otherwise $\overrightarrow{AZ} = -\overrightarrow{ZB}$ would imply that $A = B$ which is impossible.

In older textbooks on plane geometry the vector equation $\overrightarrow{AZ} = r\overrightarrow{ZB}$ is sometimes written as $AZ = rZB$ without the arrows on top, and the directed ratio $r = dr(A, B; Z)$ is written as $r = AZ/ZB$.

1.4.15 MENELAUS THEOREM *Let X, Y and Z be taken on the sides BC, AC, AB of a triangle ABC. Then X, Y and Z are collinear if*

$$dr(B, C; X)dr(C, A; Y)dr(A, B; Z) = -1$$

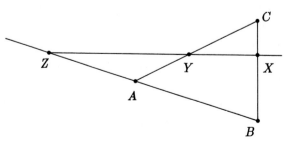

Fig 1-10

PROOF Let $r = dr(B, C; X)$, $s = dr(C, A; Y)$ $t = dr(A, B; Z)$ and $rst = -1$. It follows from the definition of directed ratio that $\overrightarrow{BX} = r\,\overrightarrow{XC}, \overrightarrow{CY} = s\,\overrightarrow{YA}$ and $\overrightarrow{AZ} = t\,\overrightarrow{ZB}$. Hence

$$(1 + r)\mathbf{x} = \mathbf{b} + r\mathbf{c}, \quad (1 + s)\mathbf{y} = \mathbf{c} + s\mathbf{a}, \quad (1 + t)\mathbf{z} = \mathbf{a} + t\mathbf{b} .$$

Using the condition $rst = -1$, we get

$$st(1 + r)\mathbf{x} + (1 + s)\mathbf{y} - s(1 + t)\mathbf{z}$$
$$= st(\mathbf{b} + r\mathbf{c}) + (\mathbf{c} + s\mathbf{a}) - s(\mathbf{a} + t\mathbf{b}) = \mathbf{0} .$$

Therefore by Theorem 1.4.13, the point X, Y and Z will be collinear if we can prove that
$$st(1 + r) + (1 + s) - s(1 + t) = 0$$

and that the scalars $st(1+r)$, $1+s$ and $s(1+t)$ are not all equal to zero. The former follows from the condition $rst = -1$. For the latter it is sufficient to show that $1 + s \neq 0$. Indeed if $s = -1$ then $\overrightarrow{CY} = -\overrightarrow{YA}$; hence $C = A$ which is impossible. The proof is now complete.

EXERCISES

1. Find the value of k so that the points $(2, 5)$, $(0, k)$, and $(-1, 1)$ are collinear.

2. Let $\mathbf{a}$ and $\mathbf{b}$ be linearly independent vectors and X, Y, Z be points whose position vectors are respectively $\mathbf{a} + 2\mathbf{b}$, $3\mathbf{a} + k\mathbf{b}$, and $2\mathbf{a} - 3\mathbf{b}$. Find the value k so that X, Y, Z are collinear.

3. Consider points A, B, C, and D on the plane. If $\overrightarrow{AB} = \mathbf{x} + 2\mathbf{y}$, $\overrightarrow{BC} = -\mathbf{x} + 7\mathbf{y}$, $\overrightarrow{CD} = 3(\mathbf{x} - \mathbf{y})$, show that A, B, D are collinear.

4. Prove Theorem 1.4.9.

5. Prove Theorem 1.4.12.

6. Given points A, B, and C with position vectors $\mathbf{a}$, $\mathbf{b}$, and $\mathbf{c}$. Suppose that $\mathbf{c} = m\mathbf{a} + n\mathbf{b}$ for some real numbers m and n, and that O, A, B are not collinear, show that A, B, and C are collinear if and only if $m + n = 1$. (Hint: Use Theorem 1.4.13)

7. Consider $\triangle ABC$. Let M be the mid-point of the median AD. Through M draw a straight line to meet AB at P and AC at Q.

(a) Suppose now A is placed at the origin O. Express the position vector of M as a linear combination of **p** and **q**.

(b) Hence deduce that $\frac{AB}{AP} + \frac{AC}{AQ}$ is a constant. What is the constant? (Hint: Use result of Question 6.)

8. Use Theorem 1.4.15 to show that if B' is the midpoint of AC of $\triangle ABC$ and that AB is extended to a point C' such that $AB = BC'$, then the straight line $B'C'$ trisects BC.

9. Let A, B be points with position vectors **a** and **b** respectively. By using Theorem 1.4.10, show that

(a) if point P divides AB internally in the ratio $m : n$, then P has position vector $(n\mathbf{a} + m\mathbf{b})/(n + m)$;

(b) if point P divides AB externally in the ratio $m : n$, then P has position vector $(n\mathbf{a} - m\mathbf{b})/(n - m)$.

10. Given two straight lines ℓ_1 and ℓ_2. A_1, B_1, C_1 and A_2, B_2, C_2 are points on ℓ_1 and ℓ_2 respectively, such that $A_1B_1 = A_2B_2$, $B_1C_1 = B_2C_2$. Let P, Q, and R be the midpoints of A_1A_2, B_1B_2, and C_1C_2 respectively.

(a) Express the position vectors of P, Q, and R in terms of position vectors of A_1, A_2, B_1, B_2, C_1 and C_2.

(b) By using Theorem 1.4.13, show that P, Q, and R are collinear.

11. Consider parallelogram $ABCD$. M and N are mid-points of BC and CD respectively. AM and AN meet the diagonal BD at E and F respectively. Show that $|BE| = |EF| = |FD| = \frac{1}{3}|BD|$.

1.5 The dot product

In the last section we have successfully employed vector algebra in the study of plane geometry. In this section we shall study the algebraic notion and the geometric meaning of dot product in the vector space $\mathbf{R}^2$, which will be used in the next section to provide us with a link to the coordinate plane geometry.

1.5.1 DEFINITION *Let* $\mathbf{a} = [a_1, a_2]$ *and* $\mathbf{b} = [b_1, b_2]$ *be vectors of the vector space* $\mathbf{R}^2$. *The dot product of* **a** *and* **b** *is defined to be the real number*

$$\mathbf{a} \cdot \mathbf{b} = a_1 b_1 + a_2 b_2 \ .$$

By definition the dot product of two vectors is no more a vector but a scalar! For example if $\mathbf{a} = [1, 2]$ and $\mathbf{b} = [-2, 3]$, then $\mathbf{a} \cdot \mathbf{b} = 1 \cdot (-2) + 2 \cdot 3 = 4$, and if $\mathbf{c} = [4, -2]$ then $\mathbf{a} \cdot \mathbf{c} = 1 \cdot 4 + 2 \cdot (-2) = 0$. In particular the dot product of a with itself $\mathbf{a} \cdot \mathbf{a} = a_1{}^2 + a_2{}^2 = |\mathbf{a}|^2$ is the square of the magnitude of the vector $\mathbf{a}$. The following fundamental properties of the dot product are readily verified.

1.5.2 THEOREM *For any vectors* $\mathbf{a}$, $\mathbf{b}$ *and* $\mathbf{c}$ *of the vector space* $\mathbf{R}^2$ *and any scalar* r, *the following statements hold.*

> (1) $\mathbf{a} \cdot \mathbf{b} = \mathbf{b} \cdot \mathbf{a}$.
> (2) $(r\mathbf{a}) \cdot \mathbf{b} = r(\mathbf{a} \cdot \mathbf{b}) = \mathbf{a} \cdot (r\mathbf{b})$.
> (3) $(\mathbf{a} + \mathbf{b}) \cdot \mathbf{c} = \mathbf{a} \cdot \mathbf{c} + \mathbf{b} \cdot \mathbf{c}$.
> (4) $|\mathbf{a}|^2 = \mathbf{a} \cdot \mathbf{a} \geq 0$, *and* $\mathbf{a} \cdot \mathbf{a} = 0$ *if and only if* $\mathbf{a} = \mathbf{0}$.

Because of (1) we say that the dot product $\mathbf{a} \cdot \mathbf{b}$ is *symmetric* in the factors $\mathbf{a}$ and $\mathbf{b}$. Properties (2) and (3) together imply that $(r\mathbf{a} + s\mathbf{b}) \cdot \mathbf{c} = r(\mathbf{a} \cdot \mathbf{c}) + s(\mathbf{b} \cdot \mathbf{c})$; thus the dot product is compatible with the formation of linear combination in the first argument. In this sense, we may say it is linear in the first argument. By (1), $\mathbf{a} \cdot (r\mathbf{b} + s\mathbf{c}) = r(\mathbf{a} \cdot \mathbf{b}) + s(\mathbf{a} \cdot \mathbf{c})$; hence it is also linear in the second argument. Therefore we say that the dot product is *bilinear* by which we mean that it is linear in both arguments. Property (4) is usually referred to as the *positive definiteness* of the dot product, expressing the fact that the dot product of a non-zero vector with itself is always positive.

1.5.3 REMARKS Some authors prefer to call $\mathbf{a} \cdot \mathbf{b}$ the *scalar product* because it is a scalar. Others call it the *inner product*. In this book it will be referred to as the dot product. The vector space $\mathbf{R}^2$ together with the addition, the scalar multiplication and now finally with the dot product consitute a complete algebraic system which is called the *2-dimensional euclidean space* still denoted by $\mathbf{R}^2$. The adjective, euclidean, comes from the euclidean measurement of length and angle which shall be presently formulated. Meanwhile we compare once more the euclidean vector space $\mathbf{R}^2$ with the complex number system $\mathbf{C}$. First of all every ordered pair of real numbers is a vector of $\mathbf{R}^2$ and a complex number of $\mathbf{C}$. Hence as sets, $\mathbf{R}^2$ and $\mathbf{C}$ are completely identical. But there are further similarities. The sum of vectors in $\mathbf{R}^2$ and the sum of complex numbers in $\mathbf{C}$ correspond to one another; so do also the scalar multiplication of a vector by a scalar and the multiplication of a complex number by a real number. Here the similarity between the two systems stops! There is no corresponding notion in **C** of the dot product of $\mathbf{R}^2$ and neither is there a corresponding notion in $\mathbf{R}^2$ of the multiplication of complex numbers of **C**.

Let us now address ourselves to the question of the euclidean measurement in the plane. If $A = (a_1, a_2)$ and $B = (b_1, b_2)$ are two points on the plane with position vectors $\mathbf{a} = [a_1, a_2]$ and $\mathbf{b} = [b_1, b_2]$, then for the euclidean lengths of segments OA and AB we have

$$|OA| = \sqrt{a_1{}^2 + a_2{}^2} \quad \text{and} \quad |AB| = \sqrt{(b_1 - a_1)^2 + (b_2 - a_2)^2}$$

while for the magnitudes of the displacement vectors $\mathbf{a}$ and $\mathbf{b} - \mathbf{a}$ we have

$$|\mathbf{a}| = \sqrt{a_1{}^2 + a_2{}^2} \quad \text{and} \quad |\mathbf{b} - \mathbf{a}| = \sqrt{(b_1 - a_1)^2 + (b_2 - a_2)^2} \ .$$

Therefore between lengths and magnitudes, we have equations

$$|OA| = |\mathbf{a}| \quad \text{and} \quad |AB| = |\mathbf{b} - \mathbf{a}| \ .$$

Next we shall study the following fundamental properties of magnitude.

1.5.4 THEOREM *Let* $\mathbf{a}$ *and* $\mathbf{b}$ *be vectors of* $\mathbf{R}^2$ *and* r *a scalar. Then the following statements hold.*
    (1) $|r\mathbf{a}| = |r||\mathbf{a}|$.
    (2) $|\mathbf{a} + \mathbf{b}|^2 + |\mathbf{a} - \mathbf{b}|^2 = 2(|\mathbf{a}|^2 + |\mathbf{b}|^2)$.
    (3) $|\mathbf{a}||\mathbf{b}| \geq |\mathbf{a} \cdot \mathbf{b}|$.
    (4) $|\mathbf{a} + \mathbf{b}| \leq |\mathbf{a}| + |\mathbf{b}|$.
    (5) $|\mathbf{a} - \mathbf{b}|^2 = |\mathbf{a}|^2 + |\mathbf{b}|^2 - 2(\mathbf{a} \cdot \mathbf{b})$.

PROOF (1) (2) and (5) are easily verified. Take (5), for example. Using 1.5.2, we get $|\mathbf{a} - \mathbf{b}|^2 = (\mathbf{a} - \mathbf{b}) \cdot (\mathbf{a} - \mathbf{b}) = \mathbf{a} \cdot \mathbf{a} - \mathbf{a} \cdot \mathbf{b} - \mathbf{b} \cdot \mathbf{a} + \mathbf{b} \cdot \mathbf{b} = |\mathbf{a}|^2 + |\mathbf{b}|^2 - 2(\mathbf{a} \cdot \mathbf{b})$.

We proceed to prove (3). If $\mathbf{a} = \mathbf{0}$, then statement (3) is trivially true. Let us assume that $\mathbf{a} \neq \mathbf{0}$ and consider, for a varying scalar x, the linear combination $x\mathbf{a} + \mathbf{b}$. By Theorem 1.5.2(4), $|x\mathbf{a} + \mathbf{b}|^2 = 0$, if and only if $x\mathbf{a} + \mathbf{b} = \mathbf{0}$. On the other hand, given two vectors $\mathbf{a} \neq \mathbf{0}$ and $\mathbf{b}$, they are either linearly independent or linearly dependent. In the former case there is no scalar x such that $x\mathbf{a} + \mathbf{b} = 0$. In the latter case there is exactly one scalar x such that $x\mathbf{a} + \mathbf{b} = 0$. Therefore $|x\mathbf{a} + \mathbf{b}|^2 = 0$ for at most one value of the scalar x. In other words, the discriminant $D = 4(\mathbf{a} \cdot \mathbf{b})^2 - 4|\mathbf{a}|^2|\mathbf{b}|^2$ of the quadratic expression

$$|x\mathbf{a} + \mathbf{b}|^2 = (x\mathbf{a} + \mathbf{b}) \cdot (x\mathbf{a} + \mathbf{b}) = x^2|\mathbf{a}|^2 + 2x(\mathbf{a} \cdot \mathbf{b}) + |\mathbf{b}|^2$$

in x is not positive. Hence $D \leq 0$ and $|\mathbf{a} \cdot \mathbf{b}| \leq |\mathbf{a}||\mathbf{b}|$.

Alternatively it follows from $(a_1 b_2 - a_2 b_1)^2 \geq 0$ that $a_1{}^2 b_2{}^2 + a_2{}^2 b_1{}^2 \geq 2a_1 a_2 b_1 b_2$. Therefore

$$a_1{}^2 b_1{}^2 + a_2{}^2 b_2{}^2 + a_1{}^2 b_2{}^2 + a_2{}^2 b_1{}^2 \geq a_1{}^2 b_1{}^2 + a_2{}^2 b_2{}^2 + 2a_1 a_2 b_1 b_2 \ .$$

Factorize both sides to get

$$(a_1{}^2 + a_2{}^2)(b_1{}^2 + b_2{}^2) \geq (a_1 b_1 + a_2 b_2)^2$$

from which (3) follows.

Finally let us prove (4). By (5) we have

$$|\mathbf{a} + \mathbf{b}|^2 = |\mathbf{a}|^2 + |\mathbf{b}|^2 + 2(\mathbf{a} \cdot \mathbf{b}) \ .$$

Therefore by (3),

$$|\mathbf{a} + \mathbf{b}|^2 \leq |\mathbf{a}|^2 + |\mathbf{b}|^2 + 2|\mathbf{a}||\mathbf{b}| = (|\mathbf{a}| + |\mathbf{b}|)^2$$

from which (4) follows.

As the magnitude of a vector is identical to the length of its corresponding segment, the five fundamental properties of magnitude may be given geometric interpretations. For example property (2) may be called the *parallelogram law*. Indeed if $OACB$ is a parallelogram on the plane, then for the sides we have $|OA| = |\mathbf{a}|$, $|OB| = |\mathbf{b}|$ and for the diagonals we have $|OC| = |\mathbf{a} + \mathbf{b}|$ and $|BA| = |\mathbf{a} - \mathbf{b}|$. Thus (2) can be taken to mean that the sum of the squares of the two diagonals of a parallelogram equals the sum of the squares of its four sides.

Property (3) is a vector formulation of the *Schwarz inequality*

$$(a_1{}^2 + a_2{}^2)(b_1{}^2 + b_2{}^2) \geq (a_1 b_1 + a_2 b_2)^2$$

and is referred to by the same name.

Property (4) can be rewritten into $|\mathbf{a} - \mathbf{b}| \leq |\mathbf{a}| + |\mathbf{b}|$ because $|\mathbf{b}| = |-\mathbf{b}|$. In this new form it says that in a triangle OAB, the sum of the two sides OA and OB is greater than or equal to the third side AB. Therefore (4) is called the *triangle inequality*.

Property (5) is to be compared with the cosine law of trigonometry which states that in a triangle OAB

$$|BA|^2 = |OA|^2 + |OB|^2 - 2|OA||OB|\cos\theta$$

where θ is the measurement of the angle at the vertex O. With $\overrightarrow{OA} = \mathbf{a}$ and $\overrightarrow{OB} = \mathbf{b}$ under conventions 1.1.4 and 1.4.1, this becomes

$$|\mathbf{a} - \mathbf{b}|^2 = |\mathbf{a}|^2 + |\mathbf{b}|^2 - 2|\mathbf{a}||\mathbf{b}|\cos\theta \ .$$

Comparing this with (5)

$$|\mathbf{a} - \mathbf{b}|^2 = |\mathbf{a}|^2 + |\mathbf{b}|^2 - 2(\mathbf{a} \cdot \mathbf{b}) \ ,$$

we obtain

$$\mathbf{a} \cdot \mathbf{b} = |\mathbf{a}||\mathbf{b}|\cos\theta \quad \text{or} \quad \cos\theta = (\mathbf{a} \cdot \mathbf{b})/|\mathbf{a}||\mathbf{b}|$$

which brings the dot product $\mathbf{a} \cdot \mathbf{b}$ and the cosine of the angle between the vectors $\mathbf{a}$ and $\mathbf{b}$ into close relationship. Therefore we call property (5) of 1.5.4 the *cosine law*.

In general for any two non-zero vectors $\mathbf{a}$ and $\mathbf{b}$ of $\mathbf{R}^2$, we have

$$|\mathbf{a}||\mathbf{b}| \geq |\mathbf{a} \cdot \mathbf{b}|$$

by the Schwarz inequality. Hence

$$-1 \leq \frac{\mathbf{a} \cdot \mathbf{b}}{|\mathbf{a}||\mathbf{b}|} \leq 1 \ .$$

These inequalities together with the above comparison permit us to put the following definition.

1.5.5 DEFINITION *Given any two non-zero vectors* $\mathbf{a}$ *and* $\mathbf{b}$ *the angle* θ *between the vectors* $\mathbf{a}$ *and* $\mathbf{b}$ *is defined by*

$$\cos\theta = \frac{\mathbf{a}\cdot\mathbf{b}}{|\mathbf{a}||\mathbf{b}|} \quad \text{and} \quad 0 \le \theta \le \pi \,.$$

For example if $\mathbf{a} = [1,0]$, $\mathbf{b} = [-1,1]$ and $\mathbf{c} = [5, -5\sqrt{3}]$, then $|\mathbf{a}| = 1$, $|\mathbf{b}| = \sqrt{2}$, $|\mathbf{c}| = 10$, $\mathbf{a}\cdot\mathbf{b} = -1$, $\mathbf{a}\cdot\mathbf{c} = 5$ and $\mathbf{b}\cdot\mathbf{c} = -5 - 5\sqrt{3}$. Therefore

$$(\mathbf{a}\cdot\mathbf{b})/|\mathbf{a}||\mathbf{b}| = -1/\sqrt{2} : \text{angle between } \mathbf{a} \text{ and } \mathbf{b} \text{ is } 3\pi/4;$$
$$(\mathbf{a}\cdot\mathbf{c})/|\mathbf{a}||\mathbf{c}| = 1/2 : \text{angle between } \mathbf{a} \text{ and } \mathbf{c} \text{ is } \pi/3;$$
$$(\mathbf{b}\cdot\mathbf{c})/|\mathbf{b}||\mathbf{c}| = (-\sqrt{2} - \sqrt{6})/4 : \text{angle between } \mathbf{b} \text{ and } \mathbf{c} \text{ is } 11\pi/12.$$

This definition of angle measurement in $\mathbf{R}^2$ provides us with a convenient criterion for perpendicularity: two non-zero vectors $\mathbf{a}$ and $\mathbf{b}$ are perpendicular if and only if $\mathbf{a}\cdot\mathbf{b} = 0$. For example $[2, 1]$ and $[1/2, -1]$ are perpendicular and so are the unit coordinate vectors $\mathbf{e}_1 = [1,0]$ and $\mathbf{e}_2 = [0,1]$. In order to extend the scope of application of the notion of perpendicularity, we shall now lift the restriction on vectors being non-zero and use the synonym orthogonality.

1.5.6 DEFINITION *Two vectors* $\mathbf{a}$ *and* $\mathbf{b}$ *of* $\mathbf{R}^2$ *are said to be orthogonal if* $\mathbf{a}\cdot\mathbf{b} = 0$. *In this case we may write* $\mathbf{a} \perp \mathbf{b}$.

We take note that the zero vector is orthogonal to all vectors of $\mathbf{R}^2$, and that the zero vector is the only vector of $\mathbf{R}^2$ that is orthogonal to itself.

As an application of the notion of orthogonality, we consider the resolution of a given vector into a sum of two orthogonal vectors. More precisely, given a vector $\mathbf{x}$ and a non-zero vector $\mathbf{a}$ we wish to write $\mathbf{x}$ into

$$\mathbf{x} = t\mathbf{a} + \mathbf{b} \quad \text{where} \quad \mathbf{a} \perp \mathbf{b}\,.$$

The problem can be solved by elementary algebra in terms of coordinates. Indeed if $\mathbf{x} = [x,y]$ and $\mathbf{a} = [a_1, a_2] \ne \mathbf{0}$, then the non-zero vector $[a_2, -a_1]$ would be orthogonal to $\mathbf{a}$ and parallel to $\mathbf{b}$. The problem at hand is to find scalars t and u such that

$$[x,y] = t[a_1, a_2] + u[a_2, -a_1]\,,$$

i.e. $\mathbf{x} = t\mathbf{a} + \mathbf{b}$ with $\mathbf{b} = u[a_2, -a_1]$. In other words, to solve for the unknowns t and u in the equations

$$a_1 t + a_2 u = x$$
$$a_2 t - a_1 u = y \ .$$

Obviously we would obtain

$$t = \frac{x a_1 + y a_2}{a_1{}^2 + a_2{}^2} = \frac{[x, y] \cdot [a_1, a_2]}{[a_1, a_2] \cdot [a_1, a_2]} = \frac{(\mathbf{x} \cdot \mathbf{a})}{(\mathbf{a} \cdot \mathbf{a})}$$

$$u = \frac{x a_2 - y a_1}{a_2{}^2 + a_1{}^2} = \frac{[x, y] \cdot [a_2, -a_1]}{[a_2, -a_1] \cdot [a_2, -a_1]} \ .$$

Alternatively we may work with vectors and reformulate the problem as follows. Given $\mathbf{x}$ and $\mathbf{a} \neq 0$. It is required to find a scalar t such that

$$\mathbf{x} - t\mathbf{a} = \mathbf{b} \quad \text{is orthogonal} \quad \mathbf{a} \ .$$

Taking dot product with $\mathbf{a}$ on both sides, we get

$$(\mathbf{x} \cdot \mathbf{a}) - t(\mathbf{a} \cdot \mathbf{a}) = (\mathbf{b} \cdot \mathbf{a}) = 0 \ .$$

Since $(\mathbf{a} \cdot \mathbf{a}) \neq 0$, we obtain solution to the problem as

$$t = (\mathbf{x} \cdot \mathbf{a})/(\mathbf{a} \cdot \mathbf{a}) \quad \text{and} \quad \mathbf{b} = \mathbf{x} - t\mathbf{a} \ .$$

Graphically the condition that $(\mathbf{x} - t\mathbf{a})$ be orthogonal to $\mathbf{a}$ means that the scalar multiple $t\mathbf{a}$ is the perpendicular projection of $\mathbf{x}$ on $\mathbf{a}$.

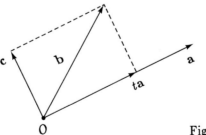

Fig 1-11

Let us put the result of our discussion into the following statement.

1.5.7 THEOREM AND DEFINITION *Let* **a** *be a non-zero vector. Then for any vector* **x** *in the plane there is a unique scalar* $t = (\mathbf{x} \cdot \mathbf{a})/(\mathbf{a} \cdot \mathbf{a})$ *and a unique vector* **b** *such that*

$$\mathbf{x} = t\mathbf{a} + \mathbf{b} \quad \text{and} \quad \mathbf{a} \perp \mathbf{b} .$$

The first summand $t\mathbf{a}$ *is called the orthogonal projection of* **x** *onto* **a** *and is denoted by* $pr_\mathbf{a}\mathbf{x}$*; thus*

$$pr_\mathbf{a}\mathbf{x} = (\mathbf{x} \cdot \mathbf{a}/\mathbf{a} \cdot \mathbf{a})\mathbf{a} .$$

The special case where any one of the summands **b** or $t\mathbf{a}$ is the zero vector gives rise to the following corollaries. **a** *and* **x** *are linearly dependent if and only if* $pr_\mathbf{a}\mathbf{x} = \mathbf{x}$. For similar reason, **x** *and* **a** *are orthogonal if and only if* $pr_\mathbf{a}\mathbf{x} = \mathbf{0}$.

1.5.8 EXAMPLE Let $X = (-1, 3)$, $Y = (3, 0)$, $A = (2, 4)$ and $B = (1, -2)$ be four points in the plane. Find the length of the perpendicular projection of the segment XY on the straight line through A and B.

SOLUTION Consider the displacement vectors

$$\mathbf{z} = \overline{XY} = [4, -3] \quad \text{and} \quad \mathbf{c} = \overrightarrow{AB} = [-1, -6] .$$

The projection of XY on the line through A and B has the same length as the projection $pr_\mathbf{c}\mathbf{z} = (\mathbf{z} \cdot \mathbf{c}/\mathbf{c} \cdot \mathbf{c})\mathbf{c}$ which is $|\mathbf{z} \cdot \mathbf{c}|/|\mathbf{c}|$. Therefore the desired length in $14/\sqrt{37}$.

EXERCISES

1. Calculate the dot product of following vectors.
   (a) $[2, -1]$, $[1, 3]$.
   (b) $[1, 2]$, $[-4, 2]$.
   (c) $[a, b]$, $[-b, a]$.
2. By expressing **a**, **b**, and **c** into their components, prove Theorem 1.5.2.
3. Let $\mathbf{a} = [1, -2]$, $\mathbf{b} = [-1, -3]$, and $\mathbf{c} = [0, -7]$. Compute
   (a) $(3\mathbf{a}) \cdot (2\mathbf{b})$,
   (b) $\mathbf{a} \cdot (\mathbf{b} + \mathbf{c})$,
   (c) $(2\mathbf{b}) \cdot (\mathbf{c} - 3\mathbf{a})$,
   (d) $(\mathbf{c} - \mathbf{a}) \cdot (2\mathbf{b} - 5\mathbf{a})$.

4. If the vectors $[k-2, 3]$ and $[k, -5]$ are orthogonal, find the values of k.

5. Let $\mathbf{x} = [2, 3]$. Show that any vector orthogonal to $\mathbf{x}$ is a scalar multiple of $[3, -2]$. What is the result when $\mathbf{x} = [a, b]$ for arbitrary real numbers a and b?

6. Given that $\mathbf{a} = [3, 2]$ and $\mathbf{b} = [-1, 5]$. Find real number k such that $\mathbf{a} + k\mathbf{b}$ is perpendicular to $\mathbf{a} - \mathbf{b}$.

7. Given that $\mathbf{a} = [\cos\alpha, \sin\alpha]$ and $\mathbf{b} = [\cos\beta, \sin\beta]$, find $\mathbf{a} \cdot \mathbf{b}$. Sketch the vectors and hence deduce that $\cos(\alpha - \beta) = \cos\alpha\cos\beta + \sin\alpha\sin\beta$.

8. By expressing $\mathbf{a}$ and $\mathbf{b}$ into their components, or otherwise, prove (1) and (2) of Theorem 1.5.4.

9. Use vector method to find the consines of the interior angles of the triangle with vertices $(0, 3)$, $(4, 0)$ and $(-2, -5)$.

10. Let $\mathbf{a} = [4, 3]$ and $\mathbf{b} = [k, 2]$. Find k so that
    (a) $\mathbf{a}$ and $\mathbf{b}$ are parallel;
    (b) $\mathbf{a}$ and $\mathbf{b}$ are orthogonal;
    (c) the angle between $\mathbf{a}$ and $\mathbf{b}$ is $\frac{\pi}{6}$.

11. Let $\mathbf{x} = [3, -4]$. Compute $pr_{\mathbf{a}}\mathbf{x}$ for each of the following vectors:
    (a) $\mathbf{a} = [1, 0]$.
    (b) $\mathbf{a} = [0, 1]$.
    (c) $\mathbf{a} = [\frac{1}{\sqrt{2}}, \frac{1}{\sqrt{2}}]$.

12. For each of the following $\mathbf{x}$ and $\mathbf{a}$, compute $pr_{\mathbf{a}}\mathbf{x}$.
    (a) $\mathbf{x} = [2, 4]$, $\mathbf{a} = [-2, 1]$.
    (b) $\mathbf{x} = [1, 101]$, $\mathbf{a} = [100, -1]$.
    (c) $\mathbf{x} = \overrightarrow{AB}$, $\mathbf{a} = \overrightarrow{CD}$, where $A = (2, 1)$, $B = (3, -2)$, $C = (1, 1)$, and $D = (2, 3)$.

13. Given vectors $\mathbf{a}$ and $\mathbf{b}$ such that $|\mathbf{a}| = 10$ and angle between $\mathbf{a}$ and $\mathbf{b}$ is $\frac{\pi}{3}$, find $pr_{\mathbf{b}}\mathbf{a}$.

14. Find a vector $\mathbf{a}$ such that $\mathbf{a}$ is parallel to $[1, 5]$ and $|pr_{[1,1]}(\mathbf{a} + [1, 5])| = 12\sqrt{2}$.

15. Show that two perpendicular non-zero vectors in $\mathbf{R}^2$ must be linearly independent.

16. Let $\mathbf{u}_1$ and $\mathbf{u}_2$ be two perpendicular unit vectors in $\mathbf{R}^2$. Show that for any $\mathbf{x}$ in $\mathbf{R}^2$,
$$\mathbf{x} = (\mathbf{x} \cdot \mathbf{u}_1)\mathbf{u}_1 + (\mathbf{x} \cdot \mathbf{u}_2)\mathbf{u}_2 .$$

17. Given two perpendicular vectors $\mathbf{a}$ and $\mathbf{b}$ such that $|\mathbf{a}| = 8$ and $|\mathbf{b}| = 15$, find $|\mathbf{a} - \mathbf{b}|$ and $|\mathbf{a} + \mathbf{b}|$.
    (Hint: Consider the geometrical meaning.)

18. Given vectors a and b such that $|\mathbf{a}| = 11$, $|\mathbf{b}| = 23$, and $|\mathbf{a} - \mathbf{b}| = 30$, find $|\mathbf{a} + \mathbf{b}|$.

19. Show that if **a** and **b** are any two vectors in $\mathbf{R}^2$ for which $|\mathbf{a}||\mathbf{b}| = |\mathbf{a} \cdot \mathbf{b}|$, then one of them must be a scalar multiple of the other.

20. Let **u** and **v** be non-zero vectors. Show that the vector $\mathbf{w} = |\mathbf{u}|\mathbf{v} + |\mathbf{v}|\mathbf{u}$ bisects the angle between **u** and **v**.

21. Let **u** and **v** be vectors in $\mathbf{R}^2$. Show that

    (a) **u** is orthogonal to **v** if and only if $|\mathbf{u} + \mathbf{v}| = |\mathbf{u} - \mathbf{v}|$.

    (b) $\mathbf{u} + \mathbf{v}$ and $\mathbf{u} - \mathbf{v}$ are orthogonal if and only if $|\mathbf{u}| = |\mathbf{v}|$.

    What do the above result mean geometrically?

22. Given non-zero vectors **a** and **b**.

    (a) Find t such that $|\mathbf{a} + t\mathbf{b}|$ is minimum.

    (b) Prove that for the value of t found in (a), **b** and $\mathbf{a} + t\mathbf{b}$ are perpendicular.

23. Let $\mathbf{a} = (\cos\alpha, \sin\alpha)$, $\mathbf{b} = (\cos\beta, \sin\beta)$, where $0 < \alpha < \beta < \pi$.

    (a) Prove that $\mathbf{a} - \mathbf{b}$ is perpendicular to $\mathbf{a} + \mathbf{b}$.

    (b) If for some non-zero real number t,

    $$|t\mathbf{a} + \mathbf{b}| = |\mathbf{a} - t\mathbf{b}|$$

    find the value of $\beta - \alpha$.

24. Given a rhombus with vertices $(0,0)$, (x_1, y_1), (x_2, y_2) and (x_3, y_3) in turn. If we let $\mathbf{x}_1 = [x_1, y_1]$, $\mathbf{x}_2 = [x_2, y_2]$ and $\mathbf{x}_3 = [x_3, y_3]$, show that

    $$|\mathbf{x}_2 - \mathbf{x}_1|^2 - |\mathbf{x}_2 - \mathbf{x}_3|^2 = 2\mathbf{x}_2 \cdot (\mathbf{x}_3 - \mathbf{x}_1) .$$

    Hence deduce that the diagonals of the rhombus must be perpendicular.

25. Given $\overrightarrow{AB} = \mathbf{x}$ and $\overrightarrow{AC} = \mathbf{y}$.

    (a) Show that the area of $\triangle ABC = \frac{1}{2}|\mathbf{x}||\mathbf{y}|\sin\theta$, where θ is the angle between **x** and **y**.

    (b) Hence deduce that the area of $\triangle ABC = \frac{1}{2}\sqrt{|\mathbf{x}|^2|\mathbf{y}|^2 - (\mathbf{x} \cdot \mathbf{y})^2}$.

    (c) Find the area of $\triangle ABC$ if $A = (1,2)$, $B = (-3,-1)$, and $C = (3,-3)$.

26. Consider $\triangle ABC$ where position vectors of the vertices are **a**, **b**, and **c** respectively. Suppose that the perpendicular bisectors of BC and AB meet at the origin.

    (a) Show that $|\mathbf{a}|^2 = |\mathbf{b}|^2 = |\mathbf{c}|^2$.

(b) Hence deduce that the perpendicular bisectors of the sides of $\triangle ABC$ are concurrent.

27. Let M be the midpoint of BC of $\triangle ABC$. If $\overrightarrow{AM} = \mathbf{x}$, $\overrightarrow{AB} = \mathbf{y}$ and $\overrightarrow{AC} = \mathbf{z}$, show that

(a) $|\mathbf{x}|^2 = \frac{1}{4}|\mathbf{y}|^2 + \frac{1}{2}(\mathbf{y} \cdot \mathbf{z}) + \frac{1}{4}|\mathbf{z}|^2$, and

(b) $AB^2 + AC^2 = 2(AM^2 + BM^2)$.

1.6 Equation of a straight line

In coordinate geometry a homogeneous linear equation

$$a_1 x + a_2 y = 0 \tag{1}$$

defines a straight line L that passes through the origin. By this we mean that a point $X = (x, y)$ lies on the line L if and only if the equation is satisfied by the coordinates x and y of the point X. Let us translate this into the language of vectors. Using the coefficients a_1, a_2 as components of a vector, we write $\mathbf{a} = [a_1, a_2]$; similarly the unknowns x, y as components of a vector $\mathbf{x} = [x, y]$. Consequently equation (1) becomes

$$\mathbf{a} \cdot \mathbf{x} = 0 \tag{2}$$

and the straight line L is seen to consist of all points X in the plane whose position vectors $\mathbf{x}$ are orthogonal to the given non-zero vector $\mathbf{a}$. As an immediate benefit of the translation, we see that the non-zero vector $\mathbf{a}$ whose components are the coefficients of (1) is a *normal vector* of the straight line L.

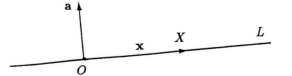

Fig 1-12

Just as (1) is an equation of the line L in terms of coordinates, (2) is an equation of the same straight line L but in terms of position vector.

More generally every straight line G in the plane is defined by a linear equation

$$a_1 x + a_2 y + c = 0 \qquad (3)$$

with an arbitrary constant term c. To write (3) in terms of vectors, we take an arbitrary point $P = (p, q)$ on G. Then the position vector $\mathbf{p} = [p, q]$ satisfies

$$\mathbf{a} \cdot \mathbf{p} + c = 0 .$$

Therefore equation (3) becomes

$$\mathbf{a} \cdot \mathbf{x} - \mathbf{a} \cdot \mathbf{p} = 0 \qquad (4)$$

or
$$\mathbf{a} \cdot (\mathbf{x} - \mathbf{p}) = 0 \qquad (5)$$

which is now an equation of the straight line G in terms of position vectors of its points.

What is the appropriate geometric interpretation of equation (5) of G? Clearly we can say that G is the straight line that is normal to the vector $\mathbf{a} = [a_1, a_2]$ and passes through the point $P = (p, q)$. Therefore (5) may be called the point-normal form of an equation of G (see Figure 1-13). Indeed if $a_2 \neq 0$, we can write (5) in term of components as

$$a_1(x - p) + a_2(y - q) = 0 \quad \text{or} \quad -\frac{a_1}{a_2} = \frac{y - q}{x - p}$$

which is in the point-slope form of an equation of G.

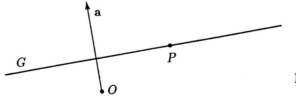

G P Fig 1-13

1.6.1 THEOREM   *The straight line that passes through a point P and has a normal vector* $\mathbf{a}$ *is defined by*

$$\mathbf{a} \cdot (\mathbf{x} - \mathbf{p}) = 0 .$$

For example

$$x + 2y + 2 = 0$$

is the straight line that passes through the point $(-2,0)$ and perpendicular to the line along $[1,2]$. Similarly the straight line which passes through the point $(3,2)$ and has a normal vector $[2,1]$ is defined by

$$[2,1] \cdot [x-3, y-2] = 0$$

or
$$2x + y - 8 = 0 .$$

1.6.2 EXAMPLE Let A and B be two distinct points in the plane. Find the straight line G that passes through a point P and perpenticular to the line L passing through A and B.

SOLUTION The line L passing through A and B is parallel to the displacement vector $\overrightarrow{BA} = \mathbf{a} - \mathbf{b}$. Therefore G, as a line perpendicular to L, has $\mathbf{a} - \mathbf{b}$ as a normal vector. Hence by 1.6.1, G is defined by the equation $(\mathbf{a} - \mathbf{b}) \cdot (\mathbf{x} - \mathbf{p}) = 0$.

For example, the perpendicular G from the point $X_0 = (1,-5)$ to the line L through the points $P = (2,3)$ and $Q = (4,-1)$ is

$$[2-4, 3+1] \cdot [x-1, y+5] = 0$$

which simplifies into

$$x - 2y - 11 = 0 .$$

The equation of L is

$$2x + y - 7 = 0 .$$

Solving these two equations, we obtain $Y_0 = (5,-3)$ as the point of intersection of the lines L and G. Therefore the distance from X_0 to the line L is the length of the segment $X_0 Y_0$ which is $\sqrt{20}$.

Let us consider an alternative method of finding the distance from X_0 to L. The segment $Y_0 X_0$ is the perpendicular projection of the segemnt PX_0 or QX_0 on the line G. On the other hand the vector $\mathbf{a} = [2,1]$ of the coefficients of the equation of L is a normal vector of L; therefore it is parallel to the line G. In terms of displacement vectors, $\overrightarrow{Y_0 X_0}$ is the orthogonal projection of $\overrightarrow{PX_0}$ on $\mathbf{a}$:

$$\overrightarrow{Y_0 X_0} = pr_{\mathbf{a}}(\overrightarrow{PX_0}) = \frac{(\mathbf{x}_0 - \mathbf{p}) \cdot \mathbf{a}}{\mathbf{a} \cdot \mathbf{a}} \mathbf{a} = \left(\frac{\mathbf{x}_0 \cdot \mathbf{a} - \mathbf{p} \cdot \mathbf{a}}{\mathbf{a} \cdot \mathbf{a}} \right) \mathbf{a}$$

and $\qquad |Y_0 X_0| = |\mathbf{x}_0 \cdot \mathbf{a} - \mathbf{p} \cdot \mathbf{a}|/|\mathbf{a}| = |-3-7|/\sqrt{5} = \sqrt{20} .$

In general, we have the following theorem.

1.6.3 THEOREM *Let L be the straight line defined by the equation*

$$a_1 x + a_2 y + c = 0 \quad \text{or} \quad \mathbf{a} \cdot \mathbf{x} + c = 0 .$$

If $X_0 = (x_0, y_0)$ is an arbitrary point in the plane then the distance from X_0 to L is given by

$$|\mathbf{a} \cdot \mathbf{x}_0 + c|/|\mathbf{a}|$$

or $\qquad |a_1 x_0 + a_2 y_0 + c|/\sqrt{a_1{}^2 + a_2{}^2} .$

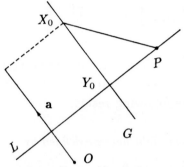

Fig 1-14

PROOF Let P be any point on L and Y_0 the foot of the perpenticular G from X_0 to L. Then the segment $Y_0 X_0$ is the perpendicular projection of the segment PX_0 on G, because P is projected to Y_0 and X_0, being on G, is projected to X_0 itself. On the other hand G is parallel to the normal vector $\mathbf{a} = [a_1, a_2]$ of L. Therefore the distance from X_0 to L is

$$|pr_{\mathbf{a}} \overrightarrow{PX_0}| = |\mathbf{a} \cdot (\mathbf{x}_0 - \mathbf{p})/\mathbf{a} \cdot \mathbf{a}||\mathbf{a}| = |\mathbf{a} \cdot \mathbf{x}_0 - \mathbf{a} \cdot \mathbf{p}|/|\mathbf{a}|$$
$$= |\mathbf{a} \cdot \mathbf{x}_0 + c|/|\mathbf{a}| .$$

The theorem leads us to consider for a given straight line L : $a_1 x_1 + a_2 y + c = 0$ or $\mathbf{a} \cdot \mathbf{x} + c = 0$ the function

$$f(x, y) = \frac{a_1 x + a_2 y + c}{\sqrt{a_1{}^2 + a_2{}^2}} \quad \text{or} \quad f(\mathbf{x}) = (\mathbf{a} \cdot \mathbf{x} + c)/|\mathbf{a}|$$

which has the property that the sum of the squares of its linear coefficients is 1. Obviously the straight line L is also defined by the equation

$$f(x, y) = 0 \quad \text{or} \quad f(\mathbf{x}) = 0$$

which is called the *normal form* of an equation of L. The advantage of the normal form is that the distance from an arbitrary point $X_0 = (x_0, y_0)$ is easily evaluated as

$$|f(x_0, y_0)| \quad \text{or} \quad |f(\mathbf{x}_0)| \ .$$

For example the normal form of

$$3x + 4y - 7 = 0$$

is $\frac{3}{5}x + \frac{4}{5}y - \frac{7}{5} = 0$. The distance from the point $(3, -2)$ to this line is therefore

$$\left| \frac{3}{5} \cdot 3 + \frac{4}{5} \cdot (-2) - \frac{7}{5} \right| = \frac{6}{5} \ .$$

Finally we observe that the angle between two given straight lines is the same as the angle between their normal vectors. In conjunction with 1.5.5 we obtain the following result.

1.6.4 THEOREM *Let two straight lines be defined by*

$$a_1 x + a_2 y + c = 0 \quad \text{or} \quad \mathbf{a} \cdot \mathbf{x} + c = 0$$

and
$$b_1 x + b_2 y + d = 0 \quad \text{or} \quad \mathbf{b} \cdot \mathbf{x} + d = 0 \ .$$

Then the angle θ between them is given by

$$\cos \theta = (\mathbf{a} \cdot \mathbf{b})/|\mathbf{a}||\mathbf{b}| \ .$$

1.6.5 EXAMPLE Let L_1 and L_2 be given by

$$4x - 2y + 7 = 0 \quad \text{and} \quad 12x + 4y - 5 = 0 \ .$$

Then the angle θ between them is given by

$$\cos \theta = (4 \cdot 12 - 2 \cdot 4)/(\sqrt{4^2 + 2^2} \cdot \sqrt{12^2 + 4^2})$$
$$= 40/(\sqrt{20} \cdot \sqrt{160}) = 1/\sqrt{2} \ .$$

Therefore L_1 and L_2 intersect each other at an angle of $45°$.

EXERCISES

1. In what follows, find a vector equation for the straight line containing the given point P and having the given vector **a** as a normal vector.
   (a) $P = (-1, 2)$, $\mathbf{a} = [1, -2]$.
   (b) $P = (3, 2)$, $\mathbf{a} = [1, 0]$.
   (c) $P = (-4, 1)$, $\mathbf{a} = [5, 2]$.
   (d) $P = (-3, -2)$, $\mathbf{a} = [2, 3]$.
   Hence express the equations in the form $ax + by + c = 0$.

2. In each of the following, find a vector equation for the straight line containing the given point P and perpendicular to the line passing through given points A and B.
   (a) $P = (0, 1)$, $A = (4, 0)$, $B = (8, 1)$.
   (b) $P = (-1, 1)$, $A = (3, 3)$, $B = (4, 4)$.
   (c) $P = (3, -2)$, $A = (1, 0)$, $B = (0, 1)$.
   (d) $P = (5, 1)$, $A = (7, 1)$, $B = (-2, 3)$.

3. Find the normal forms of the following straight lines:
   (a) $12x + 5y - 13 = 0$;
   (b) $8x + 15y + 2 = 0$;
   (c) $7x - 2y + 4 = 0$.

4. Find the distance from the point $(-4, 3)$ to each line in Question 3.

5. Use Theorem 1.6.4 to find the angle between the following pairs of straight lines.
   (a) $3x - 4y + 2 = 0$.
   (b) $7x - y - 8 = 0$.

6. Find real number k such that the point $(2, k)$ is equidistant from the lines $x + y - 2 = 0$ and $x - 7y + 2 = 0$.

7. Find the distance between the following parallel lines.
   (a) $3x - y + 12 = 0$.
   (b) $3x - y - 18 = 0$.

Given two non-parallel lines, an angle bisector of the lines is a line such that each point on the bisector is equidistant from the given lines. Note that there are two distinct angle bisectors.

8. Find the equations of the angle bisectors of
   (a) $x - y = 0$ and
   (b) $3x - y = 0$.

9. Find the equations of the angle bisectors of $4x - 3y + 1 = 0$, and $5x + 12y - 2 = 0$.

10. Consider the following non-parallel lines

$$a_1 x + b_1 y + c_1 = 0 \quad \text{and} \quad a_2 x + b_2 y + c_2 = 0 \ .$$

(a) Express the angle bisectors in the form $\mathbf{d} \cdot \mathbf{x} + c = 0$ for some $\mathbf{d}$ and c.

(b) Hence deduce that the angle bisectors are perpendicular.

CHAPTER TWO

VECTORS AND GEOMETRY IN SPACE

In this chapter we follow the pattern of last chapter to study algebra of vectors in space and solid geometry side by side. Readers will find a fairly complete treatment of the vector space $\mathbf{R}^3$ where most of the important topics are discussed. In spite of the extensive subject of geometry in space, we are only able to include some general algebraic methods in the treatment of lines and planes and a very small selection of classical theorems.

2.1 Cartesian coordinates in space

The cartesian coordinates of a point in the plane are given in relation to a pair of mutually perpendicular axes; those of a point in space are given in relation to three mutually perpendicular lines intersecting at a point. Suppose that we have such a triad of lines. The point of intersection will be called the the *origin* and is denoted by O. Suppose that each line is given a fixed direction. Then we proceed to assign the directed lines as axes of coordinates: the x-axis, the y-axis and the z-axis.

There are 6 different assignments which fall into two groups of three each. The first group is shown in Figure 2-1. Each assignment of the first group can be changed into any other of the same group by a rotation about O. Similarly each assignment of the second group (Figure 2-2) can be changed into any other of the same group. However it is impossible to change any assignment of one group into an assignment of the other group by a rotation about O.

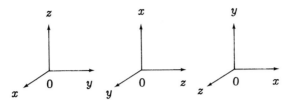

Fig 2-1

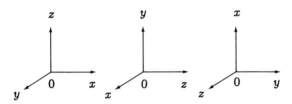

Fig 2-2

To distinguish these two groups, we shall call each assignment in the first group a *right-hand coordinate system* because the x-axis, the y-axis and the z-axis correspond respectively to the thumb, the index finger and the middle finger of the right hand when extended in mutually perpendicular directions. Similarly the other three are called *left-hand coordinate systems*. Though it is immaterial as to which coordinate system is used, we shall henceforth adopt a right-hand coordinate system and proceed to set up a one-to-one correspondence between points in space and ordered triples of real numbers.

To begin with we identify each coordinate axis with the real line in the usual manner. Given any ordered triple (a_1, a_2, a_3), each of the real numbers a_1, a_2, a_3 individually corresponds to a point on the x-axis, y-axis, z-axis respectively. Through each of these points, we erect a plane perpendicular to the corresponding coordinate axis. The three planes so erected meet at a point A in space. In this way every ordered triple (a_1, a_2, a_3) of real numbers is matched with a point A in space and we shall call the real numbers a_1, a_2, a_3 the *coordinates* of the point A and write $A = (a_1, a_2, a_3)$. In particular $O = (0, 0, 0)$.

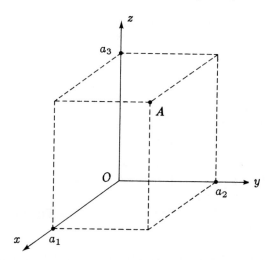

Fig 2-3

Conversely if B is a point in space, then through B we can pass three planes, each individually perpendicalur to a coordinate axis. These planes cut the x-axis, y-axis, z-axis at points corresponding to coordinates b_1, b_2, b_3 respectively. In this way the point B is matched with the ordered triple (b_1, b_2, b_3). Clearly the ordered triple (b_1, b_2, b_3) delivers back the point B by the previous process: $B = (b_1, b_2, b_3)$. Therefore we have a one-to-one coorespondence between points in space and ordered triples of real numbers. By this correspondence the cartesian space of all ordered triples of real numbers becomes a mathematical model of the three-dimensional space. This model enables us to carry out algebraic works on coordinates of points in space. As an example we give a formula for the euclidean distance between two points in space.

Let $A = (a_1, a_2, a_3)$ and $B = (b_1, b_2, b_3)$ be two points in space. Appropriate planes passing through A or B and perpendicular to the axes meet at a number of points (see Figure 2-3). Among them we find $C = (b_1, b_2, a_3)$, $A' = (a_1, a_2, 0)$ and $C' = (b_1, b_2, 0)$. A' and C' being points on the xy-plane,

$$\sqrt{(a_1 - b_1)^2 + (a_2 - b_2)^2} = |A'C'| = |AC| .$$

Moreover the triangle ABC has a right angle at C and $|BC| = |a_3 - b_3|$. Therefore by the Pythagorean theorem

$$|AB| = \sqrt{|AC|^2 + |BC|^2} = \sqrt{(a_1 - b_1)^2 + (a_2 - b_2)^2 + (a_3 - b_3)^2}$$

which is the distance between the points A and B in terms of their coordinates.

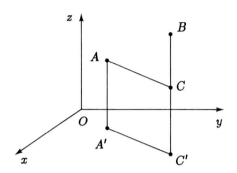

Fig 2-4

It follows from the distance formula that the sphere which has centre at $A = (a_1, a_2, a_3)$ and radius r consists of all points $X = (x, y, z)$ in space such that

$$(x - a_1)^2 + (y - a_2)^2 + (z - a_3)^2 = r^2$$

or $\quad x^2 + y^2 + z^2 - 2a_1 x - 2a_2 y - 2a_3 z + a_1{}^2 + a_2{}^2 + a_3{}^2 - r^2 = 0 \ .$

EXERCISES

1. Find the distances between the point $A = (-4, 3, 1)$ and the three axes.
2. Find the values of x, y, and z if
    (a) $A = (5, -7, 1)$, $B = (7, 2, z)$, and $|AB| = 11$;
    (b) $A = (2, 3, 3)$, $B = (x, -2, 3)$, and $|AB| = 5$;
    (c) $A = (3, y, 9)$, $B = (3, 0, 4)$, and $|AB| = 3$.
3. Find the point on the z-axis such that it is equidistant from the points $A = (6, -3, 14)$ and $B = (5, 1, -6)$.
4. Find the point on the yz-plane such that it is equidistant from the points $A = (0, 5, 1)$, $B = (3, 1, 2)$ and $C = (4, -2, -2)$.
5. Find the centre and radius of the following sphere
    $$x^2 + y^2 + z^2 + 4x - 6y + 5z + 17 = 0 \ .$$
6. Given that $A = (a, 0, 0)$, $B = (b, 0, 0)$, and $C = (c, 0, 0)$ are three points on the x-axis such that $a \geq b \geq c$. If $P = (p, 0, 0)$ is any point on the x-axis, show that the expression

$$|PA|^2 \cdot |BC| + |PB|^2 \cdot |CA| + |PC|^2 \cdot |AB|$$

is independent of p.

2.2 Vectors in space

Just as points on the cartesian plane and vectors on the plane are ordered pairs of real numbers, points in the cartesian space and vectors in space are both ordered triples.

2.2.1 DEFINITION *A vector in space is an ordered triple of real numbers. The three real numbers are the components of the vector.*

With this definition, every ordered triple of real numbers has dual meanings: it can be a vector in space or a point in space. We find ourselves in a situation very similar to the one with vectors and points in the plane. Therefore a convention in notation is necessary to distinguish the two points of view.

2.2.2 CONVENTION *An ordered triple of real numbers may either represent a point or a vector in space. As a point it will be denoted by a capital letter and the coordinates are enclosed in parentheses:*

$$A = (a_1, a_2, a_3) .$$

As a vector it will be denoted by the lower case bold-faced type of the same letter and the components are enclosed in brackets:

$$\mathbf{a} = [a_1, a_2, a_3] .$$

A vector $\mathbf{a} = [a_1, a_2, a_3]$ is represented graphically by the directed segment from the origin O to the point $A = (a_1, a_2, a_3)$. Conversely the directed segment from O to $B = (b_1, b_2, b_3)$ represents the vector $\mathbf{b} = [b_1, b_2, b_3]$. Under this association between vectors and directed segments with common initial point at O, the vector $\mathbf{a} = [a_1, a_2, a_3]$ is referred to as the *position vector* of the point $A = (a_1, a_2, a_3)$. In particular the *zero vector* $\mathbf{0} = [0, 0, 0]$ is the position vector of the origin O, and the *unit coordinate vectors* $\mathbf{e}_1 = [1, 0, 0]$ $\mathbf{e}_2 = [0, 1, 0]$, $\mathbf{e}_3 = [0, 0, 1]$ are the position vectors of the unit points on the coordinate axes. See Figure 2-5.

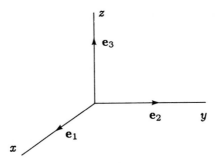

Fig 2-5

The *magnitude* of a vector $\mathbf{a} = [a_1, a_2, a_3]$ is the non-negative real number

$$|\mathbf{a}| = \sqrt{a_1{}^2 + a_2{}^2 + a_3{}^2}$$

which is also the length of the segment OA in space. Similar to Theorem 1.1.5, the zero vector $\mathbf{0} = [0, 0, 0]$ is the one and only vector in space with zero magnitude: $|\mathbf{a}| = 0$ *if and only if* $\mathbf{a} = \mathbf{0}$.

As geometric entities, points in space do not admit direct algebraic operations; such operations can only be performed on their coordinates. In contrast, the same ordered triple, when considered as a vector, becomes an algebraic entity on which algebraic operations can be carried out directly. Indeed the definitions of sum and scalar multiple of vectors in the plane are easily extended.

Let $\mathbf{a} = [a_1, a_2, a_3]$ and $\mathbf{b} = [b_1, b_2, b_3]$ be vectors in space, and r a real number. The *sum* $\mathbf{a} + \mathbf{b}$ of $\mathbf{a}$ and $\mathbf{b}$, and the *scalar multiple* $r\mathbf{a}$ of $\mathbf{a}$ by the scalar r are respectively the vectors

$$\mathbf{a} + \mathbf{b} = [a_1 + b_1, a_2 + b_2, a_3 + b_3] \,,$$

$$r\mathbf{a} = [ra_1, ra_2, ra_3] \,.$$

Like their counterparts in the plane, $\mathbf{a} + \mathbf{b}$ is the positional vector of the vertex C of the parallelogram $OACB$ and $r\mathbf{a}$ is a scaling of $\mathbf{a}$ by the scalar r.

The set of all vectors in space together with the addition and the scalar multiplication constitute an algebraic system called the *vector space* $\mathbf{R}^3$ of vectors of the 3-dimensional space. Elements of $\mathbf{R}^3$ are referred to as vectors and real numbers as scalars. The fundamental properties of the vector space $\mathbf{R}^2$ in the plane listed in 1.2.3 hold also for $\mathbf{R}^3$ without modification.

2.2.3 THEOREM *Let* a, b *and* c *be vectors of the vector space* $\mathbf{R}^3$, *and let* r *and* s *be scalars. Then the following statments hold.*
  (1) $\mathbf{a} + \mathbf{b} = \mathbf{b} + \mathbf{a}$.
  (2) $(\mathbf{a} + \mathbf{b}) + \mathbf{c} = \mathbf{a} + (\mathbf{b} + \mathbf{c})$.
  (3) *There exists a unique vector* **0** *such that* $\mathbf{a} + \mathbf{0} = \mathbf{a}$.
  (4) *For* a *there is a unique vector* $-\mathbf{a}$ *such that* $\mathbf{a} + (-\mathbf{a}) = \mathbf{0}$.
  (5) $(rs)\mathbf{a} = r(s\mathbf{a})$.
  (6) $(r + s)\mathbf{a} = r\mathbf{a} + s\mathbf{b}$.
  (7) $r(\mathbf{a} + \mathbf{b}) = r\mathbf{a} + r\mathbf{b}$.
  (8) $1\mathbf{a} = \mathbf{a}$.

These statements are readily verified in terms of components. For example

$$(rs)\mathbf{a} = [(rs)a_1, (rs)a_2, (rs)a_3]$$
$$= [r(sa_1), r(sa_2), r(sa_3)] = r(s\mathbf{a}) \ .$$

Therefore the associative law (5) of scalar multiplication holds. Many important properties of the vector space $\mathbf{R}^3$ are consequences of these eight fundamental properties. For example the entire second proof of Theorem 1.2.4 which is based on these properties can be used as a proof of the following theorem without any modification.

2.2.4 THEOREM *Let* a *be a vector of the vector space* $\mathbf{R}^3$ *and* r *a scalar. Then* $r\mathbf{a} = \mathbf{0}$ *if and only if* $r = 0$ *or* $\mathbf{a} = \mathbf{0}$.

EXERCISES

  1. Let $\mathbf{a} = [2, 2, -1]$, $\mathbf{b} = [4, 2, 5]$, $\mathbf{c} = [-1, -3, 2]$. Find
     (a) $3(\mathbf{a} + \mathbf{b}) - 2(\mathbf{b} - \mathbf{c})$,
     (b) $5(\mathbf{a} - \mathbf{c}) + 3(\mathbf{c} - \mathbf{a})$,
     (c) $2(\mathbf{a} + \mathbf{b} + \mathbf{c}) + 3(\mathbf{a} - \mathbf{b} - \mathbf{c})$.
  2. Use a, b, and c as in the above question, find x so that
     (a) $\mathbf{a} + 2\mathbf{x} = \mathbf{b} - \mathbf{c} - \mathbf{x}$;

45

(b) $2(\mathbf{a} - \mathbf{b} + \mathbf{x}) = 3(\mathbf{c} - \mathbf{x})$;

(c) $\mathbf{x} + 2(\mathbf{x} + \mathbf{a}) + 3(\mathbf{x} - \mathbf{b}) = \mathbf{0}$.

3. In general, we may define the subtraction of a vector **b** from a vector **a** by

$$\mathbf{a} - \mathbf{b} \equiv \mathbf{a} + (-\mathbf{b})$$

where $-\mathbf{b}$ is as in (4) of Theorem 2.2.3. Use this definition and Theorem 2.2.3 to prove that

(a) $\mathbf{a} - (\mathbf{b} + \mathbf{c}) = (\mathbf{a} - \mathbf{b}) - \mathbf{c}$;

(b) $r(\mathbf{a} - \mathbf{b}) = r\mathbf{a} - r\mathbf{b}$, for real number r.

4. Given two non-zero vectors **a** and **b**. By drawing suitable diagrams, find a necessary and sufficient condition for each of the following equalities to hold.

(a) $|\mathbf{a}| + |\mathbf{b}| = |\mathbf{a} + \mathbf{b}|$,

(b) $|\mathbf{a}| - |\mathbf{b}| = |\mathbf{a} + \mathbf{b}|$,

(c) $|\mathbf{b}| - |\mathbf{a}| = |\mathbf{a} - \mathbf{b}|$.

(Hint for (c): $\mathbf{a} - \mathbf{b} = \mathbf{a} + (-\mathbf{b})$.)

2.3 Linear independence

Given vectors $\mathbf{a}, \mathbf{b}, \mathbf{c}, \cdots$ of the vector space $\mathbf{R}^3$ and scalars $r, s, t, \cdots$, the vector

$$r\mathbf{a} + s\mathbf{b} + t\mathbf{c} + \cdots$$

is called a *linear combination* of the vectors $\mathbf{a}, \mathbf{b}, \mathbf{c}, \cdots$. Thus the linear combinations of a single vector **a** are all vectors of the form $r\mathbf{a}$; the linear combinations of two vectors **a** and **b** are vectors of the form $r\mathbf{a} + s\mathbf{b}$; and linear combinations of three vectors $\mathbf{a}, \mathbf{b}$ and **c** are all vectors of the form $r\mathbf{a} + s\mathbf{b} + t\mathbf{c}$, and so forth.

In section 1.3 the study of linear combinations of plane vectors led us to the important concepts of linear dependence and linear independence. For the vectors of $\mathbf{R}^3$ the same defintion applies.

2.3.1 DEFINITION *Let* **a** *and* **b** *be vectors of the vector space* $\mathbf{R}^3$. *Vectors* **a** *and* **b** *are linearly independent if* **a** *is not a scalar multiple of* **b** *and* **b** *is not a scalar multiple of* **a**. *Vectors* **a** *and* **b** *are linearly dependent if they are not linearly independent.*

For the linear dependence of two vectors of $\mathbf{R}^3$ we also have four equivalent conditions which are very similar to those in Theorem 1.3.4.

2.3.2 THEOREM *Let* $\mathbf{a} = [a_1, a_2, a_3]$ *and* $\mathbf{b} = [b_1, b_2, b_3]$ *be vectors of the vector space* $\mathbf{R}^3$. *Then the following statements are equivalent.*

(1) $\mathbf{a}$ *is a scalar multiple of* $\mathbf{b}$ *or* $\mathbf{b}$ *is a scalar multiple of* $\mathbf{a}$.

(2) *The points* $O = (0, 0, 0), A = (a_1, a_2, a_3)$ *and* $B = (b_1, b_2, b_3)$ *are collinear.*

(3) *There exist scalars* r *and* s, *not both being zero, such that* $r\mathbf{a} + s\mathbf{b} = 0$.

(4) $a_1 b_2 - a_2 b_1 = a_2 b_3 - a_3 b_2 = a_3 b_1 - a_1 b_3 = 0$.

PROOF For the equivalence of the first three conditions we may apply the same arguments that were used in section 1.3. We shall leave the details of this to the student as an exercise and proceed to prove the equivalence of conditions (1) and (4).

Let condition (1) be satisfied by $\mathbf{a}$ and $\mathbf{b}$. Without loss of generality we assume that $\mathbf{a} = r\mathbf{b}$ since the same argument can be used if $\mathbf{b} = s\mathbf{a}$. Then $a_1 = rb_1$, $a_2 = rb_2$ and $a_3 = rb_3$. Therefore $a_1 b_2 - a_2 b_1 = rb_1 b_2 - rb_2 b_1 = 0$. Similarly $a_2 b_3 - a_3 b_2 = 0$ and $a_3 b_1 - a_1 b_3 = 0$. Therefore (1) implies (4).

Conversely let condition (4) be satisfied. If $\mathbf{a} = 0$ or $\mathbf{b} = 0$ then $\mathbf{a} = 0\mathbf{b}$ or $\mathbf{b} = 0\mathbf{a}$; therefore (1) is satisfied. Otherwise suppose without loss of generality that $a_1 \neq 0$. Then it follows from $a_1 b_2 - a_2 b_1 = 0$ and $a_3 b_1 - a_1 b_3 = 0$ that $b_2 = (b_1/a_1)a_2$ and $b_3 = (b_1/a_1)a_3$. Therefore

$$\mathbf{b} = [b_1, b_2, b_3] = \left[\left(\frac{b_1}{a_1}\right)a_1, \left(\frac{b_1}{a_1}\right)a_2, \left(\frac{b_1}{a_1}\right)a_3 \right] = \left(\frac{b_1}{a_1}\right)\mathbf{a} ,$$

and condition (1) is also satisfied.

Converting each of the four conditions to its negation, we obtain four equivalent conditions for the linear independence of $\mathbf{a}$ and $\mathbf{b}$.

2.3.3 THEOREM *Let* $\mathbf{a} = [a_1, a_2, a_3]$ *and* $\mathbf{b} = [b_1, b_2, b_3]$ *be vectors of the vector space* $\mathbf{R}^3$. *Then the following statements are equivalent.*

(1) $\mathbf{a}$ *is not a scalar multiple of* $\mathbf{b}$ *and* $\mathbf{b}$ *is not a scalar multiple of* $\mathbf{a}$.

(2) *The points* $O = (0, 0, 0), A = (a_1, a_2, a_3)$ *and* $B = (b_1, b_2, b_3)$ *are distinct and not collinear.*

(3) *If* $r\mathbf{a} + s\mathbf{b} = 0$ *for scalars* r *and* s *then* $r = s = 0$.

(4) $a_1b_2 - a_2b_1 \neq 0$, $a_2b_3 - a_3b_2 \neq 0$ or $a_3b_1 - a_1b_3 \neq 0$.

The two theorems 2.3.2 and 2.3.3 give complete information on the linear dependence and the linear independence of two vectors. In each of them condition (1) is the definition, (2) is a geometric characterization, (3) is given in terms of linear combination and (4) is a condition on the components of the vectors.

Let us now look at the case of three vectors. First of all the definition.

2.3.4 DEFINITION *Three vectors in space are linearly dependent if any one of them is a linear combination of the other two. They are linearly independent otherwise. In other words, three vectors in space are linearly independent if none of them is a linear combination of the other two.*

For example the unit coordinate vectors $e_1 = [1, 0, 0]$, $e_2 = [0, 1, 0]$, $e_3 = [0, 0, 1]$ are linearly independent.

We note that this definition is a natural extension of definition 2.3.1 which can be rephrased as follows. Two vectors are linearly dependent if one of them is a linear combination of the other one. The formulation of the equivalent conditions for linear dependence is our next task. Clearly the condition in terms of linear combination should read as follows. *There exist scalars r, s and t, not all zero, such that $r\mathbf{a} + s\mathbf{b} + t\mathbf{c} = 0$.* Let us prove that this is a necessary and sufficient condition for $\mathbf{a}, \mathbf{b}$ and $\mathbf{c}$ to be linearly dependent.

Suppose that the three vectors are linearly dependent. Then one of them say $\mathbf{c}$ is a linear combination of the other two $\mathbf{a}$ and $\mathbf{b}$. Therefore for some scalar r and s, we have $\mathbf{c} = r\mathbf{a} + s\mathbf{b}$. Then $r\mathbf{a} + s\mathbf{b} + (-1)\mathbf{c} = 0$ with $t = -1$ being non-zero. Therefore the above condition is satisfied.

Conversely supposed that the condition is satisfied by $\mathbf{a}, \mathbf{b}$ and $\mathbf{c}$. Then we have $r\mathbf{a} + s\mathbf{b} + t\mathbf{c} = 0$ for some scalars r, s and t which are not all zero at the same time. Without loss of generality, assume that $t \neq 0$. Then $\mathbf{c} = (-r/t)\mathbf{a} + (-s/t)\mathbf{b}$ is a linear combination of $\mathbf{a}$ and $\mathbf{b}$. Similarly if $s \neq 0$ or $r \neq 0$, then $\mathbf{b}$ is a linear combination of $\mathbf{a}$ and $\mathbf{c}$, or $\mathbf{a}$ is a linear combination of $\mathbf{b}$ and $\mathbf{c}$. Hence $\mathbf{a}, \mathbf{b}$ and $\mathbf{c}$ are linearly dependent.

The geometric characterization of three linearly dependent vectors $\mathbf{a}, \mathbf{b}$ and $\mathbf{c}$ is also easy to write down: *the four points O, A, B and C are coplanar*, i.e. they lie on one plane in space.

Indeed if $\mathbf{a}, \mathbf{b}$ and $\mathbf{c}$ are linearly dependent, say $\mathbf{c} = r\mathbf{a} + s\mathbf{b}$ for some r and s, and if A' and B' are points in space with position vectors $r\mathbf{a}$ and $r\mathbf{b}$ respectively, then $OA'CB'$ is a parallelogram which may be degenerate or non-degenerate. In any case the four points O, A', B' and C lie on a plane. On the other hand with A on the line OA' of the plane and B on the line OB' of the plane, the four points O, A, B, C are also on the same plane. Therefore the four points are coplanar.

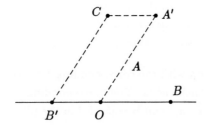

Fig 2-6

Conversely suppose that the four points O, A, B and C have the said geometric characterization of being coplanar. Then on the plane that contains these four points we may either find O, A and B collinear or OAB being a non-degenerate triangle. In the former case $\mathbf{a}$ and $\mathbf{b}$ are linearly dependent, say $\mathbf{a} = s\mathbf{b}$. Then $\mathbf{a} = s\mathbf{b} + 0\mathbf{c}$, hence the vectors $\mathbf{a}, \mathbf{b}$ and $\mathbf{c}$ are linearly dependent. In the latter case through C we draw a line parallel to OB cutting OA at A'; similiarly through C we draw a line parallel to OA cutting OB at B'. Then $OA'CB'$ is a parallelogram and $\mathbf{c} = \mathbf{a}' + \mathbf{b}'$. But O, A, A' being collinear, $\mathbf{a}' = r\mathbf{a}$ for some scalar r, and O, B, B' being collinear, $\mathbf{b}' = s\mathbf{b}$ for some scalar s, therefore $\mathbf{c} = r\mathbf{a} + s\mathbf{b}$, proving that $\mathbf{a}, \mathbf{b}$ and $\mathbf{c}$ are linearly dependent.

We have therefore proved the following two theorems:

2.3.5 THEOREM *Let $\mathbf{a} = [a_1, a_2, a_3]$, $\mathbf{b} = [b_1, b_2, b_3]$ and $\mathbf{c} = [c_1, c_2, c_3]$ be three vectors of the vector space $\mathbf{R}^3$. Then the following statements are equivalent.*

(1) *One of the vectors* **a**, **b** *and* **c** *is a linear combination of the other two vectors.*

(2) *The four points* $O = (0, 0, 0)$, $A = (a_1, a_2, a_3)$, $B = (b_1, b_2, b_3)$ *and* $C = (c_1, c_2, c_3)$ *are coplanar.*

(3) *There exist scalars* r, s *and* t, *not all being zero, such that* $r\mathbf{a} + s\mathbf{b} + t\mathbf{c} = \mathbf{0}$.

2.3.6 THEOREM Let $\mathbf{a} = [a_1, a_2, a_3]$, $\mathbf{b} = [b_1, b_2, b_3]$ *and* $\mathbf{c} = [c_1, c_2, c_3]$ *be three vectors of the vector space* $\mathbf{R}^3$. *Then the following statements are equivalent.*

(1) *None of the vectors* **a**, **b** *and* **c** *is a linear combination of the other two vectors.*

(2) *The four points* $O = (0, 0, 0)$, $A = (a_1, a_2, a_3)$, $B = (b_1, b_2, b_3)$ *and* $C = (c_1, c_2, c_3)$ *are distinct and not coplanar.*

(3) *If* $r\mathbf{a} + s\mathbf{b} + t\mathbf{c} = \mathbf{0}$ *for some scalar* r, s *and* t, *then* $r = s = t = 0$.

2.3.7 REMARKS In comparison with the corresponding 1.3.4 the absence of a condition (4) in 2.3.5 is conspicuous. The corresponding condition for the linear dependence of $\mathbf{a} = [a_1, a_2]$ and $\mathbf{b} = [b_1, b_2]$ in the plane was given in 1.3.4 as

$$a_1 b_2 - a_2 b_1 = 0 \quad \text{or} \quad \begin{vmatrix} a_1 & b_2 \\ a_1 & b_2 \end{vmatrix} = 0 .$$

A straightforward generalization of this would read: for $\mathbf{a} = [a_1, a_2, a_3]$, $\mathbf{b} = [b_1, b_2, b_3]$ and $\mathbf{c} = [c_1, c_2, c_3]$ to be linearly dependent, it is necessary and sufficient that

$$a_1 b_2 c_3 + a_2 b_3 c_1 + a_3 b_1 c_2 - a_1 b_3 c_2 - a_2 b_1 c_3 - a_3 b_2 c_1 = 0$$

or

$$\begin{vmatrix} a_1 & b_1 & c_1 \\ a_2 & b_2 & c_2 \\ a_3 & b_3 & c_3 \end{vmatrix} = 0 .$$

It is an easy exercise to see that this is a necessary condition. However to prove its sufficiency we would need some working knowledge of determinant. It is for this reason that a condition (4) is omitted in Theorems 2.3.5 and 2.3.6, and this short coming will be remedied in Chapter Six.

It is worthwhile to pick out an argument which is used implicitly in some of the previous theorems and formulate it as theorems for future reference.

2.3.8 THEOREM *Let* a, b *and* c *be three vectors. If any one of them is the zero vector or any two of them are linearly dependent, then the three vectors are linearly dependent.*

2.3.9 THEOREM *Let* a, b, *and* c *be three vectors. If they are linearly independent, then they are distinct non-zero vectors and any two of them are linearly independent.*

With these two theorems we conclude our investigation on three vectors, and it would seem natural for us to turn our attention now to the case of four or more vectors. As before we would initiate the investigation with a defintion of linear dependence which clearly should read as follows.

2.3.10 DEFINITION *Let* n *be a natural number greater than one. Then* n *vectors are said to be linearly dependent if any one of them is a linear combination of the other* n − 1 *vectors. Otherwise they are linearly independent.*

However it turns out that any four vectors of $\mathbf{R}^3$ are linearly dependent, and hence by an easy generalization of Theorem 2.3.8 any five or more vectors $\mathbf{R}^3$ are also linearly dependent. Indeed if a, b, c and d are vectors of $\mathbf{R}^3$, then the equation

$$r\mathbf{a} + s\mathbf{b} + t\mathbf{c} + u\mathbf{d} = \mathbf{0}$$

can be written in terms of components into

$$a_1 r + b_1 s + c_1 t + d_1 u = 0$$
$$a_2 r + b_2 s + c_2 t + d_2 u = 0$$
$$a_3 r + b_3 s + c_3 t + d_3 u = 0$$

which is a system of three homogeneous linear equations in the four unknowns r, s, t and u. Having more unknowns than equations, this system admits non-trivial solutions. In other words there are scalars r, s, t and u, not all zero, that satisfy the three equations. Therefore $r\mathbf{a} + s\mathbf{b} + t\mathbf{c} + u\mathbf{d} = \mathbf{0}$ where at least one of the four scalars r, s, t and u is non-zero, which means that one of the vectors is a linear combination of the other three. Hence the four vectors are linearly

51

dependent. While any four or more vectors of $\mathbf{R}^3$ are linearly dependent, there are three vectors in $\mathbf{R}^3$ that are linearly independent, e.g. the unit coordinate vectors e_1, e_2 and e_3. Therefore the natural number 3 and the vector space $\mathbf{R}^3$ have a very special relationship: 3 *is the maximum number of linearly independent vectors in* $\mathbf{R}^3$. This fact is usually expressed as: the vector space $\mathbf{R}^3$ has *dimension* 3. Similarly, 2 *is the maximum number of linearly independent vectors in* $\mathbf{R}^2$, i.e. the vector space $\mathbf{R}^2$ has *dimension* 2.

EXERCISES

1. Find the values of a and b such that $[a, -8, 6]$ is a linear combination of $[6, b, 3]$ and $[5, 4, 0]$.

2. Show that each of the following pairs of vectors are linearly independent.
   (a) $[1, 2, 3]$, $[0, 2, 1]$;
   (b) $[-3, 2, 1]$, $[5, 7, 4]$.

3. Prove that the first three conditions of Theorem 2.3.2 are equivalent.

4. Prove or give a counter-example for each of the following statements.
   (a) If points O, A, B are collinear and points O, B, C are collinear, then O, A, C are collinear.
   (b) If points O, A, B, C are coplanar and points O, C, D, E are coplanar, then O, A, C, E are coplanar.

5. Prove that for vectors $\mathbf{a} = [a_1, a_2, a_3]$, $\mathbf{b} = [b_1, b_2, b_3]$ and $\mathbf{c} = [c_1, c_2, c_3]$ to be linearly dependent, it is necessary that $a_1 b_2 c_3 + a_2 b_3 c_1 + a_3 b_1 c_2 - a_1 b_3 c_2 - a_2 b_1 c_3 - a_3 b_2 c_1 = 0$.

6. Prove Theorem 2.3.8.

7. Show that for any three vectors $\mathbf{x}$, $\mathbf{y}$, $\mathbf{z}$ of $\mathbf{R}^3$ and any scalars r, s, t, the three vectors $r\mathbf{x} - s\mathbf{y}$, $t\mathbf{y} - r\mathbf{z}$, $s\mathbf{z} - t\mathbf{x}$ are linearly dependent.

8. Given three linearly independent vectors $\mathbf{x}$, $\mathbf{y}$, $\mathbf{z}$ we construct the vectors $\mathbf{a} = 2\mathbf{x} + \mathbf{y}$, $\mathbf{b} = \mathbf{y} - \mathbf{z}$ and $\mathbf{c} = \mathbf{x} + \mathbf{y} + \mathbf{z}$. Express the vector $\mathbf{d} = \mathbf{x} + 2\mathbf{y} + \mathbf{z}$ as a linear combination of $\mathbf{a}$, $\mathbf{b}$ and $\mathbf{c}$.

2.4 Geometry in space

In this section we shall use the one-to-one correspondence between points in space and their positional vectors to study geometry in space. Like in plane geometry we shall find it convenient to use displacement vectors in this kind of work. Thus if $A = (a_1, a_2, a_3)$ and $B = (b_1, b_2, b_3)$ are two points in space and $\mathbf{a} = [a_1, a_2, a_3]$ and $\mathbf{b} = [b_1, b_2, b_3]$ their position vectors respectively, then we call the vector $\overrightarrow{AB} = \mathbf{b} - \mathbf{a} = [b_1 - a_1, b_2 - a_2, b_3 - a_3]$ the *displacement vector* of the directed segment AB. In this way every directed segment has a unique displacement vector but different directed segments may have the same displacement vector.

It is easy to see that Examples and Theorems 1.4.2 – 1.4.9 of plane geometry all hold in space geometry. Similarly the straight line L which passes through two distinct points A and B in space has a parametric representation of its points X by their position vectors

$$\mathbf{x} = (1 - t)\mathbf{a} + t\mathbf{b} \ .$$

Here again the position of the point X and the value of the parameter t have the same relationship as in plane geometry.

Clearly the same proof given for Theorem 1.4.11 can be used as a proof of the fact that the three medians of a triangle in space meet at the centroid of the triangle.

A non-degenerate triangle is a geometric configuration of three non-collinear points. It is sometimes called a *2-dimensional simplex*. A *3-dimensional simplex* is a tetrahedron. A *tetrahedron* in space is a geometric configuration of four non-coplanar points, say A, B, C and D. The points A, B, C and D are called the *vertices* of the tetrahedron $ABCD$. The four triangles ABC, BCD, CDA and DAB are the *faces* of the tetrahedron. The six line segments AB, AC, AD, BC, CD and BD are the *edges* of the tetrahedron. Two edges are said to be *opposite edges* if they do not have a vertex in common; for example AB and CD are opposite edges while CD and DA are not; similarly a vertex and a face are *opposite* if the vertex is not on the face; for example the opposite face of the vertex A is the face BCD. A line joining a vertex with the centroid of its opposite face is called a *median* of the tetrahedron. A line joining the midpoints of two opposite edges is called a *bimedian* of the tetrahedron. See Figure 2-7. We are now ready for the 3-dimensional counterpart of Theorem 1.4.11.

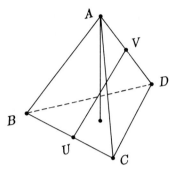

Fig 2-7

2.4.1 THEOREM *Let ABCD be a tetrahedron. Then the three bimedians and the four medians meet at a point.*

PROOF Consider the bimedian UV of the opposite edges BC and AD. With $\mathbf{u} = \frac{1}{2}(\mathbf{b} + \mathbf{c})$ and $\mathbf{v} = \frac{1}{2}(\mathbf{a} + \mathbf{d})$ we find that the vector

$$\mathbf{m} = \frac{1}{4}(\mathbf{a} + \mathbf{b} + \mathbf{c} + \mathbf{d}) = \frac{1}{2}\left\{\frac{1}{2}(\mathbf{b} + \mathbf{c}) + \frac{1}{2}(\mathbf{a} + \mathbf{d})\right\}$$

is the position vector of the midpoint M of the segment UV. Clearly M is also the midpoint of the bimedian of the opposite edges AB and CD as well as the midpoint of the bimedian of the opposite edges AC and BD, because

$$\mathbf{m} = \frac{1}{2}\left\{\frac{1}{2}(\mathbf{a} + \mathbf{b}) + \frac{1}{2}(\mathbf{c} + \mathbf{d})\right\}$$
$$= \frac{1}{2}\left\{\frac{1}{2}(\mathbf{a} + \mathbf{c}) + \frac{1}{2}(\mathbf{b} + \mathbf{d})\right\} .$$

Therefore the three bimedians of the tetrahedron $ABCD$ meet at the point M which is called the *centroid* of the tetrahedron. It remains to prove that M lies on all four medians of the tetrahedron. Let E be the centroid of the face BCD. Then

$$\mathbf{e} = \frac{1}{3}(\mathbf{b} + \mathbf{c} + \mathbf{d}) .$$

Writing $\mathbf{m}$ as a linear combination of a and e, we get

$$\mathbf{m} = \frac{1}{4}\mathbf{a} + \frac{3}{4}\mathbf{e} .$$

Since the scalars of the linear combination satisfy $\frac{1}{4} + \frac{3}{4} = 1$, we see that M lies on the median AE. Similarly M lies on all the other three medians. Therefore the seven lines meet at the centroid M of the tetradhedron $ABCD$. The proof is complete.

Through two distinct points A and B there passes one and only one straight line L. Points X on the line L are represented by their position vectors

$$\mathbf{x} = r\mathbf{a} + s\mathbf{b} \quad \text{with} \quad r + s = 1 .$$

Similarly through three non-collinear points A, B and C in space there passes one and only one plane E. Points X on the plane E are represented by their position vectors

$$\mathbf{x} = r\mathbf{a} + s\mathbf{b} + t\mathbf{c} \quad \text{with} \quad r + s + t = 1 .$$

Indeed a point X in space lies on the plane E if and only if the displacement vector $\overrightarrow{AX}$ is a linear combination of the displacement vectors $\overrightarrow{AB}$ and $\overrightarrow{AC}$:

$$\mathbf{x} - \mathbf{a} = s(\mathbf{b} - \mathbf{a}) + t(\mathbf{c} - \mathbf{a})$$

or

$$\mathbf{x} = (1 - s - t)\mathbf{a} + s\mathbf{b} + t\mathbf{c} .$$

Putting $r = 1 - s - t$ we have $\mathbf{x} = r\mathbf{a} + s\mathbf{b} + t\mathbf{c}$ with $r + s + t = 1$.

2.4.2 THEOREM *Points on the plane which passes through three non-collinear points A, B and C are represented by their position vectors in the form*

$$\mathbf{x} = r\mathbf{a} + s\mathbf{b} + t\mathbf{c} \quad \text{with} \quad r + s + t = 1 .$$

There is a well-known physical interpretation of the above theorem. If weights of r, s, t $(r + s + t = 1)$ are attached to points A, B, C respectively, then the point X would be the centre of gravity of the system. Therefore the triple r, s, t, of real numbers are sometimes called the *barycentric coordinates* of the point X relative to the triple A, B, C of non-collinear points.

As an application of this result we prove a classical theorem of geometry due to Giovanni Ceva in the 17th century.

2.4.3 CEVA'S THEOREM *Let ABC be a triangle and F, G, H points on the sides BC, CA, AB respectively. Consider the directed ratios $\lambda = dr(B,C;F)$, $\mu = dr(C,A;G)$, $\nu = dr(A,B;H)$, i.e. $\overrightarrow{BF} = \lambda\overrightarrow{FC}$, $\overrightarrow{CG} = \mu\overrightarrow{GA}$, $\overrightarrow{AH} = \nu\overrightarrow{HB}$. Then the three lines AF, BG, CH are concurrent if and only if $\lambda\mu\nu = 1$.*

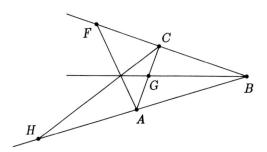

Fig 2-8

PROOF Every point X on the plane of the triangle ABC has a position vector of the form

$$\mathbf{x} = \frac{1}{r+s+t}(r\mathbf{a} + s\mathbf{b} + t\mathbf{c}) .$$

Let U be the point on the side BC with position vector

$$\mathbf{u} = \frac{1}{s+t}(s\mathbf{b} + t\mathbf{c}) .$$

Then X can be considered as a point on the line passing through A and U because

$$\mathbf{x} = \frac{1}{r+(s+t)}(r\mathbf{a} + (s+t)\mathbf{u}) .$$

On the other hand

$$\overrightarrow{BU} = \mathbf{u} - \mathbf{b} = \frac{-t}{s+t}(\mathbf{b} - \mathbf{c})$$

$$\overrightarrow{UC} = \mathbf{c} - \mathbf{u} = \frac{-s}{s+t}(\mathbf{b} - \mathbf{c}) .$$

It follows from $\overrightarrow{BU} = (t/s)\overrightarrow{UC}$ that $dr(B,C;U) = t/s$. Now if the parameters s and t of the point U are so chosen that $t/s = \lambda$, then $dr(B,C;U) = dr(B,C;F) = \lambda$ and hence $U = F$. Therefore points on the line passing through A and F are represented by

$$\mathbf{x} = \frac{1}{r+s+t}(r\mathbf{a} + s\mathbf{b} + t\mathbf{c}) \quad \text{with} \quad t/s = \lambda .$$

Similarly points on the line passing through B and G are represented by

$$\mathbf{x} = \frac{1}{r+s+t}(r\mathbf{a} + s\mathbf{b} + t\mathbf{c}) \quad \text{with} \quad r/t = \mu ,$$

and those of CH by

$$\mathbf{x} = \frac{1}{r+s+t}(r\mathbf{a} + s\mathbf{b} + t\mathbf{e}) \quad \text{with} \quad s/r = \nu .$$

Consider now the point D with position vector

$$\mathbf{d} = \frac{1}{\mu + (1/\lambda) + 1}\left(\mu\mathbf{a} + \frac{1}{\lambda}\mathbf{b} + \mathbf{c}\right) .$$

Because $1/(1/\lambda) = \lambda$, D lies on AF. Similarly D lies on BG because $\mu/1 = \mu$. Hence D is the point of intersection of the lines AF and BG. Therefore the three lines AF, BG and CH are concurrent if and only if the point D also lies on the line CH. By the parametric representation of the line CH, this is the case if and only if

$$(1/\lambda)/\mu = \nu \quad i.e. \quad \lambda\mu\nu = 1 .$$

The proof is now complete.

We note that if $\lambda = \mu = \nu = 1$, then AF, BG, CH are the medians of the triangle ABC. Therefore the concurrency of the medians is a special case of Ceva's theorem.

EXERCISES

1. Consider $\triangle ABC$ in space. D, E, and F are points on AB, BC and CA such that $\frac{AD}{DB} = \frac{BE}{EC} = \frac{CF}{FA}$, i.e. $dr(A, B; D) = dr(B, C; E) = dr(C, A; F)$. Show that $\triangle DEF$ has the same centroid as $\triangle ABC$.

2. Consider a quadrilateral $ABCD$ in space. E, F, G and H are points on AB, AC, DB and DC such that $\frac{AE}{EB} = \frac{AF}{FC} = \frac{DG}{GB} = \frac{DH}{HC}$. Show that $EFGH$ is a parallelogram.

3. Given points A, B and C with position vectors $\mathbf{a}$, $\mathbf{b}$ and $\mathbf{c}$. Suppose that the points O, A and B are not collinear and $\mathbf{c} = s\mathbf{a} + t\mathbf{b}$, prove that the points A, B and C are collinear if and only if $s + t = 1$.

4. Given points A, B and C with position vectors $\mathbf{a}$, $\mathbf{b}$ and $\mathbf{c}$. Show that a necessary and sufficient condition for the points A, B and C to be

collinear is that there exist three scalars r, s and t, not all zero, such that $r\mathbf{a} + s\mathbf{b} + t\mathbf{c} = \mathbf{0}$ and $r + s + t = 0$.

5. Given points A, B and C which are not collinear with position vectors $\mathbf{a}$, $\mathbf{b}$ and $\mathbf{c}$. Points P_1, P_2 and P_3 have respectively position vectors

$$
\begin{aligned}
r_1\mathbf{a} + m_1\mathbf{b} \quad &(r_1 + m_1 = 1) , \\
r_2\mathbf{b} + m_2\mathbf{c} \quad &(r_2 + m_2 = 1) , \\
r_3\mathbf{c} + m_3\mathbf{a} \quad &(r_3 + m_3 = 1) ,
\end{aligned}
$$

where all r_i's and m_i's are non-zero. Prove that P_1, P_2 and P_3 are collinear if and only if $m_1 m_2 m_3 = -r_1 r_2 r_3$.

6. A, B, C and D are four points in space. A' is the centroid of B, C, D; B' is the centroid of C, D, A; C' is the centroid of D, A, B; and D' is the centroid of A, B, C. Prove that if $A \neq A'$, $B \neq B'$, $C \neq C'$, and $D \neq D'$, then AA', BB', CC' and DD' all meet at the centroid G of A, B, C and D. Prove also that G is the centroid of A', B', C' and D'.

7. As an extension to Question 1, we consider four different points A, B, C and D in space. E, F, G and H are points on AB, BC, CD and DA respectively, such that $\frac{AE}{EB} = \frac{BF}{FC} = \frac{CG}{GD} = \frac{DH}{HA}$. Show that the centroids of A, B, C and D, and of E, F, G and H coincide.

2.5 The dot product

In the vector space $\mathbf{R}^3$ measurements of length and angle may be defined through the notion of dot product.

2.5.1 DEFINITION. *Let* $\mathbf{a} = [a_1, a_2, a_3]$ *and* $\mathbf{b} = [b_1, b_2, b_3]$ *be vectors of the vector space* $\mathbf{R}^3$. *The dot product of* $\mathbf{a}$ *and* $\mathbf{b}$ *is the scalar*

$$
\mathbf{a} \cdot \mathbf{b} = a_1 b_1 + a_2 b_2 + a_3 b_3 .
$$

For example if $\mathbf{a} = [1, 2, -1]$, $\mathbf{b} = [-2, 1, 0]$, $\mathbf{c} = [3, -2, 1]$, then

$$
\mathbf{a} \cdot \mathbf{b} = \mathbf{b} \cdot \mathbf{a} = 0 \; ; \; \mathbf{a} \cdot \mathbf{c} = \mathbf{c} \cdot \mathbf{a} = -2 \; ; \; \mathbf{b} \cdot \mathbf{c} = \mathbf{c} \cdot \mathbf{b} = -8 .
$$

We also notice that the dot product in $\mathbf{R}^3$ is a natural extention of the dot product in $\mathbf{R}^2$ by one more summand, $a_3 b_3$. Moreover it also behaves somewhat like the ordinary product of numbers and have the same fundamental properties as the dot product in $\mathbf{R}^2$ such as *symmetry*, *bilinearity* and *positive definiteness*.

2.5.2 THEOREM *For any vectors* **a**, **b** *and* **c** *of the vector space* $\mathbf{R}^3$ *and any scalar* r, *the following statements hold.*

(1) $\mathbf{a} \cdot \mathbf{b} = \mathbf{b} \cdot \mathbf{a}$.

(2) $(r\mathbf{a}) \cdot \mathbf{b} = r(\mathbf{a} \cdot \mathbf{b})$.

(3) $(\mathbf{a} + \mathbf{b}) \cdot \mathbf{c} = \mathbf{a} \cdot \mathbf{c} + \mathbf{b} \cdot \mathbf{c}$.

(4) $\mathbf{a} \cdot \mathbf{a} \geq 0$, *and* $\mathbf{a} \cdot \mathbf{a} = 0$ *if and only if* $\mathbf{a} = \mathbf{0}$.

Given a vector $\mathbf{a} = [a_1, a_2, a_3]$ of $\mathbf{R}^3$ we find that the dot product of **a** with itself

$$\mathbf{a} \cdot \mathbf{a} = {a_1}^2 + {a_2}^2 + {a_3}^2$$

is identitical to the square of the magnitude of **a**:

$$|\mathbf{a}|^2 = {a_1}^2 + {a_2}^2 + {a_3}^2 \ .$$

Therefore the magnitude can be expressed in terms of dot product: $|\mathbf{a}| = \sqrt{\mathbf{a} \cdot \mathbf{a}}$. On the other hand on the plane in space that contains the point $A = (a_1, a_2, a_3)$ and the z-axis we have a right-angled triangle OAA' with $A' = (a_1, a_2, 0)$. By the Pythogorean theorem

$$|OA|^2 = |OA'|^2 + |A'A|^2 = ({a_1}^2 + {a_2}^2) + {a_3}^2 \ .$$

Therefore the length of the segment OA is

$$|OA| = |\mathbf{a}| \ .$$

Similarly the length of the segment AB is

$$|AB| = \sqrt{(b_1 - a_1)^2 + (b_2 - a_2)^2 + (b_3 - a_3)^2} = |\mathbf{b} - \mathbf{a}| \ .$$

The magnitude of vector or equivalently the length of segment have the fundamental properties listed in the theorem below.

2.5.3 THEOREM *Let* **a** *and* **b** *be vectors of* $\mathbf{R}^3$ *and* r *a scalar. Then the following statements hold.*

(1) $|r\mathbf{a}| = |r||\mathbf{a}|$.

(2) $|\mathbf{a} + \mathbf{b}|^2 + |\mathbf{a} - \mathbf{b}|^2 = 2(|\mathbf{a}|^2 + |\mathbf{b}|^2)$.

(3) $|a||b| \geq |a \cdot b|$.

(4) $|a + b| \leq |a| + |b|$.

(5) $|a - b|^2 = |a|^2 + |b|^2 - 2(a \cdot b)$.

All five statements can be proved in exactly the same way as their 2-dimensional counterparts in Theorem 1.5.4. Properties (2) – (5) are called respectively the *parallelogram law*, the *Schwarz inequality*, the *triangle inequality* and the *cosine law*.

By the Schwarz inequality (3) above, there is a unique value θ, such that

$$\cos \theta = \frac{a \cdot b}{|a||b|} \quad \text{and} \quad 0 \leq \theta \leq \pi .$$

On the other hand, when applied to the triangle OAB in space, the ordinary law of consine of trigonometry yields

$$|BA|^2 = |OA|^2 + |OB|^2 - 2|OA||OB| \cos \angle AOB .$$

Noting that $|BA| = |b - a|$, $|OA| = |a|$ and $|OB| = |b|$, we conclude by (5) above that

$$\cos \angle AOB = a \cdot b / |a||b| \quad \text{and hence} \quad \theta = \angle AOB$$

which justify the following definition.

2.5.4 DEFINITION. *Given two non-zero vectors* a *and* b *of* $\mathbf{R}^3$, *the angle* θ *between the vectors* a *and* b *is defined by*

$$\cos \theta = a \cdot b / |a||b| \quad \text{and} \quad 0 \leq \theta \leq \pi .$$

Moreover the vectors a *and* b *are said to be perpendicular or orthogonal if* $a \cdot b = 0$. *For orthogonal vectors* a *and* b *we sometimes write* $a \perp b$.

Let $a = [a_1, a_2, a_3]$ be a vector in space and denote as before the mutually orthogonal unit coordinate vectors by e_1, e_2 and e_3. Then it follows from $a_i = a \cdot e_i$ for $i = 1, 2, 3$ that

$$a = (a \cdot e_1)e_1 + (a \cdot e_2)e_2 + (a \cdot e_3)e_3 ,$$

which is a resolution of the vector a into a sum of three mutually perpendicular summands.

Putting $f = (a \cdot e_2)e_2 + (a \cdot e_3) \cdot e_3$, we get

$$a = (a \cdot e_1)e_1 + f$$

where the two summands are orthogonal. This allows us to call the vector $(\mathbf{a} \cdot \mathbf{e}_1)\mathbf{e}_1$ the projection of $\mathbf{a}$ on the line along $\mathbf{e}_1$ or the projection on the x-axis, and write

$$pr_1\mathbf{a} = (\mathbf{a} \cdot \mathbf{e}_1)\mathbf{e}_1 \ .$$

Similarly the projections of $\mathbf{a}$ on the y-axis and on the z-axis are respectively the vectors

$$pr_2\mathbf{a} = (\mathbf{a} \cdot \mathbf{e}_2)\mathbf{e}_2 \quad \text{and} \quad pr_3\mathbf{a} = (\mathbf{a} \cdot \mathbf{e}_3)\mathbf{e}_3 \ .$$

More generally we have a theorem on vectors in space analogous to Theorem 1.5.7.

2.5.5 THEOREM AND DEFINITION   *Let* $\mathbf{a}$ *be a non-zero vector in space. Then for any vector* $\mathbf{b}$ *in space there are a unique scalar* $t = (\mathbf{b} \cdot \mathbf{a})/(\mathbf{a} \cdot \mathbf{a})$ *and a unique vector* $\mathbf{c}$ *such that*

$$\mathbf{b} = t\mathbf{a} + \mathbf{c} \quad \text{with} \quad \mathbf{a} \perp \mathbf{c} \ .$$

The first summand $t\mathbf{a}$ *is called the orthogonal projection of* $\mathbf{b}$ *onto* $\mathbf{a}$ *and is denoted by* $pr_\mathbf{a}\mathbf{b}$; *thus*

$$pr_\mathbf{a}\mathbf{b} = (\mathbf{b} \cdot \mathbf{a}/\mathbf{a} \cdot \mathbf{a})\mathbf{a} \ .$$

The reader will find that the proof of 1.5.7 applies without modification to the present theorem which can be illustrated in the figure below

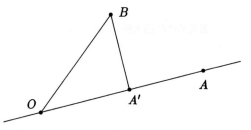

Fig 2-9

where $OA'B$ is a right-angled triangle with $\overrightarrow{OA'} = pr_\mathbf{a}\mathbf{b}$ and $\overrightarrow{A'B} = \mathbf{c}$. As OB is the hypothenuse of the right-angled triangle, we have $|OB| \geq |OA'|$, i.e.

$$|\mathbf{b}| \geq |pr_\mathbf{a}\mathbf{b}| = |\mathbf{a} \cdot \mathbf{b}/\mathbf{a} \cdot \mathbf{a}||\mathbf{a}|$$

or

$$|\mathbf{a}||\mathbf{b}| \geq |\mathbf{a} \cdot \mathbf{b}|$$

61

yielding a geometric interpretation of Schwarz's inequality. From this we note that the equality sign holds if and only if the points O, A and B are collinear.

2.5.6 COROLLARY *Let a and b be vectors of $\mathbf{R}^3$. Then $|a||b| = |a \cdot b|$ if and only if a and b are linearly dependent.*

At this juncture we shall also consider the case in which the equality sign holds in the triangle inequality.

2.5.7 THEOREM *Let a and b be vectors of $\mathbf{R}^3$. Then $|a + b| = |a| + |b|$ if and only if either vector a or b is a non-negative scalar multiple of the other.*

PROOF It follows from

$$|a + b|^2 = |a|^2 + |b|^2 + 2(a \cdot b)$$

and

$$(|a| + |b|)^2 = |a|^2 + |b|^2 + 2|a||b|$$

that $|a+b| = |a|+|b|$ if and only if $|a||b| = a \cdot b$. But this is the case if and only if $|a||b| = |a \cdot b|$ and $a \cdot b$ is non-negative, which means that either a or b is a scalar multiple of the other and $a \cdot b \geq 0$. Therefore $|a+b| = |a|+|b|$ if and only if $b = ra$ for some $r \geq 0$ or $a = sb$ for some $s \geq 0$. We may also interprete this condition as O, A and B being collinear and O not lying between A and B.

EXERCISES

1. Let **a**, **b** and **c** be the vectors $[1, 2, 0]$, $[0, -1, 2]$ and $[2, 1, 3]$ respectively. Find
   (a) $a \cdot b$,
   (b) $a \cdot (b + c)$,
   (c) $2b \cdot c$.
2. Find the cosine of the angle between **a** and **b** if
   (a) $a = [2, 3, 4]$, $b = [1, 1, 1]$;
   (b) $a = [-1, 3, -2]$, $b = [1, 1, 0]$.
3. If $a = [-2, 0, t]$ and $b = [-1, 2, 3]$ and that **a** is perpendicular to **b**, find the value of t.

4. If $\mathbf{a} = [3, -2, -3]$ and $\mathbf{b} = [r, s, t]$ such that $r^2 + s^2 + t^2 = 88$, find the maximum value of $\mathbf{a} \cdot \mathbf{b}$.

5. For vectors $\mathbf{a}$, $\mathbf{b}$ and $\mathbf{c}$ of $\mathbf{R}^3$, simplify
   (a) $(\mathbf{a} - \mathbf{b}) \cdot (\mathbf{a} + \mathbf{b})$,
   (b) $(\mathbf{a} + \mathbf{b}) \cdot \mathbf{c} - (\mathbf{a} - \mathbf{b}) \cdot \mathbf{c}$,
   (c) $(m\mathbf{a} + n\mathbf{b}) \cdot (s\mathbf{a} + t\mathbf{b})$, where m, n, s and t are real numbers.

6. Two vectors with magnitudes 15 and 32 make an angle of $\frac{\pi}{3}$ with each other. Find the value of their dot product.

7. Two vectors have magnitudes 6 and 10, and their difference has magnitude 8. Find the value of their dot product.

8. For each of the following vectors $\mathbf{a}$ and $\mathbf{b}$, find $pr_{\mathbf{a}}\mathbf{b}$.
   (a) $\mathbf{a} = [2, 2, 1]$ and $\mathbf{b} = [4, -3, 4]$.
   (b) $\mathbf{a} = [3, 4, -1]$ and $\mathbf{b} = [2, 1, -5]$.

9. Given that $\mathbf{a}$, $\mathbf{b}$, and $\mathbf{c}$ are of magnitude 1 such that $\mathbf{a} + \mathbf{b} + \mathbf{c} = \mathbf{0}$, find the value of $\mathbf{a} \cdot \mathbf{b} + \mathbf{b} \cdot \mathbf{c} + \mathbf{c} \cdot \mathbf{a}$.

10. Given vectors $\mathbf{a}$, $\mathbf{b}$ and $\mathbf{c}$ such that $|\mathbf{a}| = 1$, $|\mathbf{b}| = 2$, $|\mathbf{c}| = 3$, $\mathbf{a} \perp \mathbf{b}$, and that the angle between $\mathbf{a}$ and $\mathbf{c}$ is $\frac{\pi}{3}$ and that between $\mathbf{b}$ and $\mathbf{c}$ is $\frac{\pi}{6}$. Find $|\mathbf{a} + \mathbf{b} + \mathbf{c}|$.

11. Given non-zero vectors $\mathbf{a}$, $\mathbf{b}$ and $\mathbf{c}$, show that $\mathbf{a}$ is perpendicular to $\mathbf{b}(\mathbf{a} \cdot \mathbf{c}) - \mathbf{c}(\mathbf{a} \cdot \mathbf{b})$ and $\mathbf{b} - \frac{\mathbf{a}(\mathbf{a} \cdot \mathbf{b})}{|\mathbf{a}|^2}$ respectively.

12. Find the angle between $\mathbf{a}$ and $\mathbf{b}$ if $\mathbf{a} + 2\mathbf{b}$ is perpendicular to $5\mathbf{a} - 2\mathbf{b}$ and $\mathbf{a} - 3\mathbf{b}$ is perpendicular to $5\mathbf{a} + 3\mathbf{b}$.

13. Let $\mathbf{a} = [1, -1, 1]$, $\mathbf{b} = [3, -4, 5]$ and $\mathbf{x} = \mathbf{a} + t\mathbf{b}$ for some real number t.
    (a) Show that $|\mathbf{x}|^2 = 50t^2 + 24t + 3$.
    (b) Hence deduce that when $|\mathbf{x}|$ attains its minimum, $\mathbf{x}$ will be perpendicular to $\mathbf{b}$.

14. Given three non-zero vectors $\mathbf{a}$, $\mathbf{b}$ and $\mathbf{c}$ in which $\mathbf{b}$ is not perpendicular to $\mathbf{a}$ nor $\mathbf{c}$. Prove that $(\mathbf{a} \cdot \mathbf{b})\mathbf{c} = \mathbf{a}(\mathbf{b} \cdot \mathbf{c})$ if and only if $\mathbf{a}$ and $\mathbf{c}$ are linearly dependent.

15. In tetrahedron $ABCD$, edge AB is perpendicular to edge CD and edge AC is perpendicular to edge BD.
    (a) If $\mathbf{a}$, $\mathbf{b}$, $\mathbf{c}$ and $\mathbf{d}$ are the position vectors of points A, B, C and D respectively, show that
    $$(\mathbf{a} \cdot \mathbf{c}) + (\mathbf{b} \cdot \mathbf{d}) = (\mathbf{a} \cdot \mathbf{d}) + (\mathbf{b} \cdot \mathbf{c})$$
    $$= (\mathbf{a} \cdot \mathbf{b}) + (\mathbf{c} \cdot \mathbf{d}) .$$
    (b) Deduce that edge AD is perpendicular to edge BC.

16. Prove that for a quadrilateral in space with opposite edges being equal, the line joining the middle points of the diagonals is perpendicular to both diagonals.

2.6 Planes in space

Let

$$a_1 x + a_2 y + a_3 z = 0 \tag{1}$$

be a homogeneous linear equation in the unknowns x, y and z. Putting $\mathbf{a} = [a_1, a_2, a_3]$ and $\mathbf{x} = [x, y, z]$, we can write equation (1) into

$$\mathbf{a} \cdot \mathbf{x} = 0 . \tag{2}$$

From this we see that a point $X = (x, y, z)$ in space satisfies the equation (1) if and only if its position vector $\mathbf{x}$ is orthogonal to the vector $\mathbf{a}$ of the coefficients. Therefore both equations (1) and (2) define the same plane E in space that contains the origin O and is perpendicular to the line along $\mathbf{a}$.

More generally given a linear equation

$$a_1 x_1 + a_2 y + a_3 z + c = 0 \tag{3}$$

we pick a point $Q = (q_1, q_2, q_3)$ in space such that

$$a_1 q_1 + a_2 q_2 + a_3 q_3 + c = 0 .$$

For example we may choose $Q = (-c/a_1, 0, 0)$ if $a_1 \neq 0$. Then equation (3) can be written as

$$\mathbf{a} \cdot \mathbf{x} - \mathbf{a} \cdot \mathbf{q} = 0$$

or
$$\mathbf{a} \cdot (\mathbf{x} - \mathbf{q}) = 0. \tag{4}$$

Thus both equations (3) and (4) define the same plane F in space that is perpendicular to the line along $\mathbf{a}$ and passes through the point Q.

2.6.1 THEOREM   *A linear equation*

$$a_1 x + a_2 y + a_3 z + c = 0$$

defines a plane in space which consists of all points X whose coordinates

x, y and z satisfy the given equation. Moreover vector $\mathbf{a} = [a_1, a_2, a_3]$ is a normal vector to the plane.

2.6.2 EXAMPLE   Find the equation of the plane in space that intercepts the coordinate axes at the points $A = (a, 0, 0)$, $B = (0, b, 0)$ and $C = (0, 0, c)$. Hence find a normal vector to this plane.

SOLUTION It is known that through three non-collinear points there passes one and only one plane. Therefore it is sufficient to find a linear equation that is satisfied by the coordinates of the three given points.

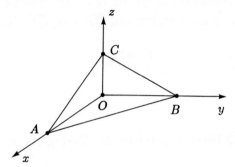

Fig 2-10

Obviously the equation

$$bcx + cay + abz = 3abc$$

is such one. By 2.6.1 the vector $[bc, ca, ab]$ is a normal vector to this plane.

We note that the problem would be somewhat more difficult if the three non-collinear points A, B and C are not the intercepts of the plane at the coordinate axes. In this more general case, we may find the coefficients and the constant term of the desired equation $a_1 x + a_2 y + a_3 z + c = 0$ by solving a system of three linear equations in the four unknowns a_1, a_2, a_3 and c. This method will be discussed in detail in Chapter Seven. However there is also an alternative method that uses the notion of cross product which we shall introduce in the following section (see Example 2.7.2).

Meanwhile we derive a few distance formulae. Recall that in the plane the distance from a point $P = (p_1, p_2)$ to a line $L : a_1 x + a_2 y + c = 0$ is given by the formula

$$\frac{|a_1 p_1 + a_2 p_2 + c|}{\sqrt{a_1{}^2 + a_2{}^2}}.$$

65

The 3-dimensional analogue of this would be

$$\frac{|a_1 p_1 + a_2 p_2 + a_3 p_3 + c|}{\sqrt{a_1{}^2 + a_2{}^2 + a_3{}^2}} = \frac{|\mathbf{a} \cdot \mathbf{p} + c|}{|\mathbf{a}|}$$

which is the distance from the point $P = (p_1, p_2, p_3)$ in space to the plane $E : a_1 x + a_2 y + a_3 z + c = 0$. Indeed if we drop a perpendicular G from P to E and if R is the foot of the perpendicular, then $|\overrightarrow{RP}|$ is the distance from P to E. On the other hand if Q is any point on E, then perpendicular projection of $\overrightarrow{QP}$ on G is $\overrightarrow{RP}$. Since G and $\mathbf{a}$ are both normal to E, they are parallel and $\overrightarrow{RP} = pr_{\mathbf{a}} \overrightarrow{QP}$. Now

$$\overrightarrow{RP} = pr_{\mathbf{a}}(\overrightarrow{QP}) = \left(\frac{(\mathbf{p} - \mathbf{q}) \cdot \mathbf{a}}{\mathbf{a} \cdot \mathbf{a}} \right) \mathbf{a} = \left(\frac{\mathbf{p} \cdot \mathbf{a} - \mathbf{q} \cdot \mathbf{a}}{\mathbf{a} \cdot \mathbf{a}} \right) \mathbf{a}$$

$$= \left(\frac{\mathbf{p} \cdot \mathbf{a} + c}{\mathbf{a} \cdot \mathbf{a}} \right) \mathbf{a}$$

because $\mathbf{q} \cdot \mathbf{a} = -c$, Q being a point on E. Taking the magnitude of $\overrightarrow{RP}$ we obtain

$$\frac{|\mathbf{a} \cdot \mathbf{p} + c|}{|\mathbf{a}|}$$

as the distance from P to E.

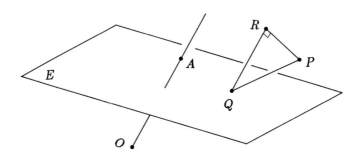

Fig 2-11

2.6.3 THEOREM *Let E be a plane in space defined by a linear equation $\mathbf{a} \cdot \mathbf{x} + c = 0$, and P a point in space. The distance from P to E is the non-negative real number $|\mathbf{a} \cdot \mathbf{p} + c|/|\mathbf{a}|$.*

It is sometimes convenient to denote the polynomial $a_1 x + a_2 y + a_3 z + c$ by $f(\mathbf{x})$. Thus

$$f(\mathbf{x}) = \mathbf{a} \cdot \mathbf{x} + c$$

is a polynomial in one vector variable $\mathbf{x}$, and the distance from the point P to the plane $f(\mathbf{x}) = 0$ is $|f(\mathbf{p})|/|\mathbf{a}|$. In particular if in $f(\mathbf{x}) = \mathbf{a} \cdot \mathbf{x} + c$ we have $|\mathbf{a}| = 1$, i.e. $\mathbf{a}$ is a unit vector, then we say that the equation $f(\mathbf{x}) = 0$ of the plane E is in *normal form*. In this case the distance from P to E is simply $|f(\mathbf{p})|$. In general, to bring an equation $f(\mathbf{x}) = \mathbf{a} \cdot \mathbf{x} + c$ of a plane into normal form, we only have to multiply the equation by the non-zero real number $1/|\mathbf{a}|$.

2.6.4 EXAMPLE   Let E be the plane defined by an equation $f(\mathbf{x}) = \mathbf{a} \cdot \mathbf{x} + c = 0$. If the straight line passing through two points P and Q intersects the plane at a point R, then $dr(P, Q; R) = -f(\mathbf{p})/f(\mathbf{q})$.

PROOF   Put $dr(P, Q; R) = \lambda$. Then $(\mathbf{r} - \mathbf{p}) = \lambda(\mathbf{q} - \mathbf{r})$ or $\mathbf{r} = \frac{1}{1+\lambda}(\mathbf{p} + \lambda \mathbf{q})$. Since R lies on E, $f(\mathbf{r}) = 0$. Therefore

$$0 = \mathbf{a} \cdot \mathbf{r} + c = \frac{1}{1+\lambda}(\mathbf{a} \cdot \mathbf{p} + c) + \frac{\lambda}{1+\lambda}(\mathbf{a} \cdot \mathbf{q} + c) .$$

Hence $\lambda = -(\mathbf{a} \cdot \mathbf{p} + c)/(\mathbf{a} \cdot \mathbf{q} + c) = -f(\mathbf{p})/f(\mathbf{q})$.

Let two planes E and F in space be defined by the equations $f(\mathbf{x}) = \mathbf{a} \cdot \mathbf{x} + c = 0$ and $g(\mathbf{x}) = \mathbf{b} \cdot \mathbf{x} + d = 0$ respectively. The non-zero vectors $\mathbf{a}$ and $\mathbf{b}$ being normal vectors to the planes E and F respectively, we see that E and F are parallel if and only if $\mathbf{b} = r\mathbf{a}$ for some non-zero scalar r. In this case if we take a point P on E and a point Q on F, then the distance between the planes is the length of the projection of $\overrightarrow{PQ}$ on the line along $\mathbf{a}$. Now

$$|pr_{\mathbf{a}}(\overrightarrow{PQ})| = |\mathbf{a} \cdot (\mathbf{q} - \mathbf{p})|/|\mathbf{a}| = \left| \left(\frac{1}{r}\mathbf{b} \right) \cdot \mathbf{q} - \mathbf{a} \cdot \mathbf{p} \right| /|\mathbf{a}|$$
$$= |(\mathbf{b} \cdot \mathbf{q})/|\mathbf{b}| - (\mathbf{a} \cdot \mathbf{p})/|\mathbf{a}|| = |c/|\mathbf{a}| - d/|\mathbf{b}|| .$$

2.6.5 THEOREM   *The distance between two parallel planes* $\mathbf{a} \cdot \mathbf{x} + c = 0$ *and* $\mathbf{a} \cdot \mathbf{x} + d = 0$ *is* $|c - d|/|\mathbf{a}|$.

On the other hand if the planes E and F are not parallel, then they intersect at a straight line L and L consists of all points $X =$

(x, y, z) whose position vectors $\mathbf{x} = [x, y, z]$ satisfy the equations

$$\mathbf{a} \cdot \mathbf{x} + c = 0 \quad \text{and} \quad \mathbf{b} \cdot \mathbf{x} + d = 0 .$$

Moreover the angle θ between the intersecting planes E and F is the same as the angle between their normal vectors $\mathbf{a}$ and $\mathbf{b}$. Therefore $\cos \theta = (\mathbf{a} \cdot \mathbf{b}) / |\mathbf{a}||\mathbf{b}|$.

In classical solid geometry, a set of planes, each passing through a given straight line L is called a *sheaf of planes* through L. In the following theorem we find a convenient expression of the planes of a sheaf.

2.6.6 THEOREM *Let two planes* $E : f(\mathbf{x}) = \mathbf{a} \cdot \mathbf{x} + c = 0$ *and* $F : g(\mathbf{x}) = \mathbf{b} \cdot \mathbf{x} + d = 0$ *intersect at a straight line* L. *Then for every pair of real numbers* r *and* s, *not both being zero, the plane defined by the equation* $rf(\mathbf{x}) + sg(\mathbf{x}) = 0$ *passes through the line* L. *Conversely every plane that passes through* L *is defined by one such equation* $rf(\mathbf{x}) + sg(\mathbf{x}) = 0$.

PROOF The first statement is self-evident. For the second statement let H defined by $h(\mathbf{x}) = \mathbf{e}\mathbf{x} + k = 0$ be a plane that passes through L. The three normal vectors $\mathbf{a}$, $\mathbf{b}$ and $\mathbf{e}$ are all orthogonal to the line L. Therefore they must be linearly dependent. On the other hand $\mathbf{a}$ and $\mathbf{b}$ being normal vectors of two non-parallel planes, they are linearly independent. Hence $\mathbf{e}$ must be linear combination of $\mathbf{a}$ and $\mathbf{b}$, say $\mathbf{e} = r\mathbf{a} + s\mathbf{b}$. Consequently H is defined by $h(\mathbf{x}) = r\mathbf{a} \cdot \mathbf{x} + s\mathbf{b} \cdot \mathbf{x} + k = 0$. Now pick an arbitrary point Q on the line L. Then $h(\mathbf{q}) = f(\mathbf{q}) = g(\mathbf{q}) = 0$, and $0 = r\mathbf{a} \cdot \mathbf{q} + s\mathbf{b} \cdot \mathbf{q} + k = -rc - sd + k$. Therefore $h(\mathbf{x}) = rf(\mathbf{x}) + sg(\mathbf{x})$.

In the next theorem we study a special property of four planes of a sheaf for which we need the notion of cross ratio.

Let A_1, A_2, A_3 and A_4 be four distinct points on a straight line. Using the first two points A_1 and A_2 as points of reference we obtain two directed ratios (see 1.4.14):

$$dr(A_1, A_2; A_3) \quad \text{and} \quad dr(A_1, A_2; A_4)$$

where $\overrightarrow{A_1 A_3} = dr(A_1, A_2; A_3)\overrightarrow{A_3 A_2}$ and $\overrightarrow{A_1 A_4} = dr(A_1, A_2; A_4)\overrightarrow{A_4 A_2}$. The quotient of these two ratios

$$dr(A_1, A_2; A_3)/dr(A_1, A_2; A_4)$$

is called the *cross-ratio* of the four collinear points A_1, A_2, A_3 and A_4 and is denoted by $cr(A_1, A_2; A_3, A_4)$. We note that the cross ratio $cr(A_1, A_2; A_3, A_4)$ is well-defined as long as A_3 is different from A_2, and A_4 is different from A_1 and A_2.

2.6.7 THEOREM *Four planes of a sheaf cut any transversal in four points of constant cross-ratio.*

PROOF Denote the four planes of the sheaf through a line by E_i, $i = 1, 2, 3, 4$. Without loss of generality we may assume that the planes E_i are defined by the equations

$$f(\mathbf{x}) + t_i g(\mathbf{x}) = 0 \quad i = 1, 2, 3, 4$$

for some fixed linear equations $f(\mathbf{x}) = 0$ and $g(\mathbf{x}) = 0$ and distinct scalars t_i. Let G be any straight line in space that intersects these planes at four distinct points A_1, A_2, A_3 and A_4. We wish to prove that the cross-ratio $cr(A_1, A_2; A_3, A_4)$ is a constant which is independent of the line G. Consider G as the straight line that passes through the points A_1, A_2 and intersects the plane E_3 at A_3. Applying the result of Example 2.6.4, we obtain

$$dr(A_1, A_2; A_3) = -\frac{f(\mathbf{a}_1) + t_3 g(\mathbf{a}_1)}{f(\mathbf{a}_2) + t_3 g(\mathbf{a}_2)} .$$

Similarly

$$dr(A_1, A_2; A_4) = -\frac{f(\mathbf{a}_1) + t_4 g(\mathbf{a}_1)}{f(\mathbf{a}_2) + t_4 g(\mathbf{a}_2)} .$$

Therefore

$$cr(A_1, A_2; A_3, A_4) = \frac{dr(A_1, A_2; A_3)}{dr(A_1, A_2; A_4)}$$

$$= \frac{(f(\mathbf{a}_1) + t_3 g(\mathbf{a}_1))(f(\mathbf{a}_2) + t_4 g(\mathbf{a}_2))}{(f(\mathbf{a}_2) + t_3 g(\mathbf{a}_2))(f(\mathbf{a}_1) + t_4 g(\mathbf{a}_1))} .$$

Since A_1 lies on E_1 and A_2 lies on E_2, we may substitute $f(\mathbf{a}_1) = -t_1 g(\mathbf{a}_1)$ and $f(\mathbf{a}_2) = -t_2 g(\mathbf{a}_2)$ into the expression above to get

$$cr(A_1, A_2; A_3, A_4) = \frac{g(\mathbf{a}_1)(t_3 - t_1)g(\mathbf{a}_2)(t_4 - t_2)}{g(\mathbf{a}_2)(t_3 - t_2)g(\mathbf{a}_1)(t_4 - t_1)}$$

$$= \frac{(t_3 - t_1)(t_4 - t_2)}{(t_3 - t_2)(t_4 - t_1)}$$

which is a constant depending only on the planes E_i.

EXERCISES

1. Find the equation of the following plane that
   (a) passes through $(-1, 5, 2)$ with normal vector $[1, -3, 2]$;
   (b) passes through $(5, 7, -6)$ and parallel to the xz-plane;
   (c) passes through $(5, 1, 4)$ and parallel to the plane $x + y - 2z = 0$.

2. In each of the following, find the perpendicular distance from the given point A to the given plane π.
   (a) $A = (0, 0, 0)$, $\pi : 3x - 4y - 12z + 2 = 0$;
   (b) $A = (1, 1, 1)$, $\pi : 5x - y - 4z + 3 = 0$;
   (c) $A = (-1, 0, 1)$, $\pi : 3x - y + 2z - 5 = 0$.

3. Find the distance between the following pairs of parallel planes.
   (a) $6x - 3y + 2z - 7 = 0$ and $6x - 3y + 2z + 14 = 0$.
   (b) $16x - 12y + 15z - 3 = 0$ and $16x - 12y + 15z - 28 = 0$.

4. In each of the following, find the cosine of the angle between the intersecting planes.
   (a) $4x - 2y + z = 0$ and $2x + 4y - z + 2 = 0$;
   (b) $x + y + 2z - 5 = 0$ and $2x - y + 3z + 7 = 0$;
   (c) $x + 3y + 4z = 5$ and $z = 0$.

5. Find the equation of a plane which is parallel to the plane $3x + 2y + 6z + 5 = 0$ and that its distance from the origin is 1.

6. Find the values of a, b and c under the following conditions.
   (a) The planes $3x + ay + 4z - 5$ and $bx - 8y - 8z + 2 = 0$ are parallel.
   (b) The planes $7x - 2y - z = 0$ and $ax + y - 3z + 7 = 0$ are perpendicular.
   (c) The planes $4x - 3y + 5z - 6 = 0$ and $2x + 3y - cz + 1 - 0$ intersect at an angle $\frac{\pi}{3}$.

7. Suppose p is the distance from the origin to the plane

$$\frac{x}{a} + \frac{y}{b} + \frac{z}{c} = 1 \qquad (abc \neq 0) ,$$

   show that

$$\frac{1}{p^2} = \frac{1}{a^2} + \frac{1}{b^2} + \frac{1}{c^2} .$$

8. Find an equation for the plane through the point $(2, 3, -1)$ and the line of intersection of the planes

$$x - y + z = 1 \quad \text{and} \quad x + y - z = 1 \ .$$

[Hint: use Theorem 2.6.6.]

9. Given the plane $Ax + By + Cz + D = 0$ and two points $M_1 = (x_1, y_1, z_1)$ and $M_2 = (x_2, y_2, z_2)$. If the line joining M_1 and M_2 intersects the plane at a point M such that $M_1 M = k M M_2$, show that $k = -\frac{Ax_1 + By_1 + Cz_1 + D}{Ax_2 + By_2 + Cz_2 + D}$.

10. Given a tetrahedron $ABCD$ with centroid G. A plane passes through G and intersects DA, DB and DC at the points P, Q and R. Show that $\frac{AP}{PD} + \frac{BQ}{QD} + \frac{CR}{RD} = 1$.

[Hint: use the result proved in Question 9.]

2.7 The cross product

Let E and F be two non-parallel planes defined by the equations $\mathbf{a} \cdot \mathbf{x} = 0$ and $\mathbf{b} \cdot \mathbf{x} = 0$ respectively. Then the planes E and F intersect in a straight line L which passes through the origin O and is perpendicular to the normal vector $\mathbf{a}$ of E and the normal vector $\mathbf{b}$ of F. Therefore if $X \neq O$ is a point on L, then its position vector $\mathbf{x}$ will be orthogonal to both $\mathbf{a}$ and $\mathbf{b}$. Moreover every point of the line L has a position vector that is a scalar multiple of $\mathbf{x}$. Let us proceed to find one such non-zero vector $\mathbf{x}$, i.e. given two linearly independent vectors $\mathbf{a}$ and $\mathbf{b}$, to find $\mathbf{x} = [x, y, z] \neq \mathbf{0}$ such that $\mathbf{a} \cdot \mathbf{x} = 0$ and $\mathbf{b} \cdot \mathbf{x} = 0$, or

$$a_1 x + a_2 y + a_3 z = 0$$
$$b_1 x + b_2 y + b_3 z = 0 \ .$$

As usual we first eliminate one of the unknowns, say z. Multiply the first equation by b_3, the second equation by $-a_3$ and add to get

$$(a_1 b_3 - a_3 b_1)x + (a_2 b_3 - a_3 b_2)y = 0 \ .$$

Put $x = a_2 b_3 - a_3 b_2$ and $y = a_3 b_1 - a_1 b_3$, and substitute these values in the original equations to get $z = a_1 b_2 - a_2 b_1$. Then the vector

$$\mathbf{x} = [a_2 b_3 - a_3 b_2, \ a_3 b_1 - a_1 b_3, \ a_1 b_2 - a_2 b_1]$$

is easily seen to be orthogonal to both $\mathbf{a}$ and $\mathbf{b}$.

71

It remains to be seen that $\mathbf{x} \neq \mathbf{0}$. But this follows immediately from 2.3.3 (4) on the hypothesis that $\mathbf{a}$ and $\mathbf{b}$ are linearly independent.

The above discussion leads us to introduce the following important definition.

2.7.1 DEFINITION *Let* $\mathbf{a} = [a_1, a_2, a_3]$ *and* $\mathbf{b} = [b_1, b_2, b_3]$ *be two arbitrary vectors of the vector space* $\mathbf{R}^3$. *The cross product* $\mathbf{a} \times \mathbf{b}$ *of* $\mathbf{a}$ *and* $\mathbf{b}$ *is the vector*

$$\mathbf{a} \times \mathbf{b} = [a_2 b_3 - a_3 b_2,\ a_3 b_1 - a_1 b_3,\ a_1 b_2 - a_2 b_1] \ .$$

There is a convenient way to memorize the components of the cross product $\mathbf{a} \times \mathbf{b}$. Recall that i, j, k is a cyclic permutation of the digits 1, 2, 3 if (i, j, k) is $(1, 2, 3)$, $(2, 3, 1)$ or $(3, 1, 2)$. Then the components of $\mathbf{x} = \mathbf{a} \times \mathbf{b}$ are $x_i = a_j b_k - a_k b_j$ where i, j, k is a cyclic permutation of 1, 2, 3.

For example if $\mathbf{a} = [1, -1, 0]$, $\mathbf{b} = [2, 1, 4]$ and $\mathbf{c} = [2, -2, 0]$, then $\mathbf{a} \times \mathbf{b} = [-4, -4, 3]$, $\mathbf{a} \times \mathbf{c} = [0, 0, 0]$, $\mathbf{c} \times \mathbf{b} = [-8, -8, 6]$. In particular for the unit coordinate vectors $\mathbf{e}_1$, $\mathbf{e}_2$, $\mathbf{e}_3$, we have $\mathbf{e}_1 \times \mathbf{e}_2 = \mathbf{e}_3$, $\mathbf{e}_2 \times \mathbf{e}_3 = \mathbf{e}_1$, $\mathbf{e}_3 \times \mathbf{e}_1 = \mathbf{e}_2$, etc. Again $\mathbf{e}_i = \mathbf{e}_j \times \mathbf{e}_k$ if i, j, k is a cyclic permutation of 1, 2, 3.

As a response to the earlier remark following Example 2.6.2, we consider the problem below.

2.7.2 EXAMPLE Find an equation of the plane E in space that passes through the three non-collinear points $A = (1, -1, 0)$, $B = (2, 3, 1)$ and $C = (0, 1, -1)$.

SOLUTION Any non-zero vector that is orthogonal to $\overrightarrow{AB}$ and $\overrightarrow{AC}$ is a normal vector $\mathbf{d}$ of the desired plane E. Therefore we may put

$$\mathbf{d} = (\mathbf{b} - \mathbf{a}) \times (\mathbf{c} - \mathbf{a}) = [-6, 0, 6] \ .$$

As the plane that passes through A perpendicular to $\mathbf{d}$, E is defined by

$$(\mathbf{x} - \mathbf{a}) \cdot \mathbf{d} = 0$$

or
$$x - z - 1 = 0 \ .$$

We have seen that following 2.3.3 (4) if **a** and **b** are linearly independent then $\mathbf{a} \times \mathbf{b} \neq 0$. Similarly it follows from 2.3.2 (4) that if **a** and **b** are linearly dependent then $\mathbf{a} \times \mathbf{b} = 0$. Therefore we have the following necessary and sufficient condition for a cross product to be non-zero.

2.7.3 THEOREM *Let* **a** *and* **b** *be vectors of* $\mathbf{R}^3$. *Then* $\mathbf{a} \times \mathbf{b} = 0$ *if and only if* **a** *and* **b** *are linearly dependent.*

The following four properties of cross product are easily verified by straightforward calculation.

2.7.4 THEOREM **Let a, b** *and* **c** *be vectors of* $\mathbf{R}^3$ *and r be a scalar. Then*
   (1) $\mathbf{a} \cdot (\mathbf{a} \times \mathbf{b}) = \mathbf{b} \cdot (\mathbf{a} \times \mathbf{b}) = 0$;
   (2) $\mathbf{a} \times \mathbf{b} = -(\mathbf{b} \times \mathbf{a})$;
   (3) $\mathbf{a} \times (\mathbf{b} + \mathbf{c}) = \mathbf{a} \times \mathbf{b} + \mathbf{a} \times \mathbf{c}$ *and* $(\mathbf{a} + \mathbf{c}) \times \mathbf{b} = \mathbf{a} \times \mathbf{b} + \mathbf{c} \times \mathbf{b}$;
   (4) $(r\mathbf{a}) \times \mathbf{b} = r(\mathbf{a} \times \mathbf{b}) = \mathbf{a} \times (r\mathbf{b})$.

Property (1) relates the fact that the cross product $\mathbf{a} \times \mathbf{b}$ is orthogonal to both its factors **a** and **b**. Property (2) states that the formation of cross product is *anticommutative*. Properties (3) and (4) mean that the cross product is linear in each factor; hence it is *bilinear*.

Now we have seen that given two vectors a and b of the vector space $\mathbf{R}^3$, we have a dot product a · b and a cross product a × b. The dot product a · b is a scalar whose properties are studies in Section 2.5. An appropriate geometric interpretation of the dot product a · b is also given in terms of the angle between a and b. While a · b is a scalar, the cross product a × b is a vector of $\mathbf{R}^3$. Some of its properties are listed in the last two theorems. As a vector, the cross product also has a definite magnitude and a definite direction. Let us first evaluate the magnitude and then interprete it geometrically. Firstly it follows from

$$(a_2b_3 - a_3b_2)^2 + (a_3b_1 - a_1b_3)^2 + (a_1b_2 - a_2b_1)^2$$
$$= (a_2b_3)^2 + (a_3b_2)^2 + (a_3b_1)^2 + (a_1b_3)^2 + (a_1b_2)^2 + (a_2b_1)^2$$
$$\quad - 2(a_2b_2)(a_3b_3) - 2(a_1b_1)(a_3b_3) - 2(a_1b_1)(a_2b_2)$$
$$= (a_1b_1)^2 + (a_1b_2)^2 + (a_1b_3)^2 + (a_2b_1)^2 + (a_2b_2)^2 + (a_2b_3)^2 + (a_3b_1)^2$$
$$\quad + (a_3b_2)^2 + (a_3b_3)^2 - (a_1b_1)^2 - (a_2b_2)^2 - (a_3b_3)^2$$
$$\quad - 2(a_1b_1)(a_2b_2) - 2(a_1b_1)(a_3b_3) - 2(a_2b_2)(a_3b_3)$$
$$= (a_1^2 + a_2^2 + a_3^2)(b_1^2 + b_2^2 + b_3^2) - (a_1b_1 + a_2b_2 + a_3b_3)^2$$

that $|a \times b|^2 = |a|^2|b|^2 - (a \cdot b)^2$. If we denote the angle between the vectors a and b by θ, then

$$|a \times b|^2 = |a|^2|b|^2 - |a|^2||b|^2 \cos^2\theta = |a|^2|b|^2 \sin^2\theta .$$

In other words $|a \times b|$ is twice the area of the triangle OAB in which one vertex is at the origin O, and the other vertices A and B have position vectors a and b.

2.7.5 THEOREM *Let* a *and* b *be vectors of* $\mathbf{R}^3$. *Then* $|a \times b|^2 = |a|^2|b|^2 - (a \cdot b)^2$, *and the magnitude of the cross product* $a \times b$ *is twice the area of the triangle spanned by* a *and* b.

Let us find the direction of $a \times b$. If a and b are linearly dependent, then $a \times b = 0$ and there is nothing more to say. Assume that a and b are linearly independent, representing points A and B respectively in space. Then $a \times b$ is a non-zero vector perpendicular to a and b; hence it should fall on the line $C'OC$ (see Fig 2-12) which is perpendicular to the plane E spanned by the rays OA and OB. In other words the direction of $a \times b$ is either the direction of the ray OC or the direction of the ray OC'. Now we proceed to rotate the positive coordinate axes of the space into the position so that the x-axis falls on the ray OA, and also both the y-axis and the ray OB lie on the same half-plane of the line OA. Then the direction of $a \times b$ is the direction of the z-axis. In particular if $a = e_1$ and $b = e_2$ then $a \times b = e_1 \times e_2 = e_3$. If we may borrow a term from mechanics, we say that the three vectors a, b and $a \times b$ form a right-hand system.

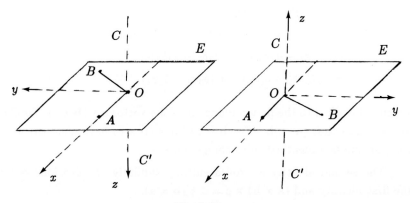

Fig 2-12

Finally as there are two different products of two vectors in space, given three vectors a, b and c of the vector space $\mathbf{R}^3$, various products can be produced formally, e.g. $a \times (b \times c)$, $(a \times b) \times c$, $(a \times b) \cdot c$, $a \cdot (b \times c)$ etc. Let us investigate these products. First of all we find that the cross product is not associative. In fact

$$e_1 \times (e_1 \times e_2) = e_1 \times e_3 = -e_2 ,$$
$$(e_1 \times e_1) \times e_2 = 0 \times e_2 = 0 .$$

The formation of cross product is therefore an *anticommutative* and *non-associative* multiplication.

Consider now the prodcut $a \times (b \times c)$. By (1) of 2.7.4, this vector is orthogonal to both a and $b \times c$. Being orthogonal to $b \times c$, it must be on the plane of b and c. Therefore $a \times (b \times c)$ is a linear combination of b and c which is orthogonal to a. More precisely we have the following theorem.

2.7.6 THEOREM *Let* a, b, c *be vectors of the vector space* $\mathbf{R}^3$. *Then*

$$a \times (b \times c) = (a \cdot c)b - (a \cdot b)c ,$$
$$(a \times b) \times c = (a \cdot c)b - (b \cdot c)a .$$

PROOF We note that the vector on the right-hand side of the first identity is a linear combination of b and c which is orthogonal to a. The first component of $(a \cdot c)b - (a \cdot b)c$ is

$$(\mathbf{a} \cdot \mathbf{c})b_1 - (\mathbf{a} \cdot \mathbf{b})c_1$$
$$= (a_1 c_1 + a_2 c_2 + a_3 c_3)b_1 - (a_1 b_1 + a_2 b_2 + a_3 b_3)c_1$$
$$= (a_2 c_2 + a_3 c_3)b_1 - (a_2 b_2 + a_3 b_3)c_1$$
$$= a_2(b_1 c_2 - b_2 c_1) - a_3(b_3 c_1 - b_1 c_3)$$

which is identical to the first component of the vector $\mathbf{a} \times (\mathbf{b} \times \mathbf{c})$. Similarly we can show that $(\mathbf{a} \cdot \mathbf{c})\mathbf{b} - (\mathbf{a} \cdot \mathbf{b})\mathbf{c}$ and $\mathbf{a} \times (\mathbf{b} \times \mathbf{c})$ have identical second components and identical third components.

The second identity is proved entirely similarly. It also follows from the first identity and $(\mathbf{a} \times \mathbf{b}) \times \mathbf{c} = \mathbf{c} \times (\mathbf{b} \times \mathbf{a})$.

An alternative proof of the first identity may be given as follows. It follows from the fact that $\mathbf{a} \times (\mathbf{b} \times \mathbf{c})$ is a linear combination of $\mathbf{b}$ and $\mathbf{c}$ which is orthogonal to $\mathbf{a}$ that

$$\mathbf{a} \times (\mathbf{b} \times \mathbf{c}) = r(\mathbf{a} \cdot \mathbf{c})\mathbf{b} - r(\mathbf{a} \cdot \mathbf{b})\mathbf{c}$$

for some scalar r. To evaluate r, we substitute $\mathbf{a} = \mathbf{b} = \mathbf{e}_1$ and $\mathbf{c} = \mathbf{e}_2$ to obtain $r = 1$. Hence the identity follows.

Meanwhile for the "mix-product" $(\mathbf{a} \times \mathbf{b}) \cdot \mathbf{c}$ we have the following theorem the proof of which is left to the reader as an exercise. The close connection between mix-product and the determinant will be discussed in section 6.6.

2.7.7 THEOREM *Let $\mathbf{a}, \mathbf{b}, \mathbf{c}$ be vectors of the vector space $\mathbf{R}^3$. The following statements hold.*

(1) *$(\mathbf{a} \times \mathbf{b}) \cdot \mathbf{c}$ is a scalar equal in magnitude to the volume of the parallelepiped of which OA, OB, OC are coterminous edges.*

(2) *$(\mathbf{a} \times \mathbf{b}) \cdot \mathbf{c} = (\mathbf{b} \times \mathbf{c}) \cdot \mathbf{a} = (\mathbf{c} \times \mathbf{a}) \cdot \mathbf{b}$.*

(3) *$(\mathbf{a} \times \mathbf{b}) \cdot \mathbf{c} = \mathbf{a} \cdot (\mathbf{b} \times \mathbf{c})$.*

Thus the product $(\mathbf{a} \times \mathbf{b}) \cdot \mathbf{c}$ is not changed by cyclic permutation of the letters $\mathbf{a}, \mathbf{b}, \mathbf{c}$; it is also unchanged if the dot and the cross are interchanged.

EXERCISES

1. If $\mathbf{a} = [1, 2, -3]$, $\mathbf{b} = [1, 0, -1]$ and $\mathbf{c} = [3, 1, 0]$. Find $\mathbf{a} \times \mathbf{b}$, $\mathbf{b} \times \mathbf{c}$ and $\mathbf{c} \times \mathbf{a}$.

2. Find the magnitude of the vector $[1, -1, 1] \times [1, 1, -1]$.

3. Find unit vectors which are perpendicular to both $[0, 1, 2]$ and $[1, -1, 3]$.

4. Find an equation of the plane which passes through the three non-collinear points $A = (1, 2, 0)$, $B = (3, 0, -3)$ and $C = (5, 2, 1)$.

5. Find an equation of the plane which passes through the points $A = (1, 2, -1)$ and $B = (-5, 2, 7)$, and is parallel to the vector $\mathbf{c} = [0, 1, 2]$.

6. Find the area of the triangle with vertices $(0, 0, 0)$, $(1, 3, 2)$, and $(2, 0, -3)$.

7. Find the area of the triangle with vertices $(1, -1, 2)$, $(3, 1, 3)$, and $(3, 3, 1)$.

8. For each of the following, calculate $(\mathbf{a} \times \mathbf{b}) \cdot \mathbf{c}$.

   (a) $\mathbf{a} = [1, 2, 3]$, $\mathbf{b} = [-1, 3, 0]$ and $\mathbf{c} = [2, 0, -3]$;

   (b) $\mathbf{a} = [2, -1, 5]$, $\mathbf{b} = [4, 1, 3]$ and $\mathbf{c} = [3, 2, -1]$.

9. Find the volume of the parallelepiped with coterminous edges OA, OB and OC, where $O = (0, 0, 0)$, $A = (1, 2, 0)$, $B = (2, 1, 0)$ and $C = (4, 3, 2)$.

10. Given that $\mathbf{a} = [3, 1, 2]$, $\mathbf{b} = [m, 0, n]$ and $\mathbf{a} \times \mathbf{b} = [2, 0, \ell]$, find m, n and ℓ.

11. By using the result, $|\mathbf{a} \times \mathbf{b}| = (\mathbf{a} \cdot \mathbf{a})(\mathbf{b} \cdot \mathbf{b}) - (\mathbf{a} \cdot \mathbf{b})^2$, show that $|\mathbf{a} \times \mathbf{b}| = |\mathbf{a}| \cdot |\mathbf{b}| \sin \theta$, where θ is the angle between $\mathbf{a}$ and $\mathbf{b}$.

12. Given that $|\mathbf{a}| = 2$, $|\mathbf{b}| = 5$ and $\mathbf{a} \cdot \mathbf{b} = -6$, find $|\mathbf{a} \times \mathbf{b}|$.

13. Prove Theorem 2.7.4.

14. Simplify $(\mathbf{a} + \mathbf{b}) \times (\mathbf{a} - \mathbf{b})$ and hence deduce that if $\mathbf{a}$ is perpendicular to $\mathbf{b}$, then $|(\mathbf{a} + \mathbf{b}) \times (\mathbf{a} - \mathbf{b})| = 2|\mathbf{a}||\mathbf{b}|$.

15. If the angle between $\mathbf{a}$ and $\mathbf{b}$ is $\frac{1}{3}\pi$, and that $|\mathbf{a}| = 2$, $|\mathbf{b}| = 3$, find the values of

   (a) $|\mathbf{a} \times \mathbf{b}|$, and

   (b) $|(2\mathbf{a} + \mathbf{b}) \times (\mathbf{a} + 2\mathbf{b})|$.

16. Let $\mathbf{a}$ be a non-zero vector in $\mathbf{R}^3$. Show that for vector $\mathbf{x}$ in $\mathbf{R}^3$, $\mathbf{x} \cdot \mathbf{a} = 0$ if and only if $\mathbf{x} = \mathbf{a} \times \mathbf{w}$ for some vector $\mathbf{w}$.

17. Prove that for the unit coordinate vectors $\mathbf{e}_1$, $\mathbf{e}_2$ and $\mathbf{e}_3$,

   (a) $\mathbf{e}_i \times (\mathbf{e}_j \times \mathbf{e}_k) = (\mathbf{e}_i \cdot \mathbf{e}_k)\mathbf{e}_j - (\mathbf{e}_i \cdot \mathbf{e}_j)\mathbf{e}_k$, and

   (b) $(\mathbf{e}_i \times \mathbf{e}_j) \times \mathbf{e}_k = (\mathbf{e}_i \cdot \mathbf{e}_k)\mathbf{e}_j - (\mathbf{e}_j \cdot \mathbf{e}_k)\mathbf{e}_i$, for any i, j and k.

   Hence prove Theorem 2.7.6.

18. Use Theorem 2.7.6 to prove the Jacobi identity: for any vectors $\mathbf{a}$, $\mathbf{b}$ and $\mathbf{c}$ in $\mathbf{R}^3$,

$$\mathbf{a} \times (\mathbf{b} \times \mathbf{c}) + \mathbf{b} \times (\mathbf{c} \times \mathbf{a}) + \mathbf{c} \times (\mathbf{a} \times \mathbf{b}) = \mathbf{0} \ .$$

19. Given that **a** and **b** are perpendicular vectors in $\mathbf{R}^3$, find $\mathbf{a} \times \{\mathbf{a} \times [\mathbf{a} \times (\mathbf{a} \times \mathbf{b})]\}$.

20. Prove Theorem 2.7.7.

21. Use Theorem 2.7.7(1) to show that the points $A = (2, 0, 1)$, $B = (5, 4, -2)$, $C = (1, -1, 2)$ and $D = (3, 2, 0)$ are coplanar.

22. For any vectors **a**, **b**, **c** and **d** in $\mathbf{R}^3$, by Theorem 2.7.6, show that
    (a) $(\mathbf{a} \times \mathbf{b}) \cdot (\mathbf{a} \times \mathbf{b}) = (\mathbf{a} \cdot \mathbf{a})(\mathbf{b} \cdot \mathbf{b}) - (\mathbf{a} \cdot \mathbf{b})(\mathbf{a} \cdot \mathbf{b})$, and
    (b) $(\mathbf{a} \times \mathbf{b}) \cdot (\mathbf{c} \times \mathbf{d}) = (\mathbf{a} \cdot \mathbf{c})(\mathbf{b} \cdot \mathbf{d}) - (\mathbf{a} \cdot \mathbf{d})(\mathbf{b} \cdot \mathbf{c})$.

23. Suppose **a**, **b** and **c** are the position vectors of the points A, B and C respectively. Prove that A, B and C are collinear if and only if $\mathbf{a} \times \mathbf{b} + \mathbf{b} \times \mathbf{c} + \mathbf{c} \times \mathbf{a} = \mathbf{0}$.

24. Given a square $ABCD$ with vertices $A = (1, 1, 1)$, $B = (-1, 1, 1)$, $C = (-1, -1, 1)$, and $D = (1, -1, 1)$. P and Q are the mid-points of AD and BC respectively. R and S are points on AB and CD respectively such that $AR = CS = a$.
    (a) Find the equations of the planes OPR and OQS where O is the origin.
    (b) If the angle between the two planes in (a) is $\frac{\pi}{3}$, find the value of a.
    (c) If area of $\triangle OPR = \frac{7}{8}$, find the value of a.

25. Consider a triangle with vertices A, B and C, and let a, b and c denote the lengths of the sides opposite A, B and C, respectively. Write $\mathbf{u} = \overrightarrow{AB}$, $\mathbf{v} = \overrightarrow{BC}$ and $\mathbf{w} = \overrightarrow{CA}$.
    (a) Deduce that $\mathbf{u} + \mathbf{v} + \mathbf{w} = \mathbf{0}$.
    (b) Show that $\mathbf{u} \times \mathbf{v} = \mathbf{w} \times \mathbf{u} = \mathbf{v} \times \mathbf{w}$.
    [Hint: Compute $\mathbf{u} \times (\mathbf{u} + \mathbf{v} + \mathbf{w})$ and $\mathbf{v} \times (\mathbf{u} + \mathbf{v} + \mathbf{w})$.]
    (c) Hence deduce the sine law:
    $$\frac{\sin A}{a} = \frac{\sin B}{b} = \frac{\sin C}{c} .$$

26. Suppose vectors **a**, **b** and **c** are edges of a triangle such that $a = |\mathbf{a}|$, $b = |\mathbf{b}|$, $c = |\mathbf{c}|$ and $s = (a + b + c)/2$.
    (a) Show that $\mathbf{a} \cdot \mathbf{b} = \frac{1}{2}(c^2 - a^2 - b^2)$.
    (b) By using Question 22(a), deduce the Heron's formula:
    $$\text{area of triangle} = \sqrt{s(s - a)(s - b)(s - c)} .$$

27. (a) Given three non-collinear points A, B and C such that $\overrightarrow{AB} = \mathbf{u}$

and $\overrightarrow{AC} = \mathbf{v}$. Show that the distance d from C to AB is given by
$d = \frac{|\mathbf{u} \times \mathbf{v}|}{|\mathbf{u}|}$.

(b) Suppose P is a point inside an equilateral triangle ABC. By (a), show that the sum of distances of P from the edges is a constant.

28. For any vectors $\mathbf{a}$, $\mathbf{b}$ and $\mathbf{c}$ in $\mathbf{R}^3$, show that
    (a) $\mathbf{a} \cdot (\mathbf{b} \times \mathbf{c}) = (\mathbf{a} \times \mathbf{b}) \cdot (\mathbf{c} + m\mathbf{a} - n\mathbf{b})$ for any real numbers m and n.
    (b) $(\mathbf{a} + \mathbf{b}) \cdot [(\mathbf{b} + \mathbf{c}) \times (\mathbf{c} + \mathbf{a})] = 2\mathbf{a} \cdot (\mathbf{b} \times \mathbf{c})$.
    (c) $\mathbf{a} \cdot (\mathbf{b} \times \mathbf{c}) = 0$ if and only if $(\mathbf{b} \times \mathbf{c}) \cdot [(\mathbf{c} \times \mathbf{a}) \times (\mathbf{a} \times \mathbf{b})] = 0$.

29. Given that $\mathbf{x}$ is a vector such that

$$\mathbf{x} \cdot \mathbf{a} = m \quad \text{and} \quad \mathbf{x} \times \mathbf{b} = \mathbf{c}, \tag{$*$}$$

where m is a given scalar and $\mathbf{a}$, $\mathbf{b}$ and $\mathbf{c}$ are given vectors such that $\mathbf{a} \cdot \mathbf{b} \neq 0$ and $\mathbf{b} \cdot \mathbf{c} = 0$.

(a) Show by considering $\mathbf{a} \times (\mathbf{x} \times \mathbf{b})$ that

$$\mathbf{x} = \left(\frac{1}{\mathbf{a} \cdot \mathbf{b}}\right) [m\mathbf{b} + \mathbf{a} \times \mathbf{c}].$$

(b) Deduce that $\mathbf{x}$ found in (a) is the unique solution of the pair of equations $(*)$.

2.8 Lines in space

The parametric representation of a straight line in space is entirely analogous to that in the plane. The straight line L in space that passes through a given point P and parallel a given non-zero vector $\mathbf{d}$ consists of all points X whose position vectors $\mathbf{x}$ are represented by

$$\mathbf{x} = \mathbf{p} + t\mathbf{d}$$

where $\mathbf{p}$ is the position vector of P and t is a parameter. Thus we may call this equation the point-direction form of L.

2.8.1 EXAMPLE   The line L that contains the point $(0, 1, 1)$ and parallel to $[1, -1, 2]$ consists of points $(t, 1 - t, 1 + 2t)$ for all possible values of t. Therefore the points $(1, 0, 3)$ and $(3, -2, 7)$ corresponding to $t = 1$ and $t = 3$ respectively lie on the line L. On the other hand the point $(2, 3, 4)$ is not a point on L because for $t = 2$ we get $1 - t = -1$ and $1 + 2t = 5$. To find the point of intersection of L with the plane $4x + 7y - z - 1 = 0$, we substitute $x = t$, $y = 1 - t$ and $z = 1 + 2t$ into the equation of the plane to get

$$4t + 7(1 - t) - (1 + 2t) - 1 = 0$$
or
$$-5t + 5 = 0 .$$

This is satisfied by $t = 1$, giving $(1, 0, 3)$ as the point of intersection.

Two distinct points in space also uniquely determine a straight line. Given two distinct points A and B, the line G passing through A and B is represented by

$$\mathbf{x} = \mathbf{a} + t(\mathbf{b} - \mathbf{a})$$
or
$$\mathbf{x} = (1 - t)\mathbf{a} + t\mathbf{b}$$
or
$$\mathbf{x} = r\mathbf{a} + s\mathbf{b} \quad \text{with} \quad r + s = 1 .$$

Each of the above is a parametric representation of the line G and may be called the two-point form of G.

Finally a line in space is also the intersection of two non-parallel planes. Therefore a straight line is also defined by a system of two linear equations

$$a_1 x + a_2 y + a_3 z + c = 0$$
$$b_1 x + b_2 y + b_3 z + d = 0$$

where the normal vectors $\mathbf{a} = [a_1, a_2, a_3]$ and $\mathbf{b} = [b_1, b_2, b_3]$ of the defining planes are linearly independent. Similarly we may call this the two-plane form of the line.

2.8.2 EXAMPLE   Let the straight line H be defined by the equations

$$2x + y - z - 1 = 0$$
$$-x + 3y + z - 8 = 0 .$$

Then H is the line of intersection of the two planes defined by the two equations separately. Therefore H is perpendicular to the normal vectors $[2, 1, -1]$ and $[-1, 3, 1]$ of these planes. It follows that the cross product of these normal vectors is a direction vector of H. To obtain a parametric representation H we still need a point on H, for example the point $(1, 2, 3)$.

Then H is represented by

$$\mathbf{x} = [1, 2, 3] + t([2, 1, -1] \times [-1, 3, 1])$$
$$= [1, 2, 3] + t[4, -1, 7]$$
$$= (1 + 4t, 2 - t, 3 + 7t) \ .$$

In many books on analytic geometry the direction of a straight line is represented by a set of real numbers called the direction cosines. These can be defined as follows. Let A and B be two distinct points in space. Then the ray AB originating from A towards B will have the same direction as the vector $\overrightarrow{AB}$ or the unit vector $\mathbf{d} = \overrightarrow{AB}/|\overrightarrow{AB}|$. Therefore the direction of the ray AB is uniquely determined by the components

$$d_1, d_2, d_3$$

of the unit vector $\mathbf{d}$. On the other hand if $\mathbf{e}_i$ $(i = 1, 2, 3)$ denotes as usual the i-th unit coordinate vector, then it follows from

$$d_i = \frac{\overrightarrow{AB} \cdot \mathbf{e}_i}{|\overrightarrow{AB}|} = \cos \theta_i \quad i = 1, 2, 3$$

that the real number d_i is the cosine of the angle between the vector $\overrightarrow{AB}$ and $\mathbf{e}_i$. Therefore the three real numbers d_1, d_2, d_3 are called the *direction cosines* of the ray AB. Obviously the opposite ray BA originating from B towards A has then the direction cosines $-d_1, -d_2, -d_3$. Since the straight line AB is constituted by the rays AB and BA, either the set d_1, d_2, d_3 or the set $-d_1, -d_2, -d_3$ may be used as the direction cosines of the line AB.

Conversely every triple d_1, d_2, d_3 of real numbers satisfying $d_1{}^2 + d_2{}^2 + d_3{}^2 = 1$ form a set of direction cosines of a line in space. Indeed if $P = (p_1, p_2, p_3)$ is a point in space then the line L with parametric representation $\mathbf{x} = \mathbf{p} + t\mathbf{d}$ has direction cosines d_1, d_2, d_3.

If none of the direction cosines d_1, d_2, d_3 is zero, (i.e. if the line L is not parallel to anyone of the coordinate planes), then we may eliminate the parameter t from the equations

$$x = p_1 + td_1, \quad y = p_2 + td_2, \quad z = p_3 + td_3$$

to get three equations

$$\frac{x - p_1}{d_1} = \frac{y - p_2}{d_2}, \quad \frac{y - p_2}{d_2} = \frac{z - p_3}{d_3}, \quad \frac{z - p_3}{d_3} = \frac{x - p_1}{d_1}.$$

Clearly any two of these three linear equations define the same straight line L. Combining these three equations into a single expression

$$\frac{x - p_1}{d_1} = \frac{y - p_2}{d_2} = \frac{z - p_3}{d_3},$$

we have what is sometimes called the *symmetric form of the equations* of L in books on analytic geometry. A detailed investigation on these three planes that intersect one another at the line L is left to the reader as an exercise.

To conclude this section, we consider two straight lines in space. Let two lines L and G be represented by

$$\mathbf{x} = \mathbf{p} + t\mathbf{d} \quad \text{and} \quad \mathbf{x} = \mathbf{q} + s\mathbf{f}$$

respectively. If the direction vectors $\mathbf{d}$ and $\mathbf{f}$ are linearly dependent, then L and G are either identical or parallel to each other. It is not difficult to see that $L = G$ if and only if $\overrightarrow{PQ}$ is a scalar multiple of $\mathbf{d}$.

If the direction vectors $\mathbf{d}$ and $\mathbf{f}$ are linearly independent then either L and G are a pair of intersecting lines or they are a pair of skew lines, according to whether L and G are coplanar or not.

We summarize the four different relative positions of two lines in space in the following theorem the proof of which is left to the reader as an exercise.

2.8.3 THEOREM *Let two straight lines L and G in space be represented by*

$$\mathbf{x} = \mathbf{p} + t\mathbf{d} \quad and \quad \mathbf{x} = \mathbf{q} + s\mathbf{f}$$

respectively. Then
   (1) *$L = G$ if and only if $\overrightarrow{PQ}$ and $\mathbf{f}$ are both scalar multiples of $\mathbf{d}$;*
   (2) *L and G are parallel lines if and only if $\mathbf{f}$ is a scalar multiple of $\mathbf{d}$ and $\overrightarrow{PQ}$ is not a scalar multiple of $\mathbf{d}$;*
   (3) *L and G intersect each other at one point if and only if $\mathbf{d}$ and $\mathbf{f}$*

are linearly independent, and $\overrightarrow{PQ}$ is a linear combination of d and f;

(4) L and G are skew lines if and only if the three vectors $\overrightarrow{PQ}$, d and f are linearly independent.

The distance between two lines L and G in space is the shortest distance between any one point of L and any one point of G. Clearly the distance between a pair of identical lines or a pair of intersecting lines is zero. For the other cases we have the following theorem whose proof is left to the reader as an exercise.

2.8.4 THEOREM *Let two lines L and G be represented by* $\mathbf{x} = \mathbf{p} + t\mathbf{d}$ *and* $\mathbf{x} = \mathbf{q} + s\mathbf{f}$ *respectively. If L and G are parallel then the distance between them is*

$$|\overrightarrow{PQ} \times \mathbf{d}|/|\mathbf{d}| \ .$$

If L and G are skew then the distance between them is

$$|\overrightarrow{PQ} \cdot (\mathbf{d} \times \mathbf{f})|/|\mathbf{d} \times \mathbf{f}| \ .$$

EXERCISES

1. Find equations of the following lines:
   (a) passes through $(3, -1, 2)$ and parallel to $[-1, 2, -3]$;
   (b) passes through $(2, 3, 4)$ and parallel to $[-2, 0, 5]$.
2. Find both parametric vector equation and symmetric equations of the line L which passes through the points $P(2, 3, -1)$ and $Q(1, -5, 1)$. Where does the line meet the three coordinate planes respectively?
3. Find the equation of the line which passes through $(2, 1, 0)$ and parallel to the planes $5x - 2y + z = 0$ and $x - z = 0$.
4. Find symmetric equations of the line of intersection of the planes $4x + 5y + z = 1$ and $2x + 3y - z = 9$.
5. If the point $P(1, -1, a)$ lies on the line $\mathbf{x} = [3, -4, 1] + t[2, -3, 4]$, find the value of a.
6. Find the coordinates of the point of intersection of the line $\mathbf{x} = [-3, -2, 0] + t[3, -2, 1]$ and the plane $x + 2y + 2z + 6 = 0$.
7. Given two straight lines $\ell_1 : \mathbf{x} = t[3, 1, 1]$ and $\ell_2 : \mathbf{x} = s[0, -1, 2]$.
   (a) Find the equation of the plane which contains ℓ_1 and ℓ_2.
   (b) Find the equation of the straight line which lies on the plane of (a) and passes through the origin such that it is perpendicular to ℓ_1.

8. Show that if the lines $\mathbf{x} = \mathbf{p} + t\mathbf{d}$ and $\mathbf{x} = \mathbf{q} + s\mathbf{f}$ intersect, then $(\mathbf{p} - \mathbf{q}) \cdot (\mathbf{d} \times \mathbf{f}) = 0$. Is the converse true? Justify your answer.

9. Show that the line $\frac{x-p_1}{d_1} = \frac{y-p_2}{d_2} = \frac{z-p_3}{d_3}$ lies on the plane $Ax+By+Cz+D = 0$ if and only if $Ad_1 + Bd_2 + Cd_3 = 0$ and $Ap_1 + Bp_2 + Cp_3 + D = 0$.

10. Prove Theorem 2.8.3.

11. Prove Theorem 2.8.4.

12. Find the distance between the following pairs of lines.
    (a) $\frac{x}{2} = \frac{y+2}{-2} = \frac{z-1}{1}$ and $\frac{x-1}{2} = \frac{y-3}{-2} = \frac{z+1}{1}$;
    (b) $\mathbf{x} = [0, 1, 1] + t[1, -1, 0]$ and $\mathbf{x} = [-1, 1, 0] + s[2, 1, 2]$.

13. Given two straight lines

$$\ell : \mathbf{x} = t[1, 1, 1]$$
$$m : \mathbf{x} = [14, 28, 42] + s[4, 2, 1] .$$

A straight line ℓ_1 intersects ℓ and m, and ℓ_1 is also perpendicular to both lines.
    (a) Find the points of intersection of ℓ_1 with ℓ and m.
    (b) Find the equation of ℓ_1.

14. Given point $A = (2, 0, 3)$ and the straight line $\ell : \mathbf{x} = [0, 3, 1] + t[1, -1, -1]$.
    (a) Find the point B on ℓ such that $AB \perp \ell$.
    (b) Find the equation of the plane which contains A and ℓ.

15. Let ℓ be the line of intersection of the planes

$$\begin{cases} x + y + z = 0 , \\ x - y + z = 0 . \end{cases}$$

    (a) Find the equation of ℓ.
    (b) Another straight line ℓ_1 which lies on the plane $x + y + z = 0$ and passes through the origin makes an angle $\frac{\pi}{3}$ with ℓ. Find its equation.

16. A straight line ℓ passes through the point $A = (0, 4, 1)$, parallel to the vector $[2, b, -1]$ and intersects the straight line $\ell_1 : \mathbf{x} = [6, -5, 4] + t[4, -3, 1]$.
    (a) Find the value of b and the point of intersection of ℓ and ℓ_1.
    (b) Find the equation of the plane which passes through A and perpendicular to $[2, b, -1]$, with the value of b found in (a).
    (c) Find the point of intersection of the line ℓ_1 and the plane in (b).

2.9 Cylindrical coordinates and spherical coordinates

Recall that every ponit X in the plane can be given a pair of polar coordinates (r, θ) with respect to a fixed point (say the origin O) and a ray originating from that fixed point (say the positive x-axis). Here r is the distance from X to O and θ is the angle of rotation in counter-clockwise direction bringing the positive x-axis into coincidence with the ray OX. The relationship between the polar coordinates (r, θ) of X and the cartesian coordinates (x, y) of X is expressed by the following identities:

$$r = \sqrt{x^2 + y^2}; \quad x = r\cos\theta; \quad y = r\sin\theta .$$

There is a natural extension of the polar coordinate system in the plane to a useful coordinate system in space. Let X be the point in space with cartesian coordinates (x, y, z). Denote by X' the point on the xy-plane with coordinate $(x, y, 0)$. Then XX' is perpendicular to the xy-plane. As a point on the xy-plane, X' has cartesian coordinate (x, y) which can be converted into polar coordinates (r, θ) by the identities above. Putting $h = z$, we obtain the triple (r, θ, h) as the *cylindrical coordinates* of the point X. The relationship between the cylindrical and the cartesian coordinates of X is therefore

$$r = \sqrt{x^2 + y^2}; \quad x = r\cos\theta, \quad y = r\sin\theta; \quad h = z .$$

We observe that the given point X lies on the cylinder which has the z-axis as its axis and the radius r.

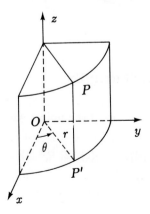

Fig 2-13

There is yet another coordinate system in space. The point X with cartesian coordinates (x, y, z) lies on the sphere of radius $r = \sqrt{x^2 + y^2 + z^2}$ with centre at O. Let X' again be the perpendicular projection of X on the xy-plane. In the xy-plane let θ again denote the angle of rotation about O in counter-clockwise direction bringing the positive x-axis into coincidence with the ray OX'. In the plane that contains the z-axis and the points X and X', let φ denote the angle of rotation about O that rotates the positive z-axis towards OX' bringing it into coincidence with the ray OX. Then the triple (r, φ, θ) will be called the *spherical coordinates* of the point X. The angles θ and φ are called the *longitude* and the *colatitude* (complement of the latitude) of the point X respectively. The relationship between the cartesian and the spherical coordinates of the point X is as follows:

$$r = \sqrt{x^2 + y^2 + z^2} \; ; x = r \cos \theta \sin \varphi \; ; y = r \sin \theta \sin \varphi \; ; z = r \cos \varphi \; .$$

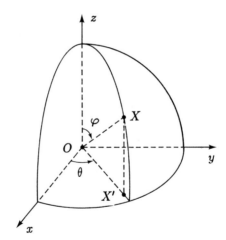

Fig 2-14

EXERCISES

1. Write the rectangular coordinates of the points whose cylindrical coordinates are

   (a) $(3, \frac{\pi}{6}, 9)$;   (b) $(4, \frac{\pi}{4}, 8)$;   (c) $(2, \frac{11\pi}{6}, 3)$.

2. Write the rectangular coordinates of the points whose spherical coordinates are

   (a) $(4,, \frac{\pi}{2}, \frac{\pi}{6})$;   (b) $(3, \frac{\pi}{2}, \frac{\pi}{2})$;   (c) $(2, \frac{\pi}{3}, -\frac{\pi}{4})$.

3. Write the cylindrical coordinates and spherical coordinates of the points whose rectangular coordinates are

   (a) $(2\sqrt{3}, 2, 4)$;   (b) $(\frac{\sqrt{3}}{2\sqrt{2}}, \frac{1}{2\sqrt{2}}, -\frac{1}{\sqrt{2}})$.

4. Derive the equations which change

   (a) cylindrical coordinates into spherical coordinates; and

   (b) spherical coordinates into cylindrical coordinates.

5. For each of the following, find a cartesian equation for the surface whose equation in cylindrical coordinates is given.

   (a) $r^2 + h^2 = 16$;   (b) $r = 6$;   (c) $h = 6 + r$.

6. For each of the following, find an equation in cylindrical coordinates.

   (a) $x - y = 0$;   (b) $x^2 + y^2 = 9$;   (c) $x^2 + y^2 = 3z$.

7. For each of the following, find a cartesian equation for the equation in spherical coordinates.

   (a) $r = a \cos \theta \sin \phi$;   (b) $r = a \sin \phi$;   (c) $r \cos \phi = a$.

CHAPTER THREE

CONIC SECTIONS

The curves known as conic sections comprise the ellipse, hyperbola and parabola. They are, after the circle, the simplest curves. This being so, it is not surprising that they have been known and studied for a long time. Their discovery is attributed to Menaechmus, a Greek geometer and astronomer of the 4th century BC. Like Hipprocrates of the 5th century BC before him, Menaechmus, in attacking the Delian problem of duplication of a cube, found himself facing the task of constructing two mean proportionals x and y between two given line segments of length a and b:

$$a : x = x : y = y : b .$$

From this continued proportion it follows that

$$x^2 = ay \qquad \text{and} \qquad xy = ab .$$

Hence
$$x^3 = axy = a^2 b$$

leading to
$$a^3 : x^3 = a^3 : a^2 b = a : b .$$

This means that if b is chosen to be $2a$, x would be the side of a cube twice the volume of the cube with side a. Menaechmus then recognized that

$$x^2 = ay \qquad \text{and} \qquad xy = ab$$

are represented by curves which are cut out of a right circular cone by plane sections. More than a hundred years later Appolonius of Perga, who is also said to be the founder of Greek mathematical astronomy, wrote a most thorough treatise on conic sections. The majority of the written works by Appolonius are now lost. Fortunately of the original eight books of Pis chef-d'oeuve – *The Conics*, the first four books remains extant in Greek. Three of the next four books also survived in Arabic translation.

In this chapter we shall use methods of analytic geometry and work with a unified definition of a conic section which is given in terms of the distance to a fixed point and the distance to a fixed straight line in the plane. Later it shall be shown in two different ways that these curves are actually cut out of a right circular cone by plane sections.

3.1 Focus, directrix and eccentricity

Instead of treating the three types of conic sections individually as it is usually done in secondary school analytic geometry, we shall take a more coherent view of our subject and begin our investigation with a unified geometric definition of the conic sections.

3.1.1 DEFINITION *A conic section or a conic is the locus of a point X which moves in a plane containing a fixed point F and a fixed line D in such a way that the distance $|XF|$ from the point F is in a constant ratio e (e $\neq$ 0) to its perpendicular distance $|XE|$ from the straight line D:*

$$|XF| = |XE|e \ .$$

The fixed point F is called a focus, the fixed straight line D is called a directrix and the constant ratio e is called the eccentricity of the conic. A conic is called an ellipse, a parabola or a hyperbola accordingly as its eccentricity e is less than, equal to or greater than unity.

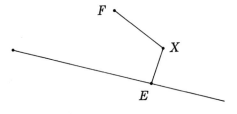

Fig 3-1

In the subsequent sections we proceed to derive equation of each type of conics relative to an appropriately chosen pair of coordinate axes.

3.2 The parabola

Given a parabola with focus F and directrix D. Then by Definition 3.1.1, the parabola in question consists of all points X in the plane such that

$$|XF| = |XE|$$

where $|XE|$ is the perpendicular distance from X to D. Our immediate task is to lay down a pair of coordinate axes so that the given parabola would be represented by an equation of the simplest form. For this purpose we drop the perpendicular FH from F on D. Then the midpoint O of the segment FH is a point on the parabola. We choose O as the origin and the ray originating from O towards F as the positive x-axis. With the origin and x-axis being chosen, the y-axis is consequently fixed in the plane (see Figure 3-2). If we denote by $2a > 0$ the distance from F to D, then $F = (a, 0)$ and the directrix D is the line $x + a = 0$. Now the distances from a point $X = (x, y)$ in the plane to the focus F and to the directrix D are respectively

$$|XF| = \sqrt{(x-a)^2 + y^2} \quad \text{and} \quad |XE| = |x+a| .$$

Therefore the parabola consists of all points $X = (x, y)$ such that

$$|x + a| = \sqrt{(x-a)^2 + y^2} . \tag{1}$$

Squaring, we get

$$x^2 + 2ax + a^2 = x^2 - 2ax + a^2 + y^2 \tag{2}$$

which simplifies to

$$y^2 = 4ax . \tag{3}$$

Thus every point X on the parabola satisfies the equation (3). Conversely if $X = (x, y)$ is a point that satisfies (3), then it also satisfies (2). Taking square roots on both sides, X must satisfy

$$\sqrt{(x-a)^2 + y^2} = x + a \quad \text{or} \quad \sqrt{(x-a)^2 + y^2} = -(x + a) .$$

But x is a non-negative number because $x = y^2/4a$ by (3). Therefore the latter alternative is impossible. The former is precisely equation (1); hence the point $X = (x, y)$ lies on the parabola. Thus equation (3) is an equation of the given parabola with respect to the xy-coordinate system in a standard position. The curve is sketched below in Figure 3-2 and a summary of the above discussion is given in the following theorem.

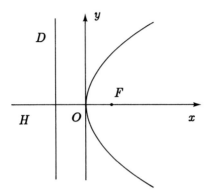

Fig 3-2

3.2.1 THEOREM *Every parabola can be so placed in the xy-plane that it is the locus of an equation of the form*

$$y^2 = 4ax$$

where $a > 0$. Conversely the locus of any such equation is a parabola.

The straight line which passes through the focus and perpendicular to the directrix of a parabola (the x-axis in the present case) is called the *axis* of the parabola. The parabola is symmetric with respect to its axis because $X = (x, y)$ lies on the parabola if and only if $X' = (x, -y)$ lies on the parabola. The point at which the parabola cuts its axis (the origin in the present case) is called the *vertex* of the parabola. The distance between the focus and the vertex is a. The straight line which passes through the focus F and perpendicular to the axis cuts the parabola at two points L and L'. The line segment LL' which is of length $4a$ is called the *latus rectum* of the parabola. Just as the parabola is completely determined by its focus and directrix or by its focus and vertex, it is also completely determined by its latus rectum alone.

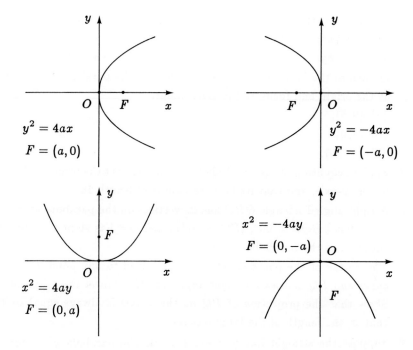

Fig 3-3

Keeping the vertex of the parabola at the origin, if we place the focus at a different position, the equation of the parabola changes its form accordingly. Figure 3-3 shows four different positions of F with the resulting equations.

It is well-known that all circles are *similar*; i.e. they have the same shape, and differ only in size. In other words one circle becomes another circle when drawn to a different scale. Like circles, all parabolas are similar because the parabola $y^2 = 4ax$ will be turned into the parabola $\bar{y}^2 = 4b\bar{x}$ by a simple scaling factor of $a/b : x = (a/b)\bar{x}$ and $y = (a/b)\bar{y}$.

EXERCISES

1. Find the coordinates of the focus and an equation for the directrix of the following parabolas with given equations:

(a) $y^2 = -8x$;

(b) $x = 2y^2$;

(c) $x^2 = 8y$.

2. In each of the following, find an equation for the parabola with vertex at the origin and focus at the given point:

   (a) $(0, 5)$;

   (b) $(2, 0)$;

   (c) $(-1, 0)$.

3. Find an equation of the parabola that has vertex at the origin and focus on the y-axis, and that its latus rectum is of length 16.

4. A right-angled triangle OBC has its vertices on the parabola $y^2 = 4ax$ such that $\angle BOC = 90°$, $|BC| = 5\sqrt{13}$ and OB has slope 2. Find the equation of the parabola.

5. The vertex of the parabola $y^2 = 4ax$ is joined to any point P on the curve and PQ is drawn at right angles to OP to meet the x-axis at Q. Show that the projection of PQ on the x-axis is always equal to $4a$, that is, the length of the latus rectum.

6. Suppose the straight line $y = mx + c$ cuts the parabola $y^2 = 4ax$ at the points $A = (x_1, y_1)$ and $B = (x_2, y_2)$, where $a > 0$.

   (a) By considering the discriminant of a suitable quadratic equation, show that $a \geq mc$.

   (b) Prove that $|x_1 - x_2| = \frac{4}{m^2}\sqrt{a(a - mc)}$.

   (c) Find the length of the chord AB.

7. OP and OQ are two chords of the parabola $y^2 = 4ax$ such that $\angle POQ = 90°$, where O is the origin.

   (a) If the straight line OP has the equation $y = mx$, find the coordinates of P and Q in terms of m.

   (b) By finding the equation of chord PQ, show that PQ always passes through a fixed point on the x-axis as m varies.

   (c) Find the locus of the midpoint of PQ as m varies. What kind of curve is it?

8. A chord of the parabola $y^2 = 4ax$ that passes through the focus is called a *focal chord*. Suppose a focal chord of slope m intersects the parabola at the points $A = (x_1, y_1)$ and $B = (x_2, y_2)$.

   (a) Show that $|AB| = 4a(m^2 + 1)/m^2$.

   (b) Let C be the midpoint of AB.

(i) A perpendicular is drawn from C to the directrix of the parabola at D, show that $|CD| = \frac{1}{2}|AB|$. What does this mean geometrically?

(ii) Show also that $DF \perp AB$, where F is the focus.

9. Let AB and CD be two focal chords of the parabola $y^2 = 4ax$. Suppose that $AB \perp CD$.

(a) If AB is of slope m, show that

$$|AF| \cdot |BF| = \frac{4a^2(1+m^2)}{m^2} \, ,$$

where F is the focus.

(b) Deduce that

$$\frac{1}{|AF| \cdot |BF|} + \frac{1}{|CF| \cdot |DF|} = \frac{1}{4a^2} \, .$$

10. Let P and Q be the points of intersection of the straight line $y = m(x+2)$ with the parabola $y = x^2$.

(a) Show that the midpoint of PQ has coordinates $(\frac{m}{2}, \frac{1}{2}m^2 + 2m)$.

(b) Find the equation of circle with PQ as diameter.

(c) If $OP \perp OQ$, where O is the origin, find the value of m.

3.3 The ellipse

Given an ellipse with focus F', directrix D' and eccentricity e $(0 < e < 1)$. Again we drop from F' the perpendicular $F'\overline{O}$ on D'. For the time being the point $\overline{O}$ on D' is chosen as the origin and the ray originating from $\overline{O}$ towards F' as the positive $\overline{x}$-axis. Consequently the directrix becomes the $\overline{y}$-axis. If we denote by $k > 0$ the distance from F' to D', then in the $\overline{x}\overline{y}$-plane, $F' = (k, 0)$ and D' is the line $\overline{x} = 0$. The distances from a point $X = (\overline{x}, \overline{y})$ in the $\overline{x}\overline{y}$-plane to the focus F' and the directrix D' are then respectively

$$|XF'| = \sqrt{(\overline{x} - k)^2 + \overline{y}^2} \quad \text{and} \quad |XE'| = |\overline{x}| \, .$$

Therefore by Definition 3.1.1, the ellipse under discussion consists of all points $X = (\overline{x}, \overline{y})$ in the $\overline{x}\overline{y}$-plane such that

$$|XF'| = |XE'|e \tag{1}$$

or in terms of coordinates in the $\overline{x}\,\overline{y}$-plane

$$\sqrt{(\overline{x} - k)^2 + \overline{y}^2} = |\overline{x}|e \ . \tag{2}$$

Squaring, we get

$$\overline{x}^2 - 2k\overline{x} + k^2 + \overline{y}^2 = e^2\overline{x}^2 \tag{3}$$

which simplifies to

$$(1 - e^2)\overline{x}^2 + \overline{y}^2 - 2k\overline{x} + k^2 = 0 \ . \tag{4}$$

Thus every point $X = (\overline{x}, \overline{y})$ on the ellipse satisfies equation (4). Using similar argument as in the case of the parabola, we can show conversely that if $X = (\overline{x}, \overline{y})$ satisfies (4) then it also satisfies (2) proving that equation (4) is the equation of an ellipse where the $\overline{y}$-axis is the directrix and the focus F' lies on the positive $\overline{x}$-axis. The curve is sketched in Figure 3-4.

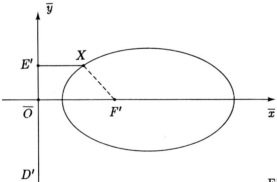

Fig 3-4

Again the straight line that passes through the focus and perpendicular to the directrix is an axis of symmetry of the ellipse. In the present case this is the $\overline{x}$-axis, and a point $X = (\overline{x}, \overline{y})$ lies on the ellipse if and only if the mirrored point $X' = (\overline{x}, -\overline{y})$ lies on the ellipse. This axis cuts the ellipse at two points $A' = (k/(1 + e), 0)$ and $A = (k/(1 - e), 0)$. The midpoint O of A and A' has the coordinates $(k/(1 - e^2), 0)$ in the $\overline{x}\,\overline{y}$-plane.

In order to better reveal the symmetries of the ellipse through its equation, we transform the $\overline{x}\,\overline{y}$-coordinate system to a new xy-

coordinate system by shifting the origin to the point O while retaining the directions of the coordinate axes (see Figure 3-5).

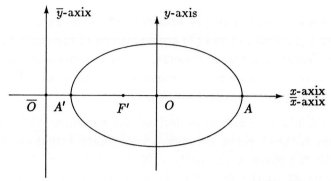

Fig 3-5

Then every point X with coordinates $(\overline{x}, \overline{y})$ in the $\overline{x}\,\overline{y}$-plane will have new coordinates (x, y) in the xy-plane where

$$x = \overline{x} - \frac{k}{1 - e^2} \quad \text{and} \quad y = \overline{y} .$$

In particular, in xy-plane, we have

$$O = (0,0), \quad A = (ek/(1 - e^2), 0), \quad A' = (-ek/(1 - e^2), 0) .$$

Substitution in the $\overline{x}\,\overline{y}$-equation (4) of the ellipse yields

$$(1 - e^2)(x + \frac{k}{1 - e^2})^2 + y^2 - 2k(x + \frac{k}{1 - e^2}) + k^2 = 0$$

which simplifies to

$$(1 - e^2)x^2 + y^2 = \frac{e^2 k^2}{1 - e^2} . \tag{5}$$

Further simplification can still be obtained. By putting $a = ek/(1 - e^2)$, we get

$$(1 - e^2)x^2 + y^2 = a^2(1 - e^2) . \tag{6}$$

The right-hand side is positive since $e < 1$; hence we may set

$$b^2 = a^2(1 - e^2) .$$

Finally we obtain

$$\frac{x^2}{a^2} + \frac{y^2}{b^2} = 1 \tag{7}$$

which is an equation of the ellipse with respect to the xy-coordinate system. Both equations (4) and (7) are equations of the same ellipse relative to different coordinate systems. Clearly because of the absence of linear terms equation (7) is preferred and we shall call it the equation of an ellipse in *standard position*.

With hindsight we could have derived the equation (7) in standard position without going through a coordinate transformation. Starting with the given eccentricity e ($0 < e < 1$) and distance k from the focus F' to the directrix D', we put $a = ek/(1 - e^2)$ and $b^2 = a^2(1 - e^2)$, and choose a pair of coordinate xy-axes so that $F' = (-ae, 0)$ and D' is defined by $x = -k - ae$ or equivalently by $x = -a/e$. Then equation (1) becomes

$$\sqrt{(x + ae)^2 + y^2} = e|x + \frac{a}{e}|$$

which simplifies into equation (7).

3.3.1 THEOREM *For any ellipse, a coordinate system in the plane can be so chosen that it is the locus of an equation of the from*

$$\frac{x^2}{a^2} + \frac{y^2}{b^2} = 1$$

where $a^2 > b^2$. *Conversely the locus of any such equation is an ellipse.*

With our ellipse placed in this convenient position, we proceed to investigate its geometric properties through algebraic methods. First of all we find that the focus F' is at the point $F' = (-ae, 0)$ and the directrix D' is defined by the equation $x = -a/e$. Furthermore the ellipse cuts the x-axis at the points $A = (a, 0)$, $A' = (-a, 0)$ and the y-axis at the points $B = (b, 0)$, $B' = (-b, 0)$ (see Figure 3-6).

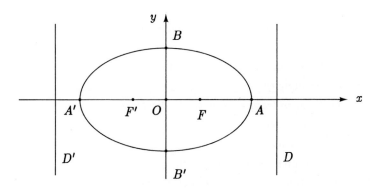

Fig 3-6

By 3.3.1, a point (x, y) lies on the ellipse if and only if the point $(x, -y)$ lies on it. Therefore the x-axis is an axis of symmetry of the ellipse. Similarly the y-axis is also an axis of symmetry: (x, y) lies on the ellipse if and only if $(-x, y)$ does. Consequently the ellipse is also symmetric with respect to the origin O in the sense that a point (x, y) is on it if and only if the point $(-x, -y)$ is on it. This leads us to call the point O the *centre* of the ellipse and an ellipse a *central conic.*

Due to the symmetry with respective to the y-axis, the point $F = (ae, 0)$ and the line D defined by $x = a/e$ which are the mirror images of F' and D' respectively have the characteristic of being a focus and its corresponding directrix of the ellipse. Therefore, as a central conic, an ellipse has a pair of foci and a pair of corresponding directrices whereas a parabola which is not a central conic has just one focus and one directrix. The axis of symmetry that passes through the foci is called the *major axis* of the ellipse; the other axis of symmetry is then the *minor axis.* Sometimes it is convenient to refer to the chord of the ellipse which falls on the major axis or its length also as the major axis of the ellipse, the context making clear in any case the meaning. Thus for the ellipse defined by (7), the major axis is the x-axis, the segment AA' or the length $|AA'| = 2a$ and the minor axis is the y-axis, the segment BB' or the length $|BB'| = 2b$.

Let us recall that an ellipse is completely determined by a pair of positive numbers, namely the eccentricity e $(0 < e < 1)$ and the distance k from the focus to the directrix. In the above discussion several other measurements on the ellipse are found. The relationship among them is given in the following theorem.

3.3.2 THEOREM *Given an ellipse. Denote by*

e = *the eccentricity;*

k = *the distance from a focus to its directrix;*

$2a$ = *the major axis;*

$2b$ = *the minor axis;*

$2c$ = *the distance between the foci;*

$2d$ = *the distance between the directrices.*

Then these six quantities are related by the four equations

$$a^2 = b^2 + c^2, \quad c = ae, \quad d = a/e, \quad k = d - c .$$

These quantities are illustrated in Figure 3-7. Moreover all six of them can be calculated by the above equations from any two among a, b, c, e, k.

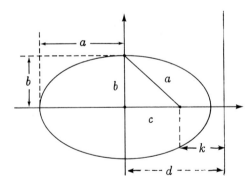

Fig 3-7

In the derivation of equation (7) of the ellipse, preference is given to the x-axis by placing on it both foci of the ellipse. If preference is given to the y-axis by placing both foci on it, then we would arrive at the equation

$$\frac{x^2}{b^2} + \frac{y^2}{a^2} = 1, \quad a^2 > b^2 ,$$

and the ellipse is turned 90° about the origin.

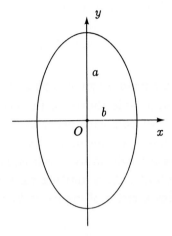

Fig 3-8

In general an equation of the form

$$\frac{x^2}{p^2} + \frac{y^2}{q^2} = 1$$

defines an ellipse which cut the x-axis at the points $(-p,\ 0)$ and $(p,\ 0)$, and the y-axis at the points $(0,\ -q)$ and $(0,\ q)$. The x-axis will be the major axis of the ellipse if $p > q > 0$ and the y-axis will be the major axis if $0 < p < q$.

3.3.3 EXAMPLE The ellipse defined by the equation

$$\frac{x^2}{9} + \frac{y^2}{25} = 1$$

has the y-axis as its major axis. Therefore the points $(0, -\sqrt{25 - 9}) = (0, -4)$ and $(0, 4)$ are the foci, and $e = 4/5$ is the eccentricity.

All circles are similar, and the same is true of the parabolas. But it is not true of the ellipses. In fact, if an ellipse is taken as an oblate circle, then its eccentricity e is a measurement of its *oblateness*. For example an ellipse of eccentricity $1/2$ is more oblate than another one of eccentricity $1/4$. This can be seen from the equation

$$e = \frac{c}{a} = \sqrt{1 - (b/a)^2} \quad \text{where} \quad a > b\ .$$

It also follows that all ellipses with the same eccentricity are *similar,* and vice versa. Thus while e and k together will completely determine both the shape and the size of an ellipse, the eccentricity e alone will only determine the shape.

Let us consider an ellipse with a fixed major axis $2a$ and a varying minor axis $2b$ $(a > b)$. As the minor axis shortens, the eccentricity will increase, the foci will move away from each other and the ellipse will become more oblate. Conversely as the minor axis lengthens, the eccentricity will decrease, the foci will move towards each other and the ellipse will become less oblate. As a limiting case, where $2b$ approaches $2a$, the ellipse becomes a circle of diameter $2a$ when the eccentricity approaches zero.

Another well-known characteristic of an ellipse is that from every point on an ellipse the sum of the distances to the foci is a constant. This property and its converse will be proved in the following theorem.

3.3.4 THEOREM *The locus of a point on the plane, the sum of whose distances from two given points F and F' is constant, is an ellipse with foci at F and F'.*

PROOF Let the said constant be conveniently denoted by $2a$ and the distance between F and F' by $2c$. Then $a > c$, otherwise the locus does not exists or consists of points on the segment FF'. We choose the line passing through F and F' as the x-axis and the midpoint O of F and F' to be the origin. Then $F' = (-c, 0)$ and $F = (c, 0)$. The said locus will then consists of all points $X = (x, y)$ such that

$$\sqrt{(x + c)^2 + y^2} + \sqrt{(x - c)^2 + y^2} = 2a . \tag{8}$$

Transpose one of the radical and square:

$$(x + c)^2 + y^2 = (x - c)^2 + y^2 - 4a\sqrt{(x - c)^2 + y^2} + 4a^2 .$$

Hence

$$a\sqrt{(x - c)^2 + y^2} = a^2 - cx . \tag{9}$$

To remove this radical, square again:

$$a^2(x^2 - 2cx + c^2 + y^2) = a^4 - 2a^2cx + c^2x^2 \tag{10}$$

or $$(a^2 - c^2)x^2 + a^2y^2 = a^2(a^2 - c^2) .$$

Put $$b^2 = a^2 - c^2$$

and hence $$b^2x^2 + a^2y^2 = a^2b^2 \tag{11}$$

or $$\frac{x^2}{a^2} + \frac{y^2}{b^2} = 1. \tag{12}$$

This is the equation of an ellipse with foci at F and F'. Therefore every point X of the said locus lies on the ellipse (12). Conversely suppose that $X = (x, y)$ is a point on the ellipse. Then (12) is satisfied. From (12) we mount up to (11) and then (10). But in each of the steps from (10) to (9) and (9) to (8), we extract a square root. Therefore we are led finally to four equations:

$$\sqrt{(x-c)^2 + y^2} + \sqrt{(x+c)^2 + y^2} = 2a ;$$
$$\sqrt{(x-c)^2 + y^2} - \sqrt{(x+c)^2 + y^2} = 2a ;$$
$$-\sqrt{(x-c)^2 + y^2} + \sqrt{(x+c)^2 + y^2} = 2a ;$$
$$-\sqrt{(x-c)^2 + y^2} - \sqrt{(x+c)^2 + y^2} = 2a .$$

Therefore any $X = (x, y)$ that satisfies (12) must satisfy one of the four equations above. The fourth equation is impossible because $2a$ is positive. The third equation would mean that $|XF'| - |XF| = 2a$. Since $2a > 2c = |FF'|$ this would mean that $|XF'| > |XF| + |FF'|$, i.e. the sum of two sides of the triangle XFF' is less than the third side, but this is impossible. Similarly the second equation is also impossible, leaving the first equation which is identical to (8) as the only possibility. Therefore every point on the ellipse (12) is a point on the locus (8). The proof is now complete.

We mentioned earlier that given the quantities a and c, the quantites e and k can be calculated by 3.3.2. Theorem 3.3.4 now gives a geometric interpretation of the quantities a and c.

We are familiar with the parametric representation

$$x = r\cos\theta, \quad y = r\sin\theta$$

of points on the circle

$$x^2 + y^2 = r^2 .$$

Similarly for points of the ellipse

$$\frac{x^2}{a^2} + \frac{y^2}{b^2} = 1 \tag{13}$$

we have a parametric representation

$$x = a \cos \theta, \quad y = b \sin \theta . \tag{14}$$

Clearly the point $X = (a \cos \theta, b \sin \theta)$ satisfies (13) and hence lies on the ellipse. Conversely if $X = (x, y)$ lies on the ellipse, then from (13) it follows that

$$\frac{x}{a} = \cos \theta \quad \text{and} \quad \frac{y}{b} = \sin \theta$$

for some θ.

The parametric representation (14) of the ellipse (13) affords a convenient geometric construction of an ellipse with given major axis $2a$ and minor axis $2b$. Draw concentric circles of radius a and radius b with the origin O as the common centre. A radius OR of the outer circle cut the inner circle at Q. Through R and Q draw lines parallel to the coordinate axes to meet at X as shown in Figure 3-9. Then X is a point on the desired ellipse. Indeed if θ is the angle between OR and the x-axis, then $X = (a \cos \theta, b \sin \theta)$. Other points of the ellipse are obtained similarly.

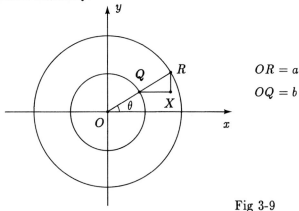

$$OR = a$$
$$OQ = b$$

Fig 3-9

The outer circle is called the *auxiliary circle* of the ellipse. The angle θ is called the *eccentric angle* of the point X of the ellipse, and we note that θ is, in general, not the angle between OX and the x-axis.

EXERCISES

1. For each of the following, find an equation of the ellipse with given conditions:
   (a) foci at $(\pm 8, 0)$, and major axis of length 20;
   (b) vertices of major axis at $(0, \pm 8)$, and minor axis of length 6;
   (c) vertices of minor axis at $(\pm 3, 0)$, and passing through $(2, 5)$.

2. Given the ellipse $\frac{x^2}{16} + \frac{y^2}{4} = 1$.
   (a) Find the coordinates of the foci.
   (b) Find an equation of the ellipse which has the same foci found in (a) and passes through $(4, \frac{2\sqrt{10}}{5})$.

3. Show that a latus rectum of the ellipse $\frac{x^2}{a^2} + \frac{y^2}{b^2} = 1$ has length $\frac{2b^2}{a}$.

4. Find the equation of the ellipse which has the same foci as the ellipse $\frac{x^2}{a^2} + \frac{y^2}{b^2} = 1$, $a > b > 0$, and that its minor axis is of length $2a$.

5. Let F be a focus and O the centre of an ellipse. A straight line is drawn from F to meet the ellipse at point P. By using Theorem 3.3.4 prove that as P moves along the ellipse, the locus of the midpoint of FP is again an ellipse with F as one focus.

6. Given the ellipse $\frac{x^2}{a^2} + \frac{y^2}{a^2 - 1} = 1$, $a^2 > 1$.
   (a) Find the coordinates of foci, $F_1 = (c, 0)$ and $F_2 = (-c, 0)$ for $c > 0$.
   (b) The straight line $y = x - 1$ intersects the ellipse at points A and B. If the circle with AB as diameter passes through F_2, show that $a^2 = 2 + \sqrt{3}$.

7. Given the ellipse $\frac{x^2}{a^2} + \frac{y^2}{b^2} = 1$, $a > b > 0$. $M = (-a, 0)$, $N = (a, 0)$ are the vertices on the major axis. If $A = (a \cos \theta, b \sin \theta)$ and $B = (a \cos \theta, -b \sin \theta)$, $0 \le \theta < 2\pi$, are two points on the ellipse such that the straight lines MA and NB are produced to meet at a point P, find the locus of P.

8. Let $P = (a \cos \theta, b \sin \theta)$ and $Q = (a \cos \phi, b \sin \phi)$ be two points on the ellipse $\frac{x^2}{a^2} + \frac{y^2}{b^2} = 1$, $a > b > 0$. If the product of the slope of OP and the slope of OQ is $-\frac{b^2}{a^2}$,
   (a) show that $\cos(\theta - \phi) = 0$;
   (b) hence deduce that $|OP|^2 + |OQ|^2 = a^2 + b^2$.
   (c) If w is the acute angle between OP and OQ, show that $\min w = \tan^{-1} \frac{2ab}{a^2 - b^2}$.

9. Given point $P = (x_1, y_1)$ on the ellipse $\frac{x^2}{a^2} + \frac{y^2}{b^2} = 1$, $a > b > 0$. Another point $Q = (s, 0)$ lies on the x-axis such that no matter how P and Q move, $|PQ| = b$. If M is the midpoint of PQ,

   (a) show that $s = \frac{(a+b)x_1}{a}$ or $s = \frac{(a-b)x_1}{a}$;

   (b) hence find the locus of M.

10. Given the ellipse $\frac{x^2}{a^2} + \frac{y^2}{b^2} = 1$, $a > b > 0$. $A = (-a, 0)$ is one of the vertices of the major axis. Suppose $P = (x_1, y_1)$ is any point on the ellipse with $x_1 \neq 0$.

    (a) A straight line AR is drawn parallel to OP. Show that AR has the equation $y = \frac{y_1}{x_1}(x + a)$.

    (b) If AR meets the ellipse at the point Q, show that $Q = \left(\frac{2x_1^2 - a^2}{a}, \frac{2x_1 y_1}{a}\right)$.

    (c) Hence deduce that $|AR| \cdot |AQ| = 2 \cdot |OP|^2$.

11. Prove Theorem 3.3.2.

12. Show that if two of the five quantities a, b, c, e, k are given, the remaining three quantities can be calculated from the following equations:
    $a^2 = b^2 + c^2$, $c = ae$, $k = \frac{a}{e} - c$.

3.4 The hyperbola

Given a hyperbola with focus F, directrix D and eccentricity e ($e > 1$). Then by Definition 3.1.1, the hyperbola under discussion consists of all points X such that

$$|XF| = |XE|e \tag{1}$$

where $|XE|$ is the perpendicular distance from X to D with the point E on D. Denote by $k > 0$ the distance from F to D and put $a = ek/(e^2 - 1)$. With minor modification on the direct derivation of the equation of an ellipse in standard position, we choose a pair of coordinate axes so that $F = (ae, 0)$ and D is the line $x = ae - k$ or equivalently $x = a/e$. Then equation (1) becomes

$$\sqrt{(x - ae)^2 + y^2} = e|x - a/e| . \tag{2}$$

Squaring, we get

$$x^2 - 2aex + a^2 e^2 + y^2 = e^2 x^2 - 2aex + a^2 \tag{3}$$

or
$$(e^2 - 1)x^2 - y^2 = a^2(e^2 - 1). \tag{4}$$

The right-hand side $a^2(e^2 - 1)$ is positive since $e > 1$; therefore we may put $b^2 = a^2(e^2 - 1)$ and write (4) as

$$\frac{x^2}{a^2} - \frac{y^2}{b^2} = 1 . \tag{5}$$

This means that every point $X = (x, y)$ on the hyperbola satisfies equation (5). Conversely if $X = (x, y)$ satisfies (5), it satisfies (4) and (3). Extracting square roots and taking absolute value X also satisfies (2). Therefore equation (5) is the equation of the hyperbola in *standard position*. The hyperbola which is a curve with two *branches* is sketched below in Figure 3-10.

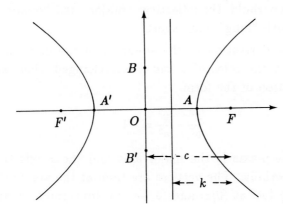

Fig 3-10

3.4.1 THEOREM *Any hyperbola in the xy-plane can be so placed that it is the locus of an equation of the form*

$$\frac{x^2}{a^2} - \frac{y^2}{b^2} = 1 .$$

Conversely the locus of any such equation is a hyperbola.

From the equation of a hyperbola in standard position we see that a hyperbola has two axes of symmetry and a centre. Like the ellipse, the hyperbola is a central conic. Consequently it has a pair of foci and a pair of corresponding directrices. The axis of symmetry that passes through the foci is called the *major axis* and the other one the *minor axis*. With the hyperbola in standard position we find the foci at $F(c,0)$ and $F' = (-c,0)$ with $c = ae$ on the major axis. The major axis cuts the curve at two points $A = (a,0)$ and $A' = (-a,0)$ called the *vertices*. Sometimes the segment AA' or its length $|AA'|$ is also referred to as the major axis of the hyperbola. The minor axis, on the other hand, has no point in common with the hyperbola and separates the two *branches* of the curve. Although the two points $B = (0,b)$ and $B' = (0,-b)$ of the minor axis are not points of the hyperbola, they turn out to be rather useful. For convenience the line segment BB' or its length $|BB'|$ is sometimes referred to as the minor axis. The major axis of an ellipse is always longer than the minor axis; this is not always true of the hyperbola. Therefore for the axes of a hyperbola, the adjectives "major" and "minor" are not synomymous with "long" and "short".

If the same derivation of the equation of the given hyperbola is repeated with the roles of x and y interchanged, then we would obtain an equation of the form

$$\frac{x^2}{b^2} - \frac{y^2}{a^2} = -1$$

In this case the y-axis is the major axis and the x-axis the minor axis of the hyperbola. The vertices are then at the points $(0,a)$ and $(0,-a)$, and the foci at $(0,c)$ and $(0,-c)$, all four points lying on the major axis.

3.4.2 EXAMPLE   The hyperbola

$$\frac{x^2}{4} - \frac{y^2}{9} = -1$$

cuts the y-axis at $A = (0,3)$ and $A' = (0,-3)$. It does not intersect the x-axis. Therefore A, A' are the vertices of the hyperbola; the y-axis is the major axis and the x-axis the minor axis.

Among the quantities a, b, c and others in connection with the hyperbola we have the relationships listed below.

3.4.3 THEOREM   *Given a hyperbola, where*

e = *the eccentricity* ;

k = *the distance from the focus to its directrix* ;

$2a$ = the major axis ;

$2b$ = the minor axis ;

$2c$ = the distance between the foci ;

$2d$ = the distance between the directrices.

Then these six quantities are related by the four equations

$$c^2 = a^2 + b^2 , \quad c = ae , \quad d = a/e , \quad k = c - d .$$

These quantities are illustrated in Figure 3-11. Moreover all six quantities can be calculated by the equations of 3.4.3 from any two among a, b, c, e, k. In particular the eccentricity $e\,(e > 1)$ alone characterizes the shape of the hyperbola; hyperbolas having the same eccentricity are all *similar*, differing only in the scale to which they are drawn, and vice versa.

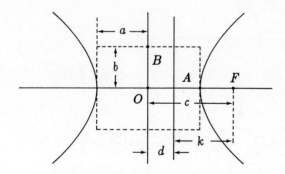

Fig 3-11

As central conics both the ellipse and the hyperbola have a major axis, a minor axes, a pair of foci and a pair of directrices. However the hyperbola

$$\frac{x^2}{a^2} - \frac{y^2}{b^2} = \pm 1$$

has an extra pair of intersecting straight lines, called the *asymptotes*, which stand in a peculiar and important relation to the hyperbola. They are the lines

$$\frac{x}{a} + \frac{y}{b} = 0 \quad \text{and} \quad \frac{x}{a} - \frac{y}{b} = 0 .$$

They intersect at the centre of the hyperbola and form the diagonals of the dotted rectangle of sides $2a$ and $2b$ in Figure 3-11.

3.4.4 THEOREM *Each branch of the hyperbola lies completely within one of the angles formed by the asymptotes. Points of the hyperbola which are far from the centre are arbitrarily close to the asymptotes.*

PROOF Clearly it is sufficient to consider a hyperbola with major axis on the x-axis, the other case being similar. By the symmetries of the hyperbola it is sufficient for us to consider points of the hyperbola in the first quadrant. Let $X_1 = (x_1, y_1)$ be such a point. For the first statement of the theorem we have to show that the point X_1 lies below the asymptote $x/a - y/b = 0$.

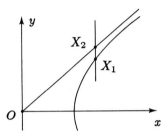

Fig 3-12

It follows from

$$1 = \frac{x_1^{\,2}}{a^2} - \frac{y_1^{\,2}}{b^2} = \left(\frac{x_1}{a} - \frac{y_1}{b}\right)\left(\frac{x_1}{a} + \frac{y_1}{b}\right)$$

that $x_1/a - y_1/b > 0$ and $x_1 b/a > y_1$. If $X_2 = (x_2, y_2)$ is the point on the asymptote with the same abscissa as that of X_1, then $x_2 = x_1$ and $y_2 = x_1 b/a$. Therefore $y_2 > y_1$, proving that the point X_2 on the asymptote lies vertically above the point X_1 on the hyperbola.

For the second statement of the theorem, consider the distance

$$f = \frac{ab}{c}\left(\frac{x_1}{a} - \frac{y_1}{b}\right) \quad \text{where} \quad c = \sqrt{a^2 + b^2}$$

from the point X_1 to the asymptote $x/a - y/b = 0$. Now it follows from

$$\frac{x_1}{a} - \frac{y_1}{b} = 1 / \left(\frac{x_1}{a} + \frac{y_1}{b}\right)$$

110

that

$$0 < f = ab/c\left(\frac{x_1}{a} + \frac{y_1}{b}\right) < ab/c\left(\frac{x_1}{a}\right) = a^2b/cx_1 .$$

Therefore $\lim_{x_1 \to \infty} f = 0$, proving the second statement.

The above properties of the asymptotes suggest a convenient way of sketching the hyperbola $x^2/a^2 - y^2/b^2 = 1$. First we mark off the major axis AA' of length $2a$ and the minor axis BB' of length $2b$ on the coordinate axes. Through their extremities, we draw horizontal and vertical lines to obtain a rectangle. The diagonals of the rectangle have the same length as the distance between the foci of the given hyperbola. Therefore the foci F and F' can be easily located. Extend the diagonals to obtain the asymptotes of the hyperbola. Now the branches can be sketched between the asymptotes passing through the vertices A and A'. See Figure 3-13. The hyperbola $x^2/a^2 - y^2/b^2 = -1$ is similarly sketched.

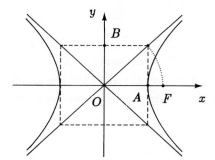

Fig 3-13

Similar to 3.3.4 we have another geometric characterization of the hyperbola.

3.4.5 THEOREM *The locus of a point on the plane, the difference of whose distances from two given points F and F' is constant is a hyperbola with foci at F and F'.*

PROOF Denote the distance between F and F' by $2c$ and the said constant by $2a$. Then $c > a > 0$, otherwise the locus does not exist or consists of two rays in opposite directions with initial points at F and F' respectively. We choose the line passing through F and F' as the x-axis and the midpoint O of F and F' as the origin. Then $F = (c, 0)$ and $F' = (-c, 0)$. The said locus will consist of all points $X = (x, y)$ such that

$$\left|\sqrt{(x-c)^2+y^2}-\sqrt{(x+c)^2+y^2}\right|=2a \qquad (6)$$

or
$$\sqrt{(x-c)^2+y^2}=\sqrt{(x+c)^2+y^2}\pm 2a. \qquad (7)$$

Squaring, we obtain

$$x^2-2cx+c^2+y^2=x^2+2cx+c^2+y^2\pm 4a\sqrt{(x+c)^2+y^2}+4a^2 \qquad (8)$$

or
$$\pm a\sqrt{(x+c)^2+y^2}=a^2+cx. \qquad (9)$$

Squaring again:

$$a^2x^2+2a^2cx+a^2c^2+a^2y^2=a^4+2a^2cx+c^2x^2 \qquad (10)$$

or
$$(a^2-c^2)x^2+a^2y^2=a^2(a^2-c^2)\ . \qquad (11)$$

Since $c>a$, we may put $-b^2=a^2-c^2$ and get

$$\frac{x^2}{a^2}-\frac{y^2}{b^2}=1\ . \qquad (12)$$

Therefore every point X on the said locus is a point of the hyperbola (12) with foci at F and F'. Conversely if $X=(x,y)$ is a point of the hyperbola (12), then equations (11) and (10) are satisfied. Taking square roots, we get (9) and (8). Taking square roots again, we get

$$\sqrt{(x-c)^2+y^2}-\sqrt{(x+c)^2+y^2}=2a$$
or
$$-\sqrt{(x-c)^2+y^2}+\sqrt{(x+c)^2+y^2}=2a$$
or
$$-\sqrt{(x-c)^2+y^2}-\sqrt{(x+c)^2+y^2}=2a$$
or
$$\sqrt{(x-c)^2+y^2}+\sqrt{(x+c)^2+y^2}=2a\ .$$

The third equation is clearly impossible since $a>0$. The fourth equation would mean that in the triangle XFF' the sum of the two sides $|XF|+|XF'|$ is $2a$ which is less than the third side $|FF'|=2c$. But this is absurd. Therefore either the first or the second equation holds, yielding equation (6). The proof is now complete.

Finally the trigonometric identity

$$\sec^2\theta-\tan^2\theta=1$$

can be used to find the parametric representation:

Conic Sections

$$x = a \sec \theta \quad \text{and} \quad y = b \tan \theta$$

of points $X = (x, y)$ of the hyperbola

$$\frac{x^2}{a^2} - \frac{y^2}{b^2} = 1 .$$

This in turn suggests the geometric construction of the hyperbola as sketched in Figure 3-14 below.

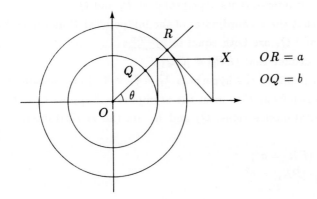

$OR = a$

$OQ = b$

EXERCISES

1. For each of the following, find the equation of hyperbola with the given conditions.
   (a) Vertices at $(0, 1)$ and $(0, -1)$, and foci at $(0, 3)$ and $(0, -3)$.
   (b) Foci at $(12, 0)$ and $(-12, 0)$, and major axis is of length 10.
   (c) Vertices at $(-6, 0)$ and $(6, 0)$, and eccentricity 2.
2. For each of the following, find the equation of hyperbola with the given conditions.
   (a) Passes through the point $(-3, 2\sqrt{3})$ and has the same asymptotes as the hyperbola $\frac{x^2}{9} - \frac{y^2}{16} = 1$.
   (b) With focus $(4, 0)$ and one of the asymptotes $3x - 4y = 0$.
3. Show that the distance between a focus and an asymptote of the hyperbola $\frac{x^2}{a^2} - \frac{y^2}{b^2} = 1$ is equal to b.
4. Given that $P = (x_1, y_1)$ be any point on the hyperbola $\frac{x^2}{a^2} - \frac{y^2}{b^2} = 1$.
   (a) Find the perpendicular distances of P to the two asymptotes.

113

(b) Hence deduce that the product of the two perpendicular distances is a constant.

5. Let $A = (a, 0)$ and $A' = (-a, 0)$ be the vertices of the hyperbola $\frac{x^2}{a^2} - \frac{y^2}{b^2} = 1$. P is any point on the hyperbola. Q is a point on the line $A'P$ such that $\angle QAP$ is a right angle. Find the locus of Q.

6. Let 2θ be the angle between the asymptotes of the hyperbola $\frac{x^2}{a^2} - \frac{y^2}{b^2} = 1$. Prove that $\cos\theta = \frac{1}{e}$.

7. A straight line $y = mx + n$ cuts the hyperbola $\frac{x^2}{a^2} - \frac{y^2}{b^2} = 1$ at points P and Q, and intersects its asymptotes at P_1 and Q_1.

   (a) Show that the x-coordinates of the midpoint of P and Q and that of P_1 and Q_1 are both equal to $\frac{a^2mn}{b^2 - a^2m^2}$.

   (b) Hence deduce that $|PP_1| = |QQ_1|$.

8. Let P be a point on the hyperbola $\frac{x^2}{a^2} - \frac{y^2}{b^2} = 1$. Lines parallel to the axes are drawn from P to meet the asymptotes. Suppose Q and R are the horizontal intersections, Q_1 and R_1 are the vertical intersections. Show that

   (a) $|PQ| \cdot |PR| = a^2$;

   (b) $|PQ_1| \cdot |PR_1| = b^2$.

3.5 Plane sections of a cone

In this section we give justification to calling an ellipse, a parabola or a hyperbola a *conic section*. In other words, we shall see that these curves are actually plane sections of a circular cone.

A circular cone is a surface of revolution, i.e. a surface obtained by rotating a curve about an axis. To obtain a cone, we take a pair of intersecting straight lines B and L and rotate L about B. The line B is called the *axis*, the point A of intersection of B and L the *vertex* and the acute angle θ between L and B the *half angle* of the cone. L or any line joining A with a point X on the cone is called a *generator* or a *ruling* of the cone (see Figure 3-15).

Next we want to derive an equation of a cone. Let K be a cone with vertex A, axis B and half angle θ. If we denote by $\mathbf{b} = [b_1, b_2, b_3]$ a unit vector along B and by $\mathbf{a} = [a_1, a_2, a_3]$ the position vector of A, then a point $X = (x, y, z)$ lies on K if and only if the angle between the vectors $\overrightarrow{AX}$ and $\mathbf{b}$ is θ or $\pi - \theta$. Therefore the cone is defined by the equation

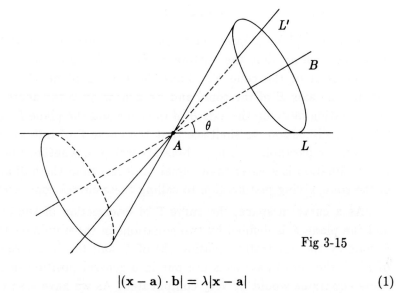

Fig 3-15

$$|(\mathbf{x} - \mathbf{a}) \cdot \mathbf{b}| = \lambda |\mathbf{x} - \mathbf{a}| \tag{1}$$

where $\lambda = \cos\theta$. The cone consists of two sets of points

$$(\mathbf{x} - \mathbf{a}) \cdot \mathbf{b} = \lambda |\mathbf{x} - \mathbf{a}| \quad \text{and} \quad (\mathbf{x} - \mathbf{a}) \cdot \mathbf{b} = -\lambda |\mathbf{x} - \mathbf{a}|$$

which have only the vertex in common and are called the two *nappes* of the cone.

Squaring both sides of (1) and putting it in terms of coordinates, we get an equation of the cone:

$$\{b_1(x - a_1) + b_2(y - a_2) + b_3(z - a_3)\}^2 =$$
$$\lambda^2 \{(x - a_1)^2 + (y - a_2)^2 + (z - a_3)^2\} \tag{2}$$

where $b_1{}^2 + b_2{}^2 + b_3{}^3 = 1$. Naturally the rather formidable equation (2) will become more manageable if the vertex A or the vector $\mathbf{b}$ of the axis takes up some special position relative to the coordinate axes. For example if the axis of the cone falls on the xz-plane, then $a_2 = b_2 = 0$ and equation (2) become

$$\{b_1(x - a_1) + b_3(z - a_3)\}^2 = \lambda^2 \{(x - a_1)^2 + y^2 + (z - a_3)^2\} .$$

Similarly a cone with its vertex at the origin and its axis on the z-axis would be defined by

115

$$z^2 = \lambda^2(x^2 + y^2 + z^2) \ .$$

Now let a plane E intersect a cone K in a curve Γ. For the time being we assume that the cutting plane E does not pass through the vertex A of the cone K. Thus the cutting plane E does not contain the axis B of the cone, and we denote by φ the acute angle of inclination between the axis B of the cone and the plane E. In the following discussion we shall see that the curve Γ of intersection is an ellipse, a parabola or a hyperbola according to whether the angle φ of inclination is greater than, equal to or less than the half angle θ of the cone, giving justification to calling these curves conic sections.

As a curve in space, the curve Γ of intersection of the cone K and the plane E is defined by two equations in three unknowns x, y, z, namely the quadratic equation (2) of K and the linear equation of E. If the coordinate axes are put in a general position in space, these equations would be very complicated. As we have seen earlier that considerable simplification on equation (2) can be achieved by special positions of the axes, we shall now try to place the coordinate axes in an advantageous position.

We have in space a line B and a plane E which incline to each other at an acute angle φ. Therefore we can put the coordinate axes in such a position that
- (a) E coincides with the xy-plane $z = 0$,
- (b) B lies on the xz-plane $y = 0$, and
- (c) φ is the angle between B and the x-axis.

Then it follows from (a) that Γ is a curve on the xy-plane and is defined by a single equation

$$\{b_1(x - a_1) + b_2(y - a_2) - b_3 a_3\}^2 = \lambda^2\{(x - a_1)^2 + (y - a_2)^2 + a_3{}^2\} \quad (3)$$

which is an equation in two unknowns x and y obtained by substituting $z = 0$ into (2).

From (b) it follows that $a_2 = b_2 = 0$ for the vectors **a** and **b**. Therefore equation (3) of Γ becomes

$$\{b_1(x - a_1) - b_3 a_3\}^2 = \lambda^2\{(x - a_1)^2 + y^2 + a_3{}^2\} \ . \quad (4)$$

Finally as $\mathbf{b} = [b_1, 0, b_3]$ is a unit vector on B, its components are the direction cosines of B. Therefore it follows from (c) that $b_1 = \cos\varphi$.

One more freedom of movement of the coordinate axes is still left to us without sacrificing the previously achieved simplification. While keeping E on the xy-plane and at the same time A and B on the xz-plane, we may still move the origin along the x-axis, so that the first coordinate a_1 of A may be adjusted to any prescribed value without altering the values of a_3, b_1 and b_3 in equation (4).

As a summary of the above discussion, we conclude that coordinate axes in space can be so chosen that the plane section Γ of the cone K is defined by an equation of the form (4), where $b_1 = \cos\varphi$, $\lambda = \cos\theta \neq 0$, $a_3 \neq 0$ and $b_3 \neq 0$ if the vertex and the axis of K do not lie on the cutting plane; moreover the value a_1 in (4) can be arbitrarily prescribed. We now propose to study the plane section Γ of the cone in four separate cases in terms of the half angle θ and the angle φ of inclination. The four cases are : (1) $\varphi = \pi/2$, (2) $\varphi = \theta$, (3) $\varphi > \theta$ and (4) $\theta > \varphi$. In each case we first look at the xz-plane and the xy-plane. These two coordinate planes are perpendicular to each other and intersect at the x-axis. The former coordinate plane contains the axis B of the cone K and intersects K in two intersecting rulings. The latter is the cutting plane E on which Γ lies.

CASE 1. $\varphi = \pi/2$.

The axis B of the cone is now perpendicular to the cutting plane (see Figure 3-15). We shall see that the plane section Γ is then a circle (see Figure 3-16). Now $b_1 = \cos\varphi = 0$. Because $b_1{}^2 + b_2{}^2 + b_3{}^2 = 1$ and $b_2 = 0$, we get $b_3{}^2 = 1$. Furthermore we move the origin O to a position so that the vertex A of the cone fall on the z-axis to get $a_1 = 0$. Then the equation (4) of the plane section becomes

$$\lambda^2 x^2 + \lambda^2 y^2 = a_3{}^2(1 - \lambda^2) . \tag{5}$$

With $\lambda = \cos\theta$ and the vertex A not lying on the cutting plane $z = 0$, we see that $a_3{}^2(1 - \lambda^2) > 0$. Therefore (5) is the equation of a circle on the xy-plane.

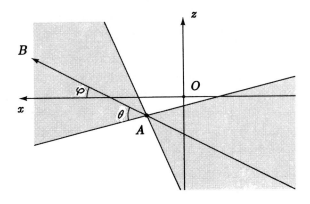

Fig 3-16

CASE 2. $\varphi = \theta$.

On the xz-plane we find exactly one generator of the cone parallel to the x-axis (see Figure 3-19). This generator is therefore parallel to the cutting plane $z = 0$. Moreover the plane $z = 0$ cuts only one nappe of the cone. We shall see that the section Γ is a parabola (see Figure 3-20).

With $b_1 = \cos\varphi = \cos\theta = \lambda$, the equation (4) of the plane section simplifies into

$$2\lambda b_3 a_3 x + \lambda^2 y^2 = 2\lambda b_3 a_1 a_3 - a_3{}^2(\lambda - b_3{}^2) . \tag{6}$$

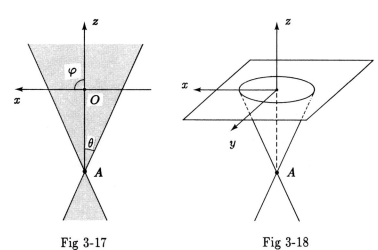

Fig 3-17   Fig 3-18

118

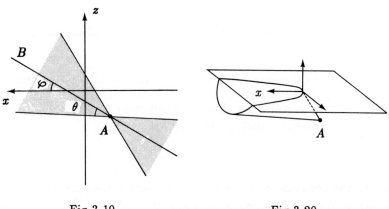

Fig 3-19   Fig 3-20

Since λ, a_3, b_3 are all non-zero, the coefficient of x is non-zero; furthermore we can adjust the value of a_1 so that the constant term becomes zero. Therefore equation (6) is of the form $y^2 = cx$ ($c \neq 0$) and the plane section must be a parabola.

CASE 3. $\varphi > \theta$.

On the xz-plane we now see that the cutting plane goes through only one nappe of the cone (see Figure 3-21). The resulting plane section Γ will be seen to be an ellipse (see Figure 3-22).

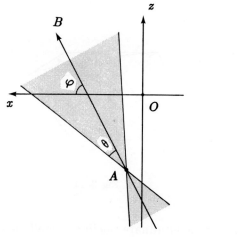

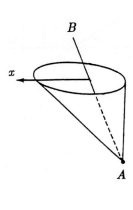

Fig 3-21   Fig 3-22

It follows from $\frac{\pi}{2} > \varphi > \theta > 0$ that $0 < b_1 < \lambda$. Therefore we may put $\mu^2 = \lambda^2 - b_1^2 \neq 0$ and write the equation (4) of the plane section into

$$\mu^2 x^2 - 2(a_1 \mu^2 - b_1 b_3 a_3)x + \lambda^2 y^2 = 2b_1 b_3 a_1 a_3 - a_1^2 \mu^2 - a_3(\lambda^2 - b_3^2) \ . \ (7)$$

Now a_1 can be so adjusted that the coefficient $2(a_1\mu^2 - b_1 b_3 a_3)$ of the linear term vanishes. Hence (7) becomes

$$\mu^2 x^2 + \lambda^2 y^2 = H \ .$$

Finally the constant term H must be positive because for any point $(x_0, y_0) \neq (0,0)$ on the curve we have $H = \mu^2 x_0^2 + \lambda^2 y_0^2$. Therefore the equation of the plane section has the form

$$\mu^2 x^2 + \lambda^2 y^2 = h^2$$

which is the equation of an ellipse.

Case 4. $\theta > \varphi$.

The plane $z = 0$ now cuts both nappes of the cone (see Figure 3-23). The resulting curve has two branches and will be seen to be a hyperbola (see Figure 3-24).

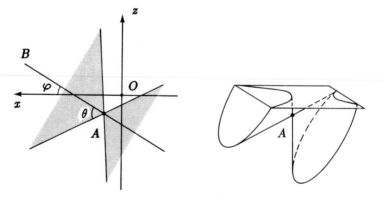

Fig 3-23   Fig 3-24

Set $-\mu^2 = \lambda^2 - b_1^2$ and adjust a_1 to eliminate the linear coefficient. Hence the equation (4) of the plane section becomes

$$-\mu^2 x^2 + \lambda^2 y^2 = K \ .$$

The curve intercepts the x-axis in two points, one of which must be distinct from the origin. Therefore K is negative and the equation has the form

$$-\mu^2 x^2 + \lambda^2 y^2 = -k^2$$

which is the equation of a hyperbola.

The long discussion above shows that circles, ellipses, parabolas and hyperbolas are actually curves of intersection of a cone and a cutting plane. We can now summarize it in the following theorem.

3.5.1 THEOREM *Given a circular cone with half angle θ. Then any plane which does not pass through the vertex of the cone cuts the cone in a conic. If φ denotes the acute angle of inclination between the axis of the cone and the plane then the conic is*

    (a) *a circle if $\varphi = \pi/2$;*
    (b) *an ellipse if $\varphi > \theta$;*
    (c) *a parabola if $\varphi = \theta$;*
    (d) *a hyperbola if $\varphi < \theta$.*

It remains for us to consider the special cases in which the cutting plane passes through the vertex A of the cone K of half angle θ ($0 < \theta < \pi/2$). Let φ ($0 \leq \varphi \leq \pi/2$) denote again the acute angle of inclination between the cutting plane E and the axis B of the cone. Let E pass through the vertex A of the cone. It is easy to see that

    (e) if $\theta < \varphi \leq \pi/2$, then the plane section of the cone consists of a single point A, in which case the point A is called a *degenerate circle* or a *degenerate ellipse*;

    (f) if $\theta = \varphi$, then the plane section of the cone consists of a single ruling of the cone, in which case we call the line a *degenerate parabola*;

    (g) if $\varphi < \theta$, then the plane section of the cone consists of a pair of intersecting rulings of the cone, in which case we call the lines a *degenerate hyperbola*.

A circular cylinder is also thought of as a cone with its vertex at infinity. In this case if the cutting plane is not parallel to the axis of the cylinder, than it cuts the cylinder in a ellipse or a circle. If the plane is parallel to the axis of the cylinder then their intersection is either

(h) empty or

(i) a ruling of the cylinder which is a degenerate parabola or

(j) a pair of parallel rulings of the cylinder which constitute a degenerate hyperbola.

3.6 The Dandelin sphere

A circular cylinder is also a surface of revolution which is obtained by rotating a straight line about an axis parallel to it. Every plane perpendicular to the axis cuts the cylinder in a circle. A plane not perpendicular nor parallel to the axis cuts the cylinder in an ellipse. This can be verified by an investigation similar to the one carried out on the plane sections of a cone in the last section. The detail of this is left to the interested reader as an exercise. But the same conclusion can be reached by a geometric argument. For this purpose, we take a sphere that just fits into the cylinder and move it within the cylinder until it touches the cutting plane. Then we do the same on the other side of the plane. Now the two spheres touch the cylinder in two circles C_1 and C_2; they also touch the cutting plane at two points F_1 and F_2. The circles C_1 and C_2 are situated on two parallel planes both perpendicular to the axis of the cylinder (see Figure 3-25). Let the distance between these planes be denoted by $2a$.

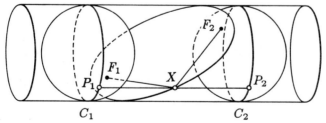

Fig 3-25

Let X be a point on the curve of intersection of the plane and the cylinder. Draw a line on the cylinder through X and parallel to the axis. Then it meets the circles C_1 and C_2 at two points P_1 and P_2. Clearly $|P_1P_2| = 2a$. On the cutting plane, draw XF_1 and XF_2. Then the segments XP_1 and XF_1 are both tangents from X to one of the spheres, while the segments XP_2 and XF_2 are both tangents from X to the other sphere. Therefore $|XP_1| = |XF_1|$ and $|XP_2| = |XF_2|$. It follows that

$$|XF_1| + |XF_2| = |XP_1| + |XP_2| = |P_1P_2| = 2a .$$

By 3.3.4 the curve is an ellipse with foci at F_1 and F_2.

The idea of the proof came from the Belgian mathematician Germinal Dandelin (1794-1847) and the two spheres are called *Dandelin spheres* in his honour. An interesting interpretation of Figure 2-25 is that the ellipse is seen as the shadow of the circle C_1 thrown onto the oblique plane when the light from the left is perpendicular to the plane of the circle. We observe that when the cutting plane is perpendicular to the axis of the cylinder then the two spheres will touch the cutting plane at the same point which is the centre of the circle of intersection of the cylinder and the plane.

The same idea of using two Dandelin spheres can be used to study the plane sections of a circular cone. If the cutting plane is perpendicular to axis of the cone then the spheres which are now of different radii r and r', but both inside the same nappe of the cone, touch the plane at the same point F (see Figure 3-26). As before they also touch the cone in two circles. If X is a point on the plane section, the line joining X and the vertex A of the cone meets the circles of tangency of the incribed spheres at two points P and P'. Being tangents to the same sphere, $|XF| = |XP|$, and similarly $|XF| = |XP'|$. Therefore $|XF| = (|PX| + |XP'|)/2 = |PP'|/2$. Since $|PP'|$ is constant, the plane section must be a circle with centre at F.

When the plane is slightly inclined , still cutting all generators on just one nappe of the cone, the spheres which are still inside the same nappe of the cone now touch the plane at two different points F and F' (see Figure 3-27). The same argument leads us to conclude that the plane section is an ellipse with foci at F and F', but the detail is left to the reader as exercise.

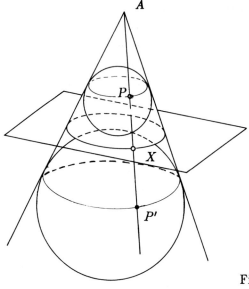

Fig 3-26

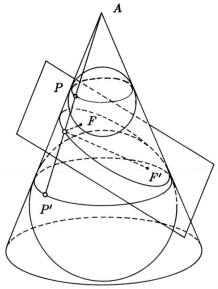

Fig 3-27

Consider one of the spheres, say the one further away from the vertex of the cone. We already know that it touches the cutting plane at a focus F of the ellipse. Let us try to locate the corresponding directrix. The plane of the circle C of tangency of the sphere and the cone is perpendicular to the axis of the cone. On the other hand the cutting plane of the ellipse is obligue to the axis of the cone. Therefore the plane of the circle and the plane of the ellipse intersect at a straight line D. We shall see that D is a directrix of the ellipse. If we denote as before by φ the angle between the axis of the cone and the plane of the ellipse, then the plane of the ellipse intersects the plane of the circle at an angle $90° - \varphi$ in the line D (see Figure 3-28).

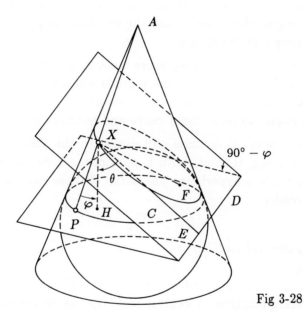

Fig 3-28

To see that D is a directrix of the ellipse, we take an arbitrary point X on the ellipse. The generator of the cone that passes through X intersects the circle C at a point P. Since XF and XP are both tangents to the sphere, we have

$$|XF| = |XP| .$$

Drop from X a perpendicular XH to the plane of the circle, and a perpendicular XE to the line D. Then

$$|XH| = |XE| \sin(90° - \varphi) = |XE| \cos\varphi .$$

On the other hand, XH is parallel to the axis of the cone, therefore

$$|XH| = |XP| \cos \theta$$

where θ is the half angle of the cone. Therefore

$$|XF| = |XE|(\cos \varphi / \cos \theta)$$

which means that from every point X of ellipse the distance $|XF|$ to the focus F and the distance $|XE|$ to the line D are in a constant ratio $\cos \varphi / \cos \theta$. Hence D must be the directrix of the ellipse corresponding to the focus F, and the constant $\cos \varphi / \cos \theta$ must be the eccentricity e of the ellipse.

As the plane is inclined more towards the axis, the angle φ decreases and the ellipse becomes more elongated. Finally when $\varphi = \theta$, the plane becomes parallel to a generator of the cone. The plane section ceases to be a closed curve and becomes a parabola.

If the plane is inclined still nearer to the axis with $\varphi < \theta$ and still decreasing, it meets both nappes of the cone. Now the curve of intersection is a hyperbola. The detail of these two cases will be left to the reader.

3.7 Parallel translation

In the previous sections we have derived the equation of a conic in standard position. A central conic (i.e. an ellipse or a hyperbola) is in a standard position if the centre coincides with the origin and the foci lie on a coordinate axis while a parabola is in standard position if the vertex is at the origin and the focus on a coordinate axis. A conic in standard position is given by an equation of the following form

$$y^2 = 4ax \quad \text{or} \quad x^2 = 4ay$$

or
$$\frac{x^2}{a^2} + \frac{y^2}{b^2} = 1$$

or
$$\frac{x^2}{a^2} - \frac{y^2}{b^2} = \pm 1.$$

In this section we consider such conics being moved out of standard position by a parallel translation of the plane. A *parallel translation* τ or simply a *translation* is a motion of the plane that takes every point $X = (x, y)$ of the plane to the point $\tau(X) = (x+h, y+k)$ for some fixed constants h and k. The translation τ takes the origin $O = (0,0)$ to the point $\tau(O) = (h, k)$ whose coordinates are the given constants h and k that determine τ. Thus the origin and every other point on the plane are displaced by the same vector $[h, k]$ in the sense that for every point X the displacement vector $\overrightarrow{X\tau(X)}$ from X to $\tau(X)$ is $[h, k]$. Therefore we may call $[h, k]$ the *displacement vector of the translation* τ. Clearly the translation τ will take every straight line to a straight line, parallel lines to parallel lines and a conic to a congruent conic.

The translation τ is reversible by its *inverse* τ^{-1} which itself is a translation that takes $X = (x, y)$ to $\tau^{-1}(X) = (x - h, y - k)$. The composite motions $\tau^{-1} \circ \tau$ and $\tau \circ \tau^{-1}$ are both the *identity motion* ι of the plane that leaves every point unchanged: $\iota(X) = X$. A curve Γ in the plane is moved by the translation τ to a congruent curve $\tau(\Gamma)$. Indeed *the curve $\tau(\Gamma)$ consists of all points X on the plane such that $\tau^{-1}(X)$ lies on* Γ. Therefore if the curve Γ is defined by an equation $f(x, y) = 0$, then the translated curve $\tau(\Gamma)$ is defined by the equation

$$f(x - h, y - k) = 0 .$$

For example with $(h, k) = (1, -2)$ the straight line

$$L : 2x + 3y = 6$$

is moved to the line

$$\tau(L) : 2(x - 1) + 3(y + 2) = 6 \quad \text{or} \quad 2x + 3y = 2 .$$

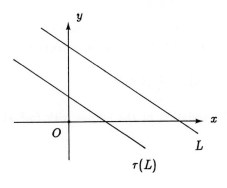

Fig 3-29

127

Similarly an ellipse Γ in standard position

$$\frac{x^2}{a^2} + \frac{y^2}{b^2} = 1$$

is taken by the translation τ to the ellipse $\tau(\Gamma)$:

$$\frac{(x-h)^2}{a^2} + \frac{(y-k)^2}{b^2} = 1$$

or $\qquad b^2 x^2 + a^2 y^2 - 2b^2 hx - 2a^2 ky + b^2 h^2 + a^2 k^2 = a^2 b^2$

which has the centre at (h, k) and the axes at $x = h$ and $y = k$. Thus if $(h, k) \neq (0,0)$ then the translated ellipse $\tau(\Gamma)$ is out of standard position and its equation has non-vanishing linear terms.

3.7.1 EXAMPLE Find the equation of the parabola with focus $F = (2, 3)$ and vertex $V = (2, -1)$.

SOLUTION The axis of the parabola is $x = 2$ and the directrix $y = -5$. Therefore by Definition 1.1.1 the parabola is defined by

$$\sqrt{(x-2)^2 + (y-3)^2} = |y + 5| \, .$$

Squaring: $\qquad\qquad x^2 - 4x - 16y - 12 = 0.$

This is the equation of the parabola. We may also make use of a translation τ to give an alternative solution to the problem.

ALTERNATIVE SOLUTION The distance between the focus and vertex of the parabola is 4. A congruent parabola Γ opening upward in standard position is therefore defined by the equation

$$x^2 = 16y \, .$$

If τ is the translation taking the origin O to the vertex $V = (2, -1)$, then the desired parabola is the curve $\tau(\Gamma)$. Therefore it is defined by

$$(x - 2)^2 = 16(y + 1) \, .$$

3.7.2 EXAMPLE Find the equation of the ellipse with foci at $(2, -1)$, $(2, 7)$ and major axis 10.

SOLUTION The distance between the foci is 8. The centre of the ellipse is the midpoint $(2, 3)$ of the foci. The major axis is parallel to the y-axis and of length 10. The minor axis is parallel to the x-axis with length $2\sqrt{5^2 - 4^2} = 6$. A congruent ellipse with major axis on the y-axis is therefore

$$\frac{x^2}{9} + \frac{y^2}{25} = 1 \ .$$

Translate this ellipse by the displacement vector $[2, 3]$ to obtain

$$\frac{(x-2)^2}{9} + \frac{(y-3)^2}{25} = 1$$

which is the equation of the given ellipse.

In the above discussion we consider a parallel translation τ of the plane with reference to a pair of fixed coordinate axes. We compare the coordinates of a point X on the plane with those of the translated point $\tau(X)$, and based on this we also compare the equation of a curve Γ on the plane with that of the translated curve $\tau(\Gamma)$.

Let us adopt another point of view. We keep the plane with its points and geometric figure fixed and move the xy-coodinate axes by a parallel translation to a new position to form a new pair of $x'y'$-coordinate axes. Then we have two coordinate systems on the same plane. Consequently every point X has two pairs of coordinates, namely (x, y) with respect to the original xy-coordinate axes and a new pair $((x', y'))$ with respect to the new $x'y'$-coordiante axes. For the time being, double parentheses are used to distinguish the new coordinates from the original ones. More precisely if the axes are shifted by a displacement vector $[h, k]$, then the new origin O' of the $x'y'$-plane has coordinates (h, k) in the original xy-plane. The new x'-axis and y'-axis are respectively the lines $y = k$ and $x = h$ in the xy-plane. This is illustrated in Figure 3-30.

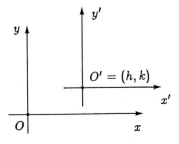

Fig 3-30

Therefore for the origins and the axes, we have

$O' = ((0,0)) = (h, k)$ $O = (0,0) = ((-h, -k))$
x'-axis : $y' = 0$ or $y = k$ x-axis : $y = 0$ or $y' = -k$
y'-axis : $x' = 0$ or $x = h$ y-axix : $x = 0$ or $x' = -h$.

More generally for every point $X = (x, y) = ((x', y'))$ on the plane we have the following relationship between the coordinates

$$\begin{cases} x' = x - h \\ y' = y - k \end{cases} \quad \text{or} \quad \begin{cases} x = x' + h \\ y = y' + k \end{cases} \tag{1}$$

Therefore the parallel shift of coordinate axes gives rise to a *transformation of coordinates* and the equations (1) are called the *equations of the transformation*.

Let us now consider the changes in the equation of a curve under such transformation of coordinates. A curve defined by an equation

$$f(x, y) = 0$$

in the xy-plane will be defined by the equation

$$g(x', y') = f(x' + h, y' + k) = 0$$

in the $x'y'$-plane which is obtained by substituting (1) into $f(x, y) = 0$. For example the straight line

$$2x + 3y - 1 = 0$$

in the xy-plane is the same straight line defined by

$$2x' + 3y' + 2h + 3k - 1 = 0$$

in the $x'y'$-plane.

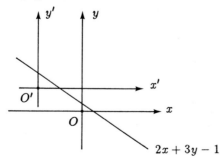

$2x + 3y - 1 = 0$ Fig 3-31

130

3.7.3 EXAMPLE   Find the equation of the parabola $y^2 = 4x$ referred to the parallel axes with the new origin at the point $(-2, 1)$.

SOLUTION   Denote the new parallel axes by x'-axis and y'-axis. With $(h, k) = (-2, 1)$ the transformation of coordinates has the following equations.

$$x = x' - 2 \quad \text{and} \quad y = y' + 1 .$$

Therefore

$$(y' + 1)^2 = 4(x' - 2)$$

or

$$y'^2 - 4x' + 2y' + 9 = 0$$

is the equation of the parabola in the $x'y'$-plane.

In the last example the parabola is in standard position in the original xy-plane and defined by an equation in x and y which is easily recognized as an equation of a parabola. After the coordinate transformation it is no more in standard position and is now defined by an equation in x' and y' which is not easily recognized as an equation of a parabola. In other words an easily recognizable equation of a curve is transformed into a more complex equation of the same curve. Clearly the procedure in the reverse direction that simplifies equations would be more useful. Let us illustrate this by the following examples.

3.7.4 EXAMPLE   What curve is defined by the equation

$$9x^2 + 4y^2 + 18x - 16y = 11 ?$$

SOLUTION   The given equation in x and y is not an equation of a conic in standard position. We shall try to transform this equation into an equation in x' and y' of an easily recognizable form. We write the equation in the form

$$9(x^2 + 2x \quad) + 4(y^2 - 4y \quad) = 11 .$$

In order to make the expression within the first parentheses a perfect square, a constant term 1 ought to be inserted. This means that 9 should be added to both sides of the equation. Similarly 16 should be so added to convert the expression within the second parentheses into a perfect square. Thus the same curve is defined by

$$9(x^2 + 2x + 1) + 4(y^2 - 4y + 4) = 11 + 9 + 16$$

or $\qquad 9(x + 1)^2 + 4(y - 2)^2 = 36 \; .$

Therefore under the coordinate transformation

$$x' = x + 1 \qquad \text{and} \qquad y' = y - 2$$

the same curve is now defined by

$$\frac{x'^2}{4} + \frac{y'^2}{9} = 1 \; .$$

The curve is now in standard position in the $x'y'$-plane. We recognize the curve as an ellipse with major axis $x' = 0$, minor axis $y' = 0$ and centre $((0,0))$. Therefore the same curve is an ellipse with major axis $x = -1$, minor axis $y = 2$ and centre $(-1, 2)$.

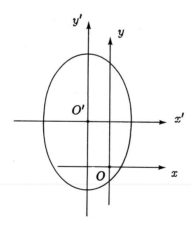

Fig 3-32

3.7.5 EXAMPLE   Show that the quadratic equation

$$4x^2 - 9y^2 - 16x + 18y - 29 = 0$$

represents a hyperbola. Hence find its foci, directrices and asymptotes.

SOLUTION   The method of completing square used in the last example gives

$$4(x^2 - 4x + 4) - 9(y^2 - 2y + 1) = 29 + 16 - 9$$

or $\qquad 4(x - 2)^2 - 9(y - 1)^2 = 36 \; .$

132

Under the coordinate transformation

$$\begin{cases} x' = x - 2 \\ y' = y - 1 \end{cases} \quad \text{or} \quad \begin{cases} x = x' + 2 \\ y = y' + 1 \end{cases}$$

the equation of the curve becomes

$$\frac{x'^2}{9} - \frac{y'^2}{4} = 1 \ .$$

As a hyperbola in standard position in the $x'y'$-plane, it has

foci : $\quad ((\sqrt{13}, 0))$ and $((-\sqrt{13}, 0))$;

directrices : $\quad x' = 9/\sqrt{13}$ and $x' = -9/\sqrt{13}$;

asymptotes : $\quad \dfrac{x'}{3} + \dfrac{y'}{2} = 0$ and $\dfrac{x'}{3} - \dfrac{y'}{2} = 0$.

In the original xy-plane it has

foci : $\quad (2 + \sqrt{13}, 1)$ and $(2 - \sqrt{13}, 1)$;

directrices : $\quad x = 2 + 9\sqrt{13}$ and $x = 2 - 9\sqrt{13}$;

asymptotes : $\quad 2x + 3y - 7 = 0$ and $2x - 3y - 1 = 0$.

After the preliminary success in solving the last two problems by the method of completing the square, it is reasonable for us to expect that by an appropriate change of coordinates we can transform any quadratic equation

$$Ax^2 + Cy^2 + Dx + Ey + F = 0 \tag{2}$$

with vanishing cross product term Bxy into an equation of a conic in standard position. This turns out to be true if we understand the term conic in its broadest sense to include both the non-degenerate and the degenerate conics. To explain the necessity of the inclusion of degenerate conics, we consider two examples. Take the quadratic equation

$$x^2 - y^2 - x - y = 0 \ .$$

After factorization we get

$$(x + y)(x - y - 1) = 0 \ .$$

Therefore the equation defines a pair of perpendicular lines $x + y = 0$ and $x - y - 1 = 0$. But these lines constitute a degenerate hyperbola which is the intersection of a cone and a plane passing the axis. The quadratic equation

$$x^2 + y^2 = -1$$

defines the empty set since no real numbers x and y satisfy the equation. Being the intersection of a cylinder and a plane, the empty set is a degenerate conic.

Let us now return to equation (2) above. Clearly the curve defined by the equation depends entirely on the values of the coefficients A, C, D, E and F. Therefore in order to determine the shape of the curve, we have to study the different configurations of the coefficients individually. These are classified into six cases. Let us begin with three simple cases.

CASE 1. $A = C = 0$.

Equation (2) becomes $Dx + Ey + F = 0$ and ceases to be a quadratic equation. Therefore this is not a genuine case of a quadratic equation and should be excluded from our consideration.

CASE 2. $A \neq 0$, $C = E = 0$.

Equation (2) becomes $Ax^2 + Dx + F = 0$ and is now a quadratic equation in one unknown x. As a quadratic equation in one unknown x, it may have no real root, a double root r or two different real roots r_1 and r_2. Taken as an equation in two unknowns x and y (with vanishing y^2-term and y-term), it defines accordingly the empty set, the straight line $x = r$ which is counted double or a pair of parallel lines $x = r_1$ and $x = r_2$. In other words the curve in this case is a degenerate parabola.

CASE 3. $C \neq 0$, $A = D = 0$.

This is the same as Case 2 with the roles of x and y interchanged. Therefore no further discussion is necessary.

The next three cases will prove to be more interesting.

CASE 4. $A \neq 0$, $C = 0$, $E \neq 0$.

Equation (2) has now the form $Ax^2 + Dx + Ey + F = 0$. By completing the square, we can absorb Ax^2 and Dx into a square $A(x - h)^2$, and rewrite the given equation into the form

$$A(x - h)^2 = -E(y - k)$$

with appropriate constants h and k. Therefore it can be further transformed into an equation of a parabola in standard position. Hence the equation defines a parabola.

CASE 5. $C \neq 0$, $A = 0$, $D \neq 0$.

This is the same as Case 4 with the roles of x and y interchanged. Therefore we have the same conclusion as in the last case.

CASE 6. $A \neq 0$, $C \neq 0$.

Both quadratic terms are now non-zero. By completing the squares, we can rewrite equation (2) into the form

$$A(x - h)^2 + C(y - k)^2 = m$$

with appropriate constants h, k and m. This gives rise to two different subcases.

CASE 6A. $AC > 0$.

A and C have the same sign. If $m = 0$ then the equation defines a curve which consists of a single point (h, k). It is therefore a degenerate ellipse. If m has the same sign as A and C, then the equation defines an ellipse as in Example 3.7.4. If m has a different sign as that of A and C, then the equation defines the empty set. Thus the curve in this case is an ellipse or a degenerate ellipse.

CASE 6B. $AC < 0$.

A and C have opposite signs. If $m \neq 0$, the equation defines a hyperbola as in Example 3.7.5. If $m = 0$, it defines a pair of intersecting lines. Thus the curve in this case is a hyperbola or a degenerate hyperbola.

Clearly the above cases exhaust all possible configurations of the coefficients A, C, D, E, F of the equation (2). Therefore a genuine quadratic equation (2) defines a conic on the xy-plane. The shape of the conic so defined will be recognized more easily if the above six genuine cases and subcases are classified into three types according to whether the expression $-4AC$ is less than zero (case 6a), equal to zero (cases 2, 3, 4, 5) or greater than zero (cases 6b). Therefore we have proved the following theorem.

3.7.6 THEOREM *The curve defined by a quadratic equation*

$$Ax^2 + Cy^2 + Dx + Ey + F = 0$$

with vanishing cross product term is either a conic or a degenerate conic. More precisely the type of the conic depends on the expression $-4AC$ as follows:

	$-4AC < 0$	$-4AC = 0$	$-4AC > 0$
non-degenerate	ellipse	parabola	hyperbola
degenerate	a point, empty set.	two parallel lines, one double line, empty set.	two intersecting lines.

We remark that a satisfactory explanation of the presence of the factor -4 in the expression $-4AC$ will be given in section 3.9 where a more general quadratic equation is examined. Meanwhile we can regard the expression $-4AC$ as a *classification index* by which we can readily recognize the type of the conic $Ax^2 + Cy^2 + Dx + Ey + F = 0$ without actually shifting the coordinate axes into standard position.

EXERCISES

1. The origin is changed to the point $(3,2)$ after a translation. Find the new coordinates of the points $(6,4)$, $(3,2)$, $(5,7)$.
2. After a translation, the points $(1,5)$, $(-2,3)$ become the points $(4,4)$, (a,b). Find the values of a and b.
3. After a translation, the points $(3,2)$, $(1,-2)$ become the points $(a,-1)$, $(-1,b)$. Find the values of a and b.

4. For each of the following, find the equation of the given curve referred to the parallel axes with the new origin at the point S.

   (a) $y^2 - 4x + 4y + 8 = 0$, $S = (1, -2)$;

   (b) $x^2 + 4y^2 - 6x - 16y - 11 = 0$, $S = (3, 2)$;

   (c) $x^2 + y^2 - 2x + 6y - 15 = 0$, $S = (1, -3)$.

5. Find the equation of each of the following parabola with given conditions.

   (a) Focus $F(3, 4)$, vertex $V(3, -2)$.

   (b) Focus $F(-2, 3)$, directrix $y = -4$.

   (c) Focus $F(\frac{3}{8}, -3)$, directrix $x = \frac{13}{8}$.

6. Find the equation of each of the following ellipse with given conditions.

   (a) Foci at $(-3, 2)$, $(-3, 8)$ and major axis of length 12.

   (b) Vertices of major axis are $(6, 2)$, $(-6, 2)$ and one focus is $(4, 2)$.

   (c) End points of major axis lie on the lines $x = -1$ and $x = 7$, and end points of minor axis lie on the lines $y = 4$ and $y = -2$.

In Question 7–10, by appropriate parallel translations, identify the curves defined by the given equations.

7. $4x^2 + y^2 - 8x + 4y + 4 = 0$.

8. $2y^2 + 5x - 4y + 17 = 0$.

9. $x^2 - y^2 + 6x + 4y + 5 = 0$.

10. $3x^2 - 2y^2 - 6x + 8y - 11 = 0$.

11. Find the foci, directrices and asymptotes of the hyperbola $9x^2 - 25y^2 + 36x - 50y + 236 = 0$.

12. Show that a curve of equation $y = ax^2 + 2bx + c$ is a parabola of the form $y' = ax'^2$ by a suitable parallel translation. What is the new origin?

13. What curve is defined by the equation

$$y = \frac{ax + b}{cx + d}, \quad \text{with } ad - bc \neq 0, \, c \neq 0?$$

14. By a parallel translation of axes, points $A = (x_1, y_1)$ and $B = (x_2, y_2)$ are given new coordinates $A = ((x_1', y_1'))$ and $B = ((x_2', y_2'))$ respectively. Show that the distance $|AB|$ does not change under the translation.

15. By a parallel translation of axes, points $A = (x_1, y_1)$, $B = (x_2, y_2)$ and $C = (x_3, y_3)$ are given new coordinates $A = ((x_1', y_1'))$, $B = ((x_2', y_2'))$

and $C = ((x'_3, y'_3))$ respectively. Let $\mathbf{a}$, $\mathbf{b}$, $\mathbf{c}$ be the positional vectors of the points A, B, C on the xy-plane respectively and $\mathbf{a}'$, $\mathbf{b}'$, $\mathbf{c}'$ be the position vectors of the points A, B, C on the $x'y'$-plane respectively. Show that

$$\frac{(\mathbf{b}' - \mathbf{a}') \cdot (\mathbf{c}' - \mathbf{a}')}{|\mathbf{b}' - \mathbf{a}'||\mathbf{c}' - \mathbf{a}'|} = \frac{(\mathbf{b} - \mathbf{a}) \cdot (\mathbf{c} - \mathbf{a})}{|\mathbf{b} - \mathbf{a}||\mathbf{c} - \mathbf{a}|} .$$

Interpret the result geometrically.

3.8 Rotation

A translation is a *rigid motion* of the plane in the sense that it does not alter the length of line segments. Another type of rigid motion is the rotation of the plane about a point. We first study rotations about the origin O.

As usual we follow the convention of measuring angles in the counter-clockwise direction. A point $X = (x, y)$ with cartesian coordinates x and y has also a pair of *polar coordinates* (r, φ) where r is the length $|OX|$ and φ is the angle between the positive x-axis and the ray OX. The relation between these two pairs of coordinates is given by the equations

$$r = \sqrt{x^2 + y^2} ; \quad x = r \cos \varphi ; \quad y = r \sin \varphi .$$

Let ρ be the rotation of the plane about the origin O by an angle θ. Under the rotation ρ, the point X with polar coordinates (r, φ) is moved to the point $\rho(X)$ with polar coordinates $(r, \varphi + \theta)$. Converting this into cartesian coordinates, we get

$$\begin{aligned}
\rho(X) &= \big(r\cos(\varphi + \theta), \ r\sin(\varphi + \theta)\big) \\
&= (r \cos \varphi \cos \theta - r \sin \varphi \sin \theta, \ r \cos \varphi \sin \theta + r \sin \varphi \cos \theta) \\
&= (x \cos \theta - y \sin \theta, x \sin \theta + y \cos \theta) .
\end{aligned}$$

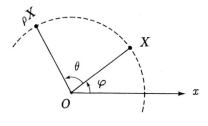

Fig 3-33

138

The rotation ρ is reversible by its inverse ρ^{-1} which is the rotation about O by $-\theta$. For $X = (x, y)$, we get

$$\rho^{-1}(X) = (x \cos\theta + y \sin\theta \, , \, -x \sin\theta + y \cos\theta) \, .$$

The composite motions $\rho^{-1} \circ \rho$ and $\rho \circ \rho^{-1}$ are both the identity motion of the plane that leaves every point of the plane unchanged.

Clearly every point in the plane except the origin O is moved by the rotation ρ if the angle θ is not a multiple of 2π. We call the fixed point O the *centre of rotation* and the angle θ the *angle of rotation*. We remark that if σ is a rotation about an arbitrary point P by an angle θ, then it is easy to see that $\sigma = \tau \circ \rho \circ \tau^{-1}$ where τ is the translation that brings the origin O to the point P. Thus it is sufficient for us to concentrate on rotations about the centre O.

Clearly a curve Γ in the plane is moved by a rotation ρ to a congruent curve $\rho(\Gamma)$. *The rotated curve $\sigma(\Gamma)$ consists of all points X such that $\rho^{-1}(X)$ lies on Γ.* Therefore if a curve Γ is defined by an equation $f(x, y) = 0$, then the rotated curve $\rho(\Gamma)$ is defined by the equation

$$f(x \cos\theta + y \sin\theta \, , \, -x \sin\theta + y \cos\theta) = 0 \, .$$

For example the parabola Γ: $y^2 = 4x$ in standard position is rotated about its vertex by $\theta = \pi/6$ to the parabola $\rho(\Gamma)$ defined by the equation

$$\left(-\frac{1}{2}x + \frac{\sqrt{3}}{2}y\right)^2 = 4\left(\frac{\sqrt{3}}{2}x + \frac{1}{2}y\right)$$

or

$$x^2 - 2\sqrt{3}xy + 3y^2 - 8\sqrt{3}x - 8y = 0.$$

Let us find the vertex, the focus and the axis of the rotated parabola $\rho(\Gamma)$. The vertex of Γ is O; therefore $\rho(O) = O$ is the vertex of $\rho(\Gamma)$. The focus $F = (1, 0)$ of Γ is moved to the foucs

$$\rho(F) = \left(1 \cdot \cos\frac{\pi}{6} - 0 \cdot \sin\frac{\pi}{6} \, , \, 1 \cdot \sin\frac{\pi}{6} + 0 \cdot \cos\frac{2\pi}{6}\right) = (\sqrt{3}/2 \, , \, 1/2)$$

of $\rho(\Gamma)$. The axis of Γ is the x-axis:

$$y = 0 \, .$$

Therefore the axis of $\rho(\Gamma)$ is the line:

$$-x \sin \frac{\pi}{6} + y \cos \frac{\pi}{6} = 0 \quad \text{or} \quad -x + \sqrt{3}y = 0 \,.$$

Thus the axis of the parabola $\rho(\Gamma)$ is no more parallel to either of the coordinate axes. Consequently the equation of the parabola $\sigma(\Gamma)$ has a non-vanishing cross product term $2\sqrt{3}xy$.

Clearly if the parabola $\rho(\Gamma)$ is rotated about O by $-\frac{\pi}{6}$, it will be brought back to the original Γ in standard position. This suggests that rotations can be used to bring a conic whose equation contains a non-vanishing cross product term into a standard position.

For example, consider the curve defined by

$$xy = 1$$

being rotated about O through an angle ψ. Then the rotated curve is defined by the equation

$$(x \cos \psi + y \sin \psi)(-x \sin \psi + y \cos \psi) = 1$$

which simplifies into

$$-(\sin \psi \cos \psi)x^2 + (\cos^2 \psi - \sin^2 \psi)xy + (\sin \psi \cos \psi)y^2 = 1$$

or $\qquad (-\tfrac{1}{2} \sin 2\psi)x^2 + (\cos 2\psi)xy + (\tfrac{1}{2} \sin 2\psi)y^2 = 1.$

In order to eliminate the cross product term in this equation, we select the angle of rotation ψ to be $-\pi/4$. Then the equation of the rotated curve is

$$x^2 - y^2 = 2$$

which is recognized as a hyperbola in standard position. Therefore the original curve $xy = 1$ is a hyperbola with the perpendicular axes $x - y = 0$ and $x + y = 0$. Readers are asked to find its vertex, foci, eccentricity and asymptotes.

Let us try to use the above idea in conjunction with a coordinate transformation. Instead of rotating the plane, we shall now keep the points and geometric figures on the plane fixed and rotate the xy-

coordinate axes about the origin O through an angle θ to a new position to become a new pair of $x'y'$-coordinate axes. Similar to the shifting of coordinate axes, every point in the plane now has two pairs of coordinates as a consequence of the rotation, namely (x, y) with respect to the original xy-axes and a new pair $((x', y'))$ with respect to the new $x'y'$-axes. If (r, φ) are the polar coordinate of the point $X = (x, y)$ in the xy-plane, then $((r, \varphi - \theta))$ will be the polar coordinates of the same points $X = ((x', y'))$ on the $x'y'$-plane.

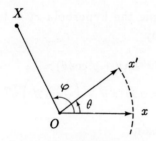

Fig 3-34

Therefore

$$x' = r\cos(\varphi - \theta) = r\cos\varphi\cos\theta + r\sin\varphi\sin\theta$$
$$= x\cos\theta + y\sin\theta$$
$$y' = r\sin(\varphi - \theta) = r\sin\varphi\cos\theta - r\cos\varphi\sin\theta$$
$$= -x\sin\theta + y\cos\theta \ .$$

Hence the rotation of coordinate axes gives rise to a transformation of coordinates which are represented by the equations:

$$\begin{cases} x' = x\cos\theta + y\sin\theta \\ y' = -x\sin\theta + y\cos\theta \end{cases} \text{or} \begin{cases} x = x'\cos\theta - y'\sin\theta \\ y = x'\sin\theta + y'\cos\theta \end{cases} . \qquad (1)$$

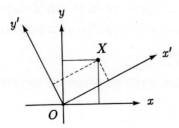

Fig 3-35

141

Consequently the equation of a curve also undergoes certain changes under the above transformation of coordinates. If a curve is defined by the equation

$$f(x, y) = 0$$

in the xy-plane, then the same curve is defined by the transformed equation

$$g(x', y') = f(x' \cos \theta - y' \sin \theta, \ x' \sin \theta + y' \cos \theta) = 0$$

in the $x'y'$-plane. For example the hyperbola $xy = 1$ in the xy-plane is defined in the $x'y'$-plane by

$$(x' \cos \theta - y' \sin \theta)(x' \sin \theta + y' \cos \theta) = 1$$

or
$$\left(\frac{1}{2} \sin 2\theta\right) x'^2 + (\cos 2\theta)x'y' - \left(\frac{1}{2} \sin 2\theta\right) y'^2 = 1 \ .$$

If we choose $\theta = \pi/4$, we obtain

$$x'^2 - y'^2 = 2$$

which is an equation of a hyperbola in standard position in that particular $x'y'$-plane.

Similarly we can use the equations of the coordinate transformation to convert the equation of a curve in the $x'y'$-plane into an equation of the same curve in the xy-plane. For example the curve

$$x'^2 - 2\sqrt{3}\ x'y' + 3y'^2 - 8\sqrt{3}\ x' - 8y' = 0$$

in the $x'y'$-plane is defined by

$$(x \cos \theta + y \sin \theta)^2 - 2\sqrt{3}\ (x \cos \theta + y \sin \theta)(-x \sin \theta + y \cos \theta)$$
$$+ 3(-x \sin \theta + y \cos \theta)^2 - 8\sqrt{3}\ (x \cos \theta + y \sin \theta)$$
$$- 8(-x \sin \theta + y \cos \theta) = 0$$

or
$$(\cos^2 \theta + \sqrt{3} \sin 2\theta + 3 \sin^2 \theta)x^2 - (2 \sin 2\theta + 2\sqrt{3} \cos 2\theta)xy$$
$$+ (\sin^2 \theta - \sqrt{3} \sin 2\theta + 3 \cos^2 \theta)y^2 - (8\sqrt{3} \cos \theta - 8 \sin \theta)x$$
$$- (8\sqrt{3} \sin \theta + 8 \cos \theta)y = 0$$

in the xy-plane. Choosing $2\theta = -\pi/3$ to eliminate the cross product term, we obtain

$$4y^2 = 16x$$

which defines a parabola in standard position in that particular xy-plane.

3.8.1 EXAMPLE   Identity the curve defined by the equation

$$5x^2 - 6xy + 5y^2 - 8 = 0 .$$

SOLUTION Transform the given equation by a rotation of axes by substituting

$$x = x' \cos\theta - y' \sin\theta \quad \text{and} \quad y = x' \sin\theta + y' \cos\theta$$

to get

$$(5 - 6\sin\theta\cos\theta)x'^2 - 6(\cos^2\theta - \sin^2\theta)x'y'$$
$$+ (5 + 6\sin\theta\cos\theta)y'^2 - 8 = 0 .$$

Now we may select $\theta = \pi/4$ to get rid of the cross product term and obtain:

$$2x'^2 + 8y'^2 - 8 = 0$$

or

$$\frac{x'^2}{4} + \frac{y'^2}{1} = 1 .$$

Therefore the original equation is the equation of an ellipse in the xy-plane. The centre of the ellipse is at the origin O and its major axis is $x - y = 0$ (or $y' = 0$).

3.8.2 EXAMPLE   Identify the curve defined by the equation

$$6x^2 - xy - 2y^2 + 4x + 9y - 10 = 0 .$$

SOLUTION In order to lighten the burden of working with sines and cosines, we first eliminate the linear terms by a shift of axes. Subtitute

$$x = x' + h \quad \text{and} \quad y = y' + k$$

into the given equation to get

$$6x'^2 - x'y' - 2y'^2 + (12h - k + 4)x' + (-h - 4k + 9)y' + F' = 0$$

where $F' = 6h^2 - hk - 2k^2 + 4h + 9k - 10$. Put $h = -1/7$ and $k = 16/7$ to get

$$6x'^2 - x'y' - 2y'^2 = 0 .$$

Factorize the left-hand side to get

$$(3x' - 2y')(2x' + y') = 0$$

which represents a pair of straight lines passing through the origin in the $x'y'$-plane. Substitute back

$$x' = x + 1/7 \quad \text{and} \quad y' = y - 16/7$$

to get

$$(3x - 2y + 5)(2x + y - 2) = 0$$

which is the original given equation. Therefore the given equation represents a pair of straight lines $3x - 2y = -5$ and $2x + y = 2$ intersecting at $(-1/7, 16/7)$ in the xy-plane.

3.8.3 EXAMPLE   Identify the curve defined by the equation

$$5x^2 - 6xy + 5y^2 - 4x - 4y - 4 = 0 \ .$$

SOLUTION Obviously the correct way of attack is to simplify the equation by a series of coordinate transformations. The equation has both non-vanishing x^2- and y^2-terms. Therefore it is possible to get rid of its linear terms by a shift of coordinate axes. Put

$$x = x' + h \quad \text{and} \quad y = y' + k$$

and substitute to get

$$5x'^2 - 6x'y' + 5y'^2 + (10h - 6k - 4)x' + (-6h + 10k - 4)y'$$
$$+ (5h^2 - 6hk + 5k^2 - 4h - 4k - 4) = 0 \ .$$

Hence we select $(h, k) = (1, 1)$ to get

$$5x'^2 - 6x'y' + 5y'^2 - 8 = 0 \ .$$

To get rid of the cross product term, we rotate the $x'y'$-axes about $O' = ((0,0)) = (h, k) = (1, 1)$ through θ. Subtitute

$$x' = \overline{x} \cos \theta - \overline{y} \sin \theta$$
$$y' = \overline{x} \sin \theta + \overline{y} \cos \theta$$

into the last equation of the curve in the $x'y'$-plane to get

144

$$(5 - 3\sin 2\theta)\bar{x}^2 - (6\cos 2\theta)\bar{x}\,\bar{y} + (5 + 3\sin 2\theta)\bar{y}^2 - 8 = 0 \ .$$

Hence we select

$$\cos 2\theta = 0 \quad \text{or} \quad \theta = \pi/4$$

to obtain

$$2\bar{x}^2 + 8\bar{y}^2 - 8 = 0$$

which is the equation of the given curve in the $\bar{x}\,\bar{y}$-plane. The curve is now recognized as an ellipse with

centre : $\quad (((0,0))) = ((0,0)) = (1,1)$;

major axis : $\quad \bar{y} = 0 \quad$ or $\quad -x' + y' = 0 \quad$ or $\quad x - y = 0$.

3.8.4 EXAMPLE   Identify the curve defined by

$$7x^2 - 8xy + y^2 + 14x - 8y - 2 = 0 \ .$$

SOLUTION Rewrite the given equation into

$$7(x^2 + 2x + 1) - 8(x + 1)y + y^2 - 9 = 0 \ .$$

Put $x' = x + 1$ and $y' = y$. Then we have the equation

$$7x'^2 - 8x'y' + y'^2 - 9 = 0$$

of the curve in the $x'y'$-plane. Now rotate the $x'y'$-axes through θ to get the equation

$$7(\bar{x}\cos\theta - \bar{y}\sin\theta)^2 - 8(\bar{x}\cos\theta - \bar{y}\sin\theta)(\bar{x}\sin\theta + \bar{y}\cos\theta)$$
$$+ (\bar{x}\sin\theta + \bar{y}\cos\theta)^2 - 9 = 0$$

of the curve in the $\bar{x}\,\bar{y}$-plane. This simplifies into

$$(1 + 6\cos^2\theta - 4\sin 2\theta)\bar{x}^2 - (8\cos 2\theta + 6\sin 2\theta)\bar{x}\,\bar{y}$$
$$+ (1 + 6\sin^2\theta + 4\sin 2\theta)\bar{y}^2 - 9 = 0$$

or

$$(4 + 3\cos 2\theta - 4\sin 2\theta)\bar{x}^2 - (8\cos 2\theta + 6\sin 2\theta)\bar{x}\,\bar{y}$$
$$+ (4 - 3\cos 2\theta + 4\sin 2\theta)\bar{y}^2 - 9 = 0 \ .$$

145

Select a value of the angle θ of rotation so that

$$0 \le 2\theta \le \pi \quad \text{and} \quad \tan 2\theta = -8/6 .$$

Then $\sin 2\theta = 8/10$ and $\cos 2\theta = -6/10$. Substitute into the equation of curve in the $\overline{x}\,\overline{y}$-plane to get

$$-\overline{x}^2 + 9\overline{y}^2 = 9 .$$

We recognize the curve to be a hyperpola.

EXERCISES

1. Find the new coordinates of $(6, 12)$ after the xy-axes are rotated through
   (a) $\frac{\pi}{4}$;
   (b) $-\frac{\pi}{6}$.
2. For each of the following, find the angle θ $(-\pi < \theta < \pi)$ the xy-axes are to be rotated so that
   (a) $P = (-1, 2) = ((2, -1))$;
   (b) $P = (4, 12) = ((-4\sqrt{2}, 8\sqrt{2}))$.
3. Show that the equation $x^2 + y^2 = r^2$ is invariant under a rotation of axes.

In Question 4–6, transform the given equations by rotating the axes in the specified angles.

4. $13x^2 - 10xy + 13y^2 = 72$, $\theta = \frac{\pi}{4}$.
5. $3x^2 - 4\sqrt{3}xy - y^2 = 6$, $\theta = \frac{\pi}{3}$.
6. $2x^2 - 3xy - 2y^2 = -20$, $\theta = \tan^{-1}\left(\frac{1}{2}\right)$ and $0 < \theta < \frac{\pi}{2}$.

In Question 7–9, identify the curves defined by the given equations by performing a suitable rotation of the axes.

7. $8x^2 + 4xy + 5y^2 = 144$.
8. $x^2 - 4xy + y^2 = 36$.
9. $9x^2 - 2\sqrt{3}xy + 11y^2 = 24$.

In Questions 10 and 11, identify the curve defined by the given equation by performing a suitable parallel translation and a rotation of the axes.

10. $2x^2 + 4xy + 5y^2 - 4x - 22y + 7 = 0$.
11. $x^2 - 2xy + y^2 - 2x - 2y + 1 = 0$.

12. Find the angle through which the axes may be turned so that the equation $Ax + By + C = 0$ is transformed to $x' = h$. Find also the value of h.

13. Given two straight lines $y = mx$ and $y = m_1 x$. The axes is now rotated so that $y = mx$ becomes the new x'-axis. Find the equation of $y = m_1 x$ in the new coordinate system.

14. Refer to Question 14 of section 3.7. Show that $|AB|$ is also invariant under a rotation.

15. Refer to Question 15 of section 3.7. If we replace "a parallel translation of axes" by "a rotation of axes" show that

$$\frac{(\mathbf{b'} - \mathbf{a'}) \cdot (\mathbf{c'} - \mathbf{a'})}{|\mathbf{b'} - \mathbf{a'}||\mathbf{c'} - \mathbf{a'}|} = \frac{(\mathbf{b} - \mathbf{a}) \cdot (\mathbf{c} - \mathbf{a})}{|\mathbf{b} - \mathbf{a}||\mathbf{c} - \mathbf{a}|} .$$

Interpret the result geometrically.

3.9 General quadratic equation

Now that we have learnt the method of shifting and rotating coordinates axes, we are able to carry out a careful study of the general quadratic equation

$$Ax^2 + Bxy + Cy^2 + Dx + Ey + F = 0 \tag{1}$$

in two variables x and y. From the several examples of the last section we know that by a coordinate transformation

$$x = x' \cos \theta - y' \sin \theta$$
$$y = x' \sin \theta + y' \cos \theta \tag{2}$$

the equation (1) is transformed into the equation

$$A'x'^2 + B'x'y' + C'y'^2 + D'x' + E'y' + F' = 0 \tag{3}$$

whose coefficients are expressed in terms of the coefficients of (1) as

$$A' = A \cos^2 \theta + B \cos \theta \sin \theta + C \sin^2 \theta \ ;$$
$$B' = 2(C - A) \cos \theta \sin \theta + B(\cos^2 \theta - \sin^2 \theta) \ ;$$
$$C' = A \sin^2 \theta - B \cos \theta \sin \theta + C \cos^2 \theta \ ; \tag{4}$$

$$D' = D \cos \theta + E \sin \theta \; ;$$
$$E' = -D \sin \theta + E \cos \theta \; ;$$
$$F' = F \; .$$

Special attention is given first to the coefficient B' of the cross product term which we wish to eliminate by an appropriate rotation of coordinate axes. Using double angles we can write it as

$$B' = (C - A) \sin 2\theta + B \cos 2\theta \; .$$

If $(C - A)^2 + B^2 \neq 0$, we can eliminate B' by selecting a value of the angle θ of rotation such that

$$\sin 2\theta = B/\sqrt{(C - A)^2 + B^2} \, , \cos 2\theta = (A - C)/\sqrt{(C - A)^2 + B^2} \; . \quad (5)$$

Using the well-known trigonometric identities

$$\cos 2\theta = 2 \cos^2 \theta - 1 \quad \text{and} \quad \cos 2\theta = 1 - 2 \sin^2 \theta \; ,$$

we can calculate the corresponding values of A', C', D', E' and F'. Therefore by an appropriate coordinate transformation (2) the general quadratic equation (1) is transformed into a quadratic equation

$$A'x'^2 + C'y'^2 + D'x' + E'y' + F' = 0 \quad (6)$$

with vanishing cross product term.

If $(C - A)^2 + B^2 = 0$, then $B = 0$. Hence the equation (1) is already a quadratic equation with vanishing cross product term. Appealing to Theorem 3.7.6, we have now proved the following theorem.

3.9.1 THEOREM. *Every quadratic equation*

$$Ax^2 + Bxy + Cy^2 + Dx + Ey + F = 0 \quad (1)$$

can be transformed by an appropriate rotation of coordinate axes into a quadratic equation

$$A'x'^2 + C'y'^2 + D'x' + E'y' + F' = 0 \quad (6)$$

with vanishing cross product term. Hence every quadratic equation defines a conic or a degenerate conic.

It follows from 3.9.1 and 3.7.6 that once equation (1) is transformed into equation (6) by an appropriate rotation of axes, we can identify the type of conic defined by (1) according to whether the value of $-4A'C'$ is less than, equal to or greater than zero. Here the expression $-4A'C'$ serves as a quick classification index for the type of conics without further shifting the $x'y'$-axes into standard position. But in order to evaluate $-4A'C'$, we have to find θ by equations (5) and evaluate A' and C' by equations (4). As we have seen in the last section, this usually means a great deal of calculation. Therefore we would like to have an expression in the coefficients $A, B, C, \ldots$ of (1) as a classification index for the conic (1) without actually rotating the xy-axes. In other words, we look for an expression in $A, B, C \ldots$ whose value would not change under any rotation of the xy-axes but would become $-4A'C'$ under the particular rotation that elimates B. We shall see shortly that $B^2 - 4AC = B^2 + (C - A)^2 - (C + A)^2$ is one such expression.

Taken individually each coefficient of (1) except F changes under a rotation of coordinate axes according to equations (4). So does the expression $-4AC$. Take for example the equation $2xy = 1$. We have $-4AC = 0$. Rotating the axes by $\pi/4$, we obtain $x'^2 - y'^2 = 1$ and $-4A'C' = 1$. But certain combinations of them are indeed invariant under any rotation of coordinate axes. Take for instance the expression $C + A$. Indeed $C + A$ becomes

$$C' + A' = A\sin^2\theta - B\cos\theta\sin\theta + C\cos^2\theta + A\cos^2\theta + B\cos\theta\sin\theta + C\sin^2\theta$$
$$= C(\sin^2\theta + \cos^2\theta) + A(\cos^2\theta + \sin^2\theta)$$
$$= C + A .$$

Therefore $C + A$ is an invariant of the equation (1) under any rotation of coordinate axes. Hence its square $(C + A)^2$ is also an invariant. But so is the expression $B^2 + (C - A)^2$. Because

$$B'^2 + (C' - A')^2 = [(C - A)\sin 2\theta + B\cos 2\theta]^2$$
$$+ [(C - A)\cos 2\theta - B\sin 2\theta]^2$$
$$= B^2(\cos^2 2\theta + \sin^2 2\theta) + (C - A)^2(\sin^2 2\theta + \cos^2 2\theta)$$
$$= B^2 + (C - A)^2 ,$$

therefore the difference of the last two invariants is also an invariant:

$$B'^2 + (C' - A')^2 - (C' + A')^2 = B^2 + (C - A)^2 - (C + A)^2$$

from which follows that

$$B'^2 - 4A'C' = B^2 - 4AC .$$

This means that the expression $B^2 - 4AC$ is an invariant under any rotation of coordinate axes. Now under the particular rotation that eliminates B', the expression becomes $-4A'C'$. Therefore the expression $B^2 - 4AC$ is the desired classification index. Applying 3.9.1 and 3.7.6, we have therefore proved the following theorem.

3.9.2 THEOREM *The conic defined by the quadratic equation*

$$Ax^2 + Bxy + Cy^2 + Dx + Ey + F = 0$$

is of elliptic type if $B^2 - 4AC < 0$. It is of parabolic type if $B^2 - 4AC = 0$. It is of hyperbolic type if $B^2 - 4AC > 0$. More precisely:

	$B^2 - 4AC < 0$	$B^2 - 4AC = 0$	$B^2 - 4AC > 0$
non-degenerate	ellipse	parabola	hyperbola
degenerate	a point, empty set.	two parallel lines, one double line, empty set.	two inter-secting lines.

The expression $B^2 - 4AC$ is sometimes called the *discriminant* of the quadratic equation $Ax^2 + Bxy + Cy^2 + Dx + Ey + F = 0$ in two unknowns. The discriminant of $Ax^2 + Cy^2 + Dx + Ey + F = 0$ is therefore $-4AC$ which explains the presence of the factor -4 in 3.7.6.

The present discriminant $B^2 - 4AC$ shares the name and the form with the discriminant $b^2 - 4ac$ of a quadratic equation $az^2 + bz + c = 0$ in one unknown. Actually they have more than name and appearance in common. Recall that if $b^2 - 4ac = 0$, then $az^2 + bz + c = 0$ has equal real roots and can be rewritten as $(2az + b)^2/4a = 0$. Similarly if the curve defined by an equation $Ax^2 + Bxy + Cy^2 + Dx + Ey + F = 0$ has

$B^2 - 4AC = 0$ and $A > 0$, the curve is a parabola and the equation can be rewritten into $(\sqrt{A}x \pm \sqrt{C}y)^2 = -(Dx + Ey + F)$. If it can be further rewritten into $(ax + by + c)^2 = k(ax - by + d)$, then we recognize the axis of the parabola to be the line $ax + by + c = 0$ and the vertex as the intersection of the perpendicular lines $ax + by + c = 0$ and $ax - by + d = 0$. This means that in the parabolic case we have an alternative method in identifying the vertex, focus and axis of the conic without using equations (4) and (5).

3.9.3 EXAMPLE   For the equation $x^2 - 4xy + 4y^2 - 4x - 2y = 0$ the discriminant $B^2 - 4AC = 4^2 - 4 \cdot 1 \cdot 4 = 0$. Therefore the curve is of parabolic type. After factorization, we may write the given equation as

$$(-x + 2y)^2 = 2(2x + y) .$$

The linear polynomials within the parentheses defined two lines

$$2x + y = 0 \quad \text{and} \quad -x + 2y = 0$$

which are perpendicular to each other and intersect at the origin O. Therefore they can be used as a pair of new coordinate axes. In terms of coordinate transformation we put

$$x' = (2/\sqrt{5})x + (1/\sqrt{5})y \quad \text{and} \quad y' = (-1/\sqrt{5})x + (2/\sqrt{5})y .$$

We observe that the multiplication by $1/\sqrt{5}$ is necessary to put them into the form of a coordinate transformation. The given equation is now transformed into

$$y'^2 = (2/\sqrt{5})x'$$

under a rotation of the axes about O through an angle θ, such that $0 < \theta < \pi/2$ and $\cos\theta = 2/\sqrt{5}$. The curve is now seen to be a parabola with

vertex : $O' = ((0,0)) = (0,0)$;

focus : $F' = ((1/2\sqrt{5}, 0)) = (2/10, 1/10)$;

axis : $y' = 0 \quad \text{or} \quad x - 2y = 0$.

3.9.4 EXAMPLE   For the equation $8x^2 + 24xy + 18y^2 - 14x - 21y + 3 = 0$, the discriminant is $B^2 - 4AC = 24^2 - 4 \cdot 8 \cdot 18 = 0$. Therefore the curve is of parabolic type. We notice here that the quadratic terms form a perfect square. Therefore the given equation is

$$2(2x + 3y)^2 - 7(2x + 3y) + 3 = 0 .$$

Treated as a polynomial in $2x + 3y$, it can be factored:

$$[2(2x + 3y) - 1][(2x + 3y) - 3] = 0 .$$

Therefore the curve is a degenate parabola consisting of a pair of parallel lines

$$4x + 6y - 1 = 0 \quad \text{and} \quad 2x + 3y - 3 = 0 .$$

In comparison with the method used in the last section our new method used in the last examples has the advantage of being simple and direct. However it is not always possible to obtain a pair of perpendicular lines right away. Sometimes a slight modification of the method will be necessary. Such modification is used in the following example.

3.9.5 EXAMPLE   For the equation $x^2 + 2xy + y^2 - 2x - 10y + 5 = 0$, $B^2 - 4AC = 2^2 - 4 \cdot 1 \cdot 1 = 0$. Therefore the curve is of parabolic type. Following the method we write it as

$$(x + y)^2 = 2x + 10y - 5 .$$

The lines $x + y = 0$ and $2x + 10 - 5 = 0$ are not perpendicular; therefore they cannot serve as new coordinate axes. To overcome this difficulty, we introduce a constant p and rewrite the equation as

$$(x + y + p)^2 = (x + y)^2 + 2p(x + y) + p^2 = (2p + 2)x + (2p + 10)y + p^2 - 5 .$$

A value of p such that the lines

$$x + y + p = 0 \quad \text{and} \quad (2p + 2)x + (2p + 10)y + p^2 - 5 = 0$$

become perpendicular to each other is $p = -3$. After substitution the given equation is now in the form

$$(x + y - 3)^2 = 4(-x + y + 1) .$$

The perpendicular lines $x + y - 3 = 0$ and $-x + y + 1 = 0$ intersect each other at the point $(2, 1)$. Now use the point $O' = ((0, 0)) = (2, 1)$ as the new origin and shift the coordinate axes by

$$\begin{cases} x' = x - 2 \\ y' = y - 1 \end{cases} \quad \text{or} \quad \begin{aligned} x &= x' + 2 \\ y &= y' + 1 \end{aligned}$$

and transform the given equation into

$$(x' + y')^2 = 4(-x' + y') .$$

Rotate the $x'y'$-axes about the new origin O' by

$$\begin{cases} \bar{x} = \dfrac{1}{\sqrt{2}}(x' + y') \\ \bar{y} = \dfrac{1}{\sqrt{2}}(-x' + y') \end{cases} \quad \text{or} \quad \begin{cases} x' = \dfrac{1}{\sqrt{2}}(\bar{x} - \bar{y}) \\ y' = \dfrac{1}{\sqrt{2}}(\bar{x} + \bar{y}) \end{cases}$$

and transform the equation into

$$\bar{x}^2 = 2\sqrt{2}\,\bar{y} .$$

Therefore the curve is a parabola in standard position in the $\bar{x}\,\bar{y}$-plane with

$$\text{vertex} : \overline{O} = (((0,0))) = ((0,0)) = (2,1) ;$$
$$\text{focus} : \overline{F} = (((0, 1/\sqrt{2}))) = ((-1/2, 1/2)) = (3/2, 3/2) ;$$
$$\text{axis} : \bar{x} = 0 \quad \text{or} \quad x' + y' = 0 \quad \text{or} \quad x + y = 3 .$$

Though the method used in the last two examples on conics of parabolic type delivers the desired result more quickly, it is not generally applicable to treat conics of elliptic or hyperbolic type.

EXERCISES

1. By using Theorem 3.9.2, name the type of graph of
   (a) $x^2 + 4y^2 - 2x - 16y + 1 = 0$;
   (b) $x^2 + 3xy - 5y^2 - 11 = 0$;
   (c) $9x^2 - 24xy + 16y^2 - 20x + 110y - 50 = 0$;
   (d) $14x^2 + 24xy + 21y^2 - 4x + 18y - 139 = 0$.
2. Show that the equation $2x^2 + 3xy - 2y^2 + 5y - 2 = 0$ represents a pair of straight lines and write down the corresponding equations.

3. Identify the curve $144x^2 - 120xy + 25y^2 - 65x - 156y - 169 = 0$ and find its equation in standard position by a suitable transformation.

3.10 Tangents

Given a point P on a circle with centre O. We draw through P a line particular to OP to obtain the tangent at P to the circle. Such simple geometric construction cannot be applied to drawing tangents to other conics.

Let Γ be an arbitrary non-degenerate conic and P a point on Γ at which it is required to find a tangent T. Let P' be a second point on Γ and draw the secant PP'. When P' moves along the conic Γ, the secant rotates about P. As P' approaches P along Γ, the secant approaches a limiting position. The secant at this limiting position is then the *tangent T* to Γ at P.

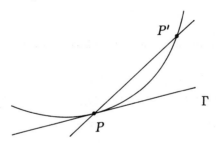

Fig 3-36

If the slope of the secant PP' is denoted by λ', then the limit λ of λ' as P' approaches P is the slope of the tangent T at P. The slope λ of the tangent to Γ at P is also called the *slope of the curve Γ at P.* We observe that with the slope λ and the point P, the equation of the tangent at P to Γ is readily written.

In the following examples, we shall see that the limit λ does exist for all types of non-degenerate conics. For this purpose it is clearly sufficient to put the conic in standard position.

3.10.1 EXAMPLE   Let $P = (x_0, y_0)$ be a fixed point on the parabola $x^2 = 4ay$, and let $P' = (x_0 + h, y_0 + k)$ with $h \neq 0$ be a second point on the parabola. Then the slope of the secant PP' is

$$\lambda' = k/h .$$

Both P and P' are points on the given parabola; therefore

$$x_0{}^2 = 4ay_0 \quad \text{and} \quad x_0{}^2 + 2hx_0 + h^2 = 4a(y_0 + k) \ .$$

Hence $\qquad k = (2hx_0 + h^2)/4a \quad \text{and} \quad \lambda' = (2x_0 + h)/4a.$

As P' approaches P, h tends to zero. Therefore

$$\lambda = \lim_{P' \to P} \lambda' = \lim_{h \to 0} (2x_0 + h)/4a = x_0/2a \ .$$

The slope λ of the parabola at P does exists and the tangent at P is the line

$$2a(y - y_0) = x_0(x - x_0) \quad \text{or} \quad 2a(y + y_0) = x_0 x \ .$$

Take for instance the parabola $y = x^2$. The tangent at $(-1, 1)$ is $2x + y + 1 = 0$ and the normal at $(-1, 1)$ is $x - 2y + 3 = 0$. As an exercise, the reader may wish to carry out a similar investigation on the tangent and the normal of the parabola $y^2 = 4ax$ at $P = (x_0, y_0)$ with special attention being paid to $P = O$.

3.10.2 EXAMPLE Let $P = (x_0, y_0)$ and $P' = (x_0 + h, y_0 + k)$ be two points on the ellipse

$$\frac{x^2}{a^2} + \frac{y^2}{b^2} = 1 \ .$$

Then $\qquad\qquad b^2 x_0{}^2 + a^2 y_0{}^2 = a^2 b^2$

and $\quad b^2 x_0{}^2 + a^2 y_0{}^2 + 2b^2 hx_0 + 2a^2 ky_0 + b^2 h^2 + a^2 k^2 = a^2 b^2.$

Subtracting, we get

$$2b^2 hx_0 + 2a^2 ky_0 + b^2 h^2 + a^2 k^2 = 0 \ .$$

Therefore the slope of the secant PP' is

$$\lambda' = \frac{k}{h} = \frac{-2b^2 x_0 - b^2 h}{2a^2 y_0 + a^2 k} \ .$$

As P' approaches P, both h and k tend to zero at the same time. Therefore

$$\lambda = \lim_{P' \to P} \lambda' = \frac{-b^2 x_0}{a^2 y_0}$$

is the slope of the tangent to the ellipse at P. The tangent itself is

$$\frac{x_0 x}{a^2} + \frac{y_0 y}{b^2} = 1 \ .$$

We observe that though the above equation of the tangent at P is formed with a definite value of the slope λ, it is valid also for tangents that are perpendicular to the x-axis at the points $(a, 0)$ and $(-a, 0)$.

In particular, the tangent to the circle

$$x^2 + y^2 = r^2$$

at $P = (x_0, y_0)$ is therefore the line

$$x_0 x + y_0 y = r^2$$

which is perpendicular to the radius $OP : y_0 x - x_0 y = 0$.

3.10.3 EXAMPLE Similarly the slope of the tangent to the hyperbola

$$\frac{x^2}{a^2} - \frac{y^2}{b^2} = 1$$

at $P = (x_0, y_0)$ is

$$\lambda = \frac{b^2 x_0}{a^2 y_0}$$

and the tangent itself is

$$\frac{x_0 x}{a^2} - \frac{y_0 y}{b^2} = 1 \ .$$

Here again the above equation also delivers the tangents at the vertices $(a, 0)$ and $(-a, 0)$ which are perpendicular to the x-axis.

Let us summarize the above discussion in the table below.

conic	slope at $P = (x_0, y_0)$	tangent at $P = (x_0, y_0)$
$y^2 = 4ax$	$\lambda = 2a/y_0$	$y_0 y = 2a(x + x_0)$
$x^2 = 4ay$	$\lambda = x_0/2a$	$2a(y + y_0) = x_0 x$
$\frac{x^2}{a^2} + \frac{y^2}{b^2} = 1$	$\lambda = -b^2 x_0/a^2 y_0$	$\frac{x_0 x}{a^2} + \frac{y_0 y}{b^2} = 1$
$\frac{x^2}{a^2} - \frac{y^2}{b^2} = 1$	$\lambda = b^2 x_0/a^2 y_0$	$\frac{x_0 x}{a^2} - \frac{y_0 y}{b^2} = 1$

A tangent to a non-degenerate conic meets the conic at a single point which is called the *point of tangency* or the *point of contact*. Conversely, with the exception of the lines parallel to the axis of a parabola, if a straight line intersects a non-degenerate conic at a single point, then the line is tangent to the conic at that point. Adopting this point of view we shall develop a method of finding tangents with a given slope.

3.10.4 EXAMPLE   Find a tangent to the parabola $y^2 = 6x$ with slope 1/2.

SOLUTION   A line of slope 1/2 is represented by an equation of the form

$$y = \frac{1}{2}x + t$$

where t is the intercept of the line on the y-axis. The intersection of this line with the parabola consists of points whose coordinates satisfy the equations

$$y^2 = 6x \quad \text{and} \quad y = \frac{1}{2}x + t .$$

Thus they satisfy

$$y^2 = 6(2y - 2t) \quad \text{or} \quad y^2 - 12y + 12t = 0 .$$

For certain values of t, e.g. $t = 4$, this equation has no-real solution. This means that the lines corresponding to these values of t do not intersect the given parabola. For other values of t, e.g. $t = 0$, the equation has two solutions, which means that there are two distinct points of intersection. For the present problem we are only concerned with the value of t for which the equation has exactly one solution. The discriminant of the equation being $12^2 - 4 \cdot 1 \cdot 12t$, the desired value is $t = 3$. Hence the corresponding straight line

$$x - 2y + 6 = 0$$

is a tangent to the parabola $y^2 = 6x$. The point of intersection is $P = (6, 6)$ which is then the point of tangency. Indeed, by using the previous result, we find that tangent to $y^2 = 6x$ at $P = (6, 6)$ is

$$6y = 3(x + 6) \quad \text{or} \quad x - 2y + 6 = 0 .$$

3.10.5 EXAMPLE   Find the tangents to the ellipse

$$4x^2 + y^2 = 5$$

which has slope 4.

SOLUTION   All straight lines with slope 4 are defined by

$$y = 4x + t \ .$$

Substitute to get $4x^2 + (4x + t)^2 = 5$ or

$$20x^2 + 8tx + t^2 - 5 = 0 \ .$$

Put the discriminant equal to 0:

$$64t^2 - 80(t^2 - 5) = 0$$

to get $t = 5$ or $t = -5$. Hence the ellipse have two tangents

$$T_1 : y = 4x + 5 \quad \text{and} \quad T_2 : y = 4x - 5$$

with slope 4. The points of tangency are respectively $(-1, 1)$ and $(1, -1)$. By our previous result the tangents at these point are respectively

$$-4x + y = 5 \quad \text{and} \quad 4x - y = 5 \ .$$

ALTERNATIVE SOLTUION   Consider the equation

$$4x_0 x + y_0 y - 5 = 0$$

of the tangent to the given ellipse at the point $P = (x_0, y_0)$ together with the equation

$$4x - y + t = 0$$

of a line of slope 4. These two lines will define the same straight line if and only if

$$4x_0 : 4 = y_0 : -1 = -5 : t$$

which means

$$x_0 = -5/t \quad \text{and} \quad y_0 = 5/t \ .$$

Since $P = (x_0, y_0)$ lies on the ellipse, we get

$$4(-5/t)^2 + (5/t)^2 = 5 \ .$$

Thus we obtain $t = 5$ and $t = -5$ as before.

From the method of solution to the two examples, it is easy to see that the parabola $y^2 = 4ax$ has a unique tangent with slope $\lambda \neq 0$, while the ellipse $b^2x^2 + a^2x^2 = a^2b^2$ has two tangents with arbitrary slope λ. With a hyperbola, it is somewhat more complicated.

Conic Sections

3.10.6 EXAMPLE Consider the intersection of the hyperbola

$$b^2x^2 - a^2y^2 = a^2b^2 \tag{1}$$

and the straight line

$$y = \lambda x + t \tag{2}$$

with given slope λ. By substitution and simplification, we see that the abscissa x of the point of the intersection must satisfy

$$(b^2 - a^2\lambda^2)x^2 - 2a^2\lambda tx - a^2(b^2 + t^2) = 0 .$$

Following the method of the last examples, we put the discriminant of this quadratic equation equal to zero, to get

$$4a^4\lambda^2t^2 + 4a^2(b^2 - a^2\lambda^2)(b^2 + t^2) = 0$$

or

$$t^2 = a^2\lambda^2 - b^2 . \tag{3}$$

Our task is therefore to find real numbers t which satisfy the condition (3) above. Depending on the value λ of the given slope, we have three cases to consider.

CASE 1. $|\lambda| > |b/a|$

In this case we obtain two real numbers

$$t = \pm\sqrt{a^2\lambda^2 - b^2}$$

that satisfy (3). Hence we have two tangents of slope λ:

$$y = \lambda x \pm \sqrt{a^2\lambda^2 - b^2}$$

to the given hyperbola.

CASE 2. $|\lambda| = |b/a|$

In this case, we get $t = 0$ and $\lambda = b/a$ or $\lambda = -b/a$. Therefore the straight line (2) of slope λ is the asymptote

$$bx + ay = 0 \quad \text{or} \quad bx - ay = 0$$

of the hyperbola (1). In other words, each asymptote can be thought of as a tangent to the hyperbola (1) with the point of contact at infinity because points of the hyperbola far away from the centre are arbitrarily close to the asymptotes. Hence we conclude that the hyperbola (1) has a tangent of slope λ which is an asymptote.

CASE 3. $|\lambda| < |b/a|$

In this case, $a^2\lambda^2 - b^2 < 0$ and we do not have real numbers t that satisfy condition (3). Hence we conclude that the hyperbola (1) does not have tangent of slope λ.

We can summarize the results of the above examples in the following theorem

3.10.7 THEOREM *The parabola $y^2 = 4ax$ has one tangent $y = \lambda x + (a/\lambda)$ with slope $\lambda \neq 0$. The parabola $x^2 = 4ay$ has one tangent $y = \lambda x - a\lambda^2$ with arbitrary slope λ. The ellipse $b^2x^2 + a^2y^2 = a^2b^2$ has two parallel tangents $y = \lambda x \pm \sqrt{a^2\lambda^2 + b^2}$ with arbitrary slope λ. The hyperbola $b^2x^2 - a^2y^2 = a^2b^2$ has two targents $y = \lambda x \pm \sqrt{a^2\lambda^2 - b^2}$ of slope λ if $|\lambda| > |b/a|$. It has one tangent $y = \lambda x$ which is an asymptote if $\lambda = \pm b/a$, and it has no tangent of slope λ such that $|\lambda| < |b/a|$.*

Finally we can use the above methods to find equations of tangents from an external point.

3.10.8 EXAMPLE Find equations of the tangents to the ellipse

$$x^2 + 2y^2 = 3$$

that pass through the given point $Q = (-1, 2)$.

SOLUTION Consider the tangent to the given ellipse at a point $P = (x_0, y_0)$:

$$x_0 x + 2y_0 y = 3 \ .$$

For this tangent to pass through the external point Q, we must have

$$-x_0 + 4y_0 = 3$$

which is one condition on the coordinates of the point P of tangency. The second condition on these coordinates must clearly be

$$x_0{}^2 + 2y_0{}^2 = 3$$

160

because P is a point on the ellipse. To find the point P of tangency we solve

$$\begin{cases} x_0{}^2 + 2y_0{}^2 = 3 \\ -x_0 + 4y_0 = 3 \end{cases}$$

to obtain $(x_0, y_0) = (1,1)$ and $(x_0, y_0) = (-5/3, 1/3)$. Therefore we have two tangents

$$x + 2y - 3 = 0 \quad \text{and} \quad 5x - 2y + 9 = 0$$

to the given ellipse from the given external point $Q = (-1, 2)$.

SECOND SOLUTION The tangents to the ellipse with slope λ are given by

$$y = \lambda x \pm \sqrt{3\lambda^2 + (3/2)} \ .$$

For them to pass through the given point Q, we must have

$$2 = -\lambda \pm \sqrt{3\lambda^2 + (3/2)} \ .$$

After squaring we obtain

$$4\lambda^2 - 8\lambda - 5 = 0 \ .$$

Solve this equation to get $\lambda = -1/2$ and $\lambda = 5/2$ which deliver two tangents

$$2(y - 2) = -(x + 1) \quad \text{and} \quad 2(y - 2) = 5(x + 1)$$

or

$$x + 2y - 3 = 0 \quad \text{and} \quad 5x - 2y + 9 = 0$$

to the ellipse passing through $Q = (-1, 2)$.

On the other hand if we substitute $\lambda = -1/2$ into $y = \lambda x \pm \sqrt{3\lambda^2 + (3/2)}$, we obtain two tangents to the ellipse with slope $-1/2$. Similarly we get two tangents with slope $5/2$. It is easy to see that the two tangents through $Q = (-1, 2)$ are obtainable by substituting the two values of λ to

$$y = \lambda x + \sqrt{3\lambda^2 + (3/2)} \ .$$

However if we substitute these values to the other equation

$$y = \lambda x - \sqrt{3\lambda^2 + (3/2)} \ ,$$

we will get two symmetric tangents

$$x + 2y + 3 = 0 \quad \text{and} \quad 5x - 2y - 9 = 0$$

with the same slopes but passing the point $Q' = (1, -2)$.

THIRD SOLUTION Lines that pass through Q are defined by equations of the form

$$(y - 2) = \lambda(x + 1)$$

where λ is the slope. Bring them to intersection with the given ellipse, we obtain the following condition on the abscissa of the point of intersection

$$x^2 + 2(\lambda x + \lambda + 2)^2 = 3 .$$

For this equation to have a single solution in x, the condition on the discriminant is

$$4\lambda^2 - 8\lambda - 5 = 0$$

which yields $\lambda = -1/2$ and $\lambda = 5/2$ as before. The desired tangents are therefore

$$2(y - 2) = -(x + 1) \quad \text{and} \quad 2(y - 2) = 5(x + 1) .$$

We observe that the second and the third methods are not applicable to find tangents that are perpendicular to the x-axis.

The line segment joining a point on a conic and a focus of the conic is called a *focal radius* of the conic. If the conic is a circle, then a focal radius is just a usual radius of the circle. A well-known relation between the tangent at a point P of a circle and the radius drawn pass P is that they are perpendicular to each other. For the other types of non-degenerate conics the relation between tangents and focal radii is given in the theorems below.

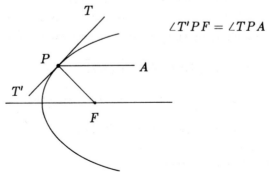

$$\angle T'PF = \angle TPA$$

Fig 3-37

3.10.9 THEOREM *The tangent at any point P of a parabola makes the same angle with the focal radius draw to P as it does with the axis of the parabola.*

3.10.10 THEOREM *The tangent to an ellipse at any point makes equal angles with the focal radii drawn to that point.*

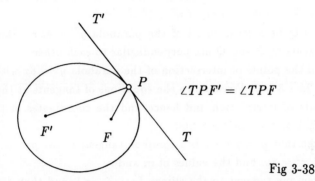

$$\angle TPF' = \angle TPF$$

Fig 3-38

3.10.11 THEOREM *The tangent of a hyperbola at any point bisects the angle between the focal radii drawn to that point.*

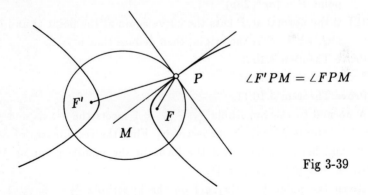

$$\angle F'PM = \angle FPM$$

Fig 3-39

The proof of these theorems is left to the reader as an exercise. We take note that these properties of conics find many applications in daily life. For example the property 3.10.9 of the parabola is put to good use in the reflector of the headlight of a car which is made in the form a paraboloid of revolution, i.e. the surface generated by revolving a parabola about its axis.

EXERCISES

1. Show that the straight line $mx + ny + \ell = 0$ is a tangent to the parabola $y^2 = 4ax$ if and only if $m\ell = an^2$.

2. Show that the straight line $mx + ny + \ell = 0$ is a tangent to the ellipse $\frac{x^2}{a^2} + \frac{y^2}{b^2} = 1$ if and only if $a^2m^2 + b^2n^2 = \ell^2$.

3. Show that the straight line $y = mx + n$ is a tangent to the hyperbola $\frac{x^2}{a^2} - \frac{y^2}{b^2} = 1$ if and only if $n^2 + b^2 = a^2m^2$.

4. Let PQ be a focal chord of the parabola $y^2 = 4ax$. Show that the tangents at P and Q are perpendicular to each other.

5. Find the points of intersection of the parabola $y^2 = 8x$ with the chord $y - 2x + 3 = 0$. Write down the equations of tangents at the respective points of intersection and hence find the coordinates of the point of intersection of the tangents.

6. Given that $y = mx + c$ is a common tangent to the parabolas $y^2 = 8x$ and $x^2 = 8y$, find the values of m and c.

7. Given any tangent to the ellipse $\frac{x^2}{a^2} + \frac{y^2}{b^2} = 1$, and that d_1, d_2 are the perpendicular distances of the foci F, F' to the tangent, show that $d_1 \cdot d_2 = b^2$.

8. (a) Find the equation of the normal to the parabola $y^2 = 4ax$ at the point $P = (ap^2, 2ap)$.

   (b) If the normal at P cuts the curve again at the point Q and $OP \perp OQ$, where O is the origin, then deduce that $p^2 = 2$.

9. Prove Theorem 3.10.9.

10. Prove Theorem 3.10.10.

11. Prove Theorem 3.10.11.

12. A normal to the hyperbola $\frac{x^2}{a^2} - \frac{y^2}{b^2} = 1$ meets the x-axis and y-axis at the points M and N respectively. Find the coordinates of M and N, and hence deduce that the locus of the middle point of MN is $4(a^2x^2 - b^2y^2) = (a^2 + b^2)^2$.

13. Given the point $P = (x_1, y_1)$ on the hyperbola $\frac{x^2}{a^2} - \frac{y^2}{b^2} = 1$. The tangent at P to the hyperbola cuts the asymptotes at points Q and R respectively.

    (a) Show that the x-coordinates of Q and R are the roots of the equation $x^2 - 2x_1x + a^2 = 0$.

    (b) Hence deduce that $|PQ| = |PR|$.

14. Given the parabola $y^2 = 4ax$.

(a) Show that the point $P = (ap^2, 2ap)$ lies on the parabola for any real p.

(b) Find the equation of the tangent to the parabola at the point $P = (ap^2, 2ap)$.

(c) Let QR be a focal chord of the ellipse which is parallel to the tangent in (b). Show that $QR = 4PF$ where F is the focus.

15. Let $P(x_1, y_1)$ and $Q(x_2, y_2)$ be two points on the ellipse $\frac{x^2}{a^2} + \frac{y^2}{b^2} = 1$. Suppose ℓ is the straight line passing through the origin O and the middle point of PQ.

   (a) Show that $\frac{y_1 + y_2}{x_1 + x_2} = -\frac{b^2(x_1 - x_2)}{a^2(y_1 - y_2)}$.

   (b) Deduce that the tangents at P and Q meet at ℓ.

16. Let $P = (t, \frac{1}{t})$ be a point on the hyperbola $xy = 1$. Suppose the normal at P cuts the hyperbola at another point Q.

   (a) Show that Q has coordinates $(-\frac{1}{t^3}, -t^3)$.

   (b) Write down the equations of the tangents at P and Q and hence deduce that the two tangents are parallel if and only if $t = \pm 1$.

   (c) Find the point of intersection of the two tangents in (b) if $t \neq \pm 1$.

17. Given the equation of the curve $xy = m^2$ $(m \neq 0)$.

   (a) By a suitable rotation of axes, show that the curve is a hyperbola.

   (b) Show that if $y = ax + b$ is a tangent to the hyperbola $xy = m^2$, then $a = -\frac{b^2}{4m^2}$.

   (c) By (b), or otherwise, deduce that the area of the triangle formed by the tangent $y = ax + b$ and the two axes is a constant.

18. Let $P = (x_1, y_1)$ be a point on the hyperbola $xy = c^2$ $(c \neq 0)$.

   (a) Find the equation of the normal to the hyperbola at P.

   (b) If the normal at P meets the curve again at Q, show that $|OP|^3 = c^2|PQ|$, where O is the origin.

19. Given a point $P = (2\sqrt{2}\cos\theta, 2\sin\theta)$ on the ellipse $\frac{x^2}{8} + \frac{y^2}{4} = 1$.

   (a) Find the equations of the tangent and the normal at P to the ellipse.

   (b) If the tangent and the normal in (a) meet the y-axis at Q and R respectively, show that $|OQ| \cdot |OR|$ is a constant, where O is the origin.

   (c) Find the coordinates of the centre of the circle passing through point P, Q and R.

20. Given an ellipse $E : b^2 x^2 + a^2 y^2 = a^2 b^2$.

    (a) Find the tangents to E with slope λ.

    (b) Consider the circle $C : x^2 + y^2 = a^2 + b^2$.

        (i) If (α, β) is a point on C which lies on a tangent to E with slope λ, prove that

        $$(\alpha^2 - a^2)\lambda^2 - 2\alpha\beta\lambda + (\beta^2 - b^2) = 0.$$

        (ii) Hence, or otherwise, prove that if two tangents from (α, β) are drawn to E, they are perpendicular to each other.

Chapter Four

QUADRIC SURFACES

On the plane a linear equation in two variables defines a line. In space a linear equation in three variables defines a plane. A line in space is the intersection of two planes; it is therefore defined by two linear equations in three variables. A quadratic equation in two variables defines a quadratic curve on the plane. Quadratic curves on the plane are called conics because they are plane sections of a circular cone in space. In Chapter Three we have seen that there are only three types of conics. A quadratic equation in three variables defines a surface in spaces. These surfaces are called *quadric surfaces*. While there are only three types of conics, we shall see in this chapter that there are numerous, in fact no less than nine, types of quadric surfaces. Plane sections of a quadric surface are defined by a quadratic equation in three variables and a linear equation in three variables. Eliminating one variable, we see that they are defined by a quadratic equation in two variables and a linear equation in three variables. Therefore plane sections of quadric surfaces are conics. This consideration provides us with a convenient way to study the shape of a quadric surface.

The various types of quadric surfaces together with their equations and plane sections are individually studied in the following sections.

4.1 Surfaces of revolution

We have seen that a circular cylinder is generated by rotating a straight line about an axis parallel to it. If the axis of revolution is chosen to be the z-axis then

$$x^2 + y^2 = r^2 \quad \text{or} \quad x^2 + y^2 + 0z^2 = r^2$$

is an equation of the circular cylinder. A circular cone is generated by rotating a straight line about an axis that intersects it. If the axis is chosen to be the z-axis and the vertex to be the origin, then

$$x^2 + y^2 = az^2$$

is an equation of the cone.

In general the surface obtained by rotating a curve about an axis is called a *surface of revolution*. The circular cone and the circular cylinder are the simplest examples of such surfaces. In this section we only consider surfaces of revolution generated by rotating conics about their axes of symmetry. Together with the circular cones and the circular cylinder, they are known as *quadric surfaces of revolution*. Since the axis of rotation is an axis of symmetry of the generating curve, every such surface is symmetric with respect to a plane that passes through the axis of revolution.

Take a circle and rotate it about its diameter to obtain a sphere. If the centre of the circle is at the origin, then an equation of the sphere is

$$x^2 + y^2 + z^2 = r^2 \ .$$

Every plane section of a sphere is either a point or a circle.

Next we consider an ellipse being rotated about one of its axes. The resulting quadric surface of revolution is called an *ellipsoid of revolution*. As there are two axes of symmetry to an ellipse there are two different kinds of ellipsoids of revolution. We consider first the ellipsoid obtained by rotating an ellipse Γ about its major axis. To obtain an equation of the ellipsoid we place the z-axis on the major axis and the y-axis on the minor axis of the given ellipse Γ. Then the ellipse Γ on the yz-plane is defined by

$$\frac{y^2}{b^2} + \frac{z^2}{a^2} = 1 \quad (a > b) \ .$$

Consider an arbitrary rotation ρ about the z-axis. Under ρ the yz-plane is taken to a plane E which contains the z-axis, and the ellipse Γ to an ellipse $\rho(\Gamma)$ on E. Points on $\rho(\Gamma)$ are therefore points of the ellipsoid of revolution under consideration, and conversely every point of the ellipsoid is on one such rotating ellipse $\rho(\Gamma)$ for some ρ. On the rotating plane E we choose a pair of auxilary coordinate axes; one of them is taken to be the original z-axis and the other one labelled as the r-axis to be the image of the y-axis under ρ (see Figure 4-1). Then on the rz-plane E, the rotating ellipse $\rho(\Gamma)$ is defined by

$$\frac{r^2}{b^2} + \frac{z^2}{a^2} = 1 \quad (a > b) \ . \tag{1}$$

Now a point $X = (x, y, z)$ on E will have coordinates (r, z) on the rz-plane E where

$$r^2 = x^2 + y^2 \ . \tag{2}$$

If furthermore X lies on $\rho(\Gamma)$, then its coordinates (r, z) also satisfy equation (1). Hence the coordinates of points $X = (x, y, z)$ of $\rho(\Gamma)$ must satisfy the equation

$$\frac{x^2}{b^2} + \frac{y^2}{b^2} + \frac{z^2}{a^2} = 1 \quad (a > b) \ . \tag{3}$$

This is now an equation of the ellipsoid of revolution under discussion.

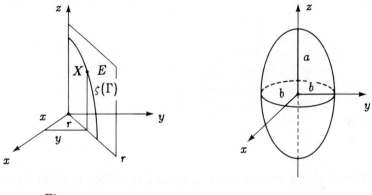

Fig 4-1

Fig 4-2

The present ellipsoid of revolution (see Figure 4-2) has the shape of an elongated sphere and is called a *prolate spheroid*. An egg or a rugby ball is an approximation to this type of spheroid.

We observe that the same rotating rz-plane E with $r^2 = x^2 + y^2$ can be used to obtain equations of other surfaces of revolution. Indeed if a curve Γ on the yz-plane is defined by an equation

$$f(y, z) = 0 \ ,$$

then the surface obtained by rotating Γ about the z-axis consists of points $X = (x, y, z)$ such that

$$f\left(\sqrt{x^2 + y^2}, z\right) = 0 \quad \text{or} \quad f\left(-\sqrt{x^2 + y^2}, z\right) = 0 \ .$$

It is easily seen that the equations of the circular cylinder and of the cone can be obtained this way.

Similarly the equation of an ellipsoid of revolution obtained by rotating an ellipse on the yz-plane about its minor axis (e.g. the z-axis) is found to be

$$\frac{x^2}{a^2} + \frac{y^2}{a^2} + \frac{z^2}{b^2} = 1 \quad (a > b) \ . \tag{4}$$

In this case the ellipsoid (see Figure 4-3) has a shape somewhat like a flattened sphere and is called an *oblate spheroid*. The earth or a volley ball being sat on is a familiar example of this type of spheroid.

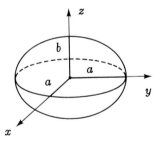

Fig 4-3

Every plane section of a spheroid is either a point or a non-degenerate closed quadratic curve. Therefore a plane section of a spheroid is either a point, a circle or an ellipse.

In a similar manner we rotate the hyperbolas

$$\frac{y^2}{a^2} - \frac{z^2}{b^2} = -1 \quad \text{and} \quad \frac{y^2}{a^2} - \frac{z^2}{b^2} = 1$$

on the yz-plane about the z-axis to obtain two *hyperboloids of revolution*

$$\frac{x^2}{a^2} + \frac{y^2}{a^2} - \frac{z^2}{b^2} = -1 \quad \text{and} \quad \frac{x^2}{a^2} + \frac{y^2}{a^2} - \frac{z^2}{b^2} = 1 \ . \tag{3}$$

For the first hyperboloid (Figure 4-4) the axis of rotation is the major axis of the revolving hyperbola. The resulting surface has two separate parts and is called a *hyperboloid of revolution of two sheets.*

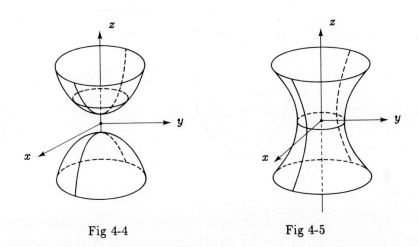

Fig 4-4   Fig 4-5

For the second hyperboloid (Figure 4-5) the axis of rotation is the minor axis of the revolving hyperbola and the resulting surface is called a *hyperboloid of revolution of one sheet.* Many cooling towers are built in the form of such hyperboloid.

All spheroids and all hyperboloids of revolution are symmetric to the centre O of the generating conic. Accordingly we call O the *centre* of these quadric surfaces, and all such quadric surfaces *central quadrics.*

Since a parabola has only one axis of symmetry, there can be only one quadric surface of revolution obtained by rotating a parabola about its axis. This is called a *paraboloid of revolution* (Figure 4-6). If the generating parabola is

$$y^2 = 4az$$

on the yz-plane and the z-axis is the axis of revolution, the equation of the paraboloid of revolution is

$$x^2 + y^2 = 4az \ .$$

171

In 1672, the French astronomer Guillaume Cassegrain designed a reflecting telescope with parabolic mirrors. Nowadays parabolic reflectors can be found everywhere. For example in the auto headlights and the telecommunication antennae.

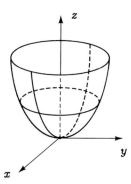

Fig 4-6

We have thus identified eight different types of quadric surfaces of revolution, all defined by quadratic equations in three variables. We shall later see that they are all special cases of more general quadric surfaces, and in Section 4.9 it shall be shown that no other type of quadric surface of revolution exists. Finally we remark that if a conic is rotated about a line which is not an axis of symmetry of the conic, then the surface of revolution so obtained will fail to be a quadric surface.

Take for example the circle

$$(y - b)^2 + z^2 = a^2 , \quad (b \neq 0)$$

on the yz-plane and rotate it about the z-axis. Then the resulting surface of revolution is not defined by a single quadratic equation any more. If $b > a$ then the resulting ring-shape surface is called a *torus* (Figure 4-7). If $b = a$, then the hole in the middle disappears and the surface tonches itself at the origin. If $b < a$, it becomes a self-penetrating surface of rather complicated shape. None of these is a quadric surface. Readers may be astonished to know that the surface obtained by rotating a line which is skew to the rotation axis is actually a hyperboloid of rotation of one sheet (see Section 4.5).

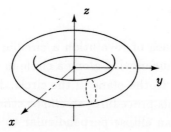

Fig 4-7

EXERCISES

1. Write down the name for each of the following surfaces:
   (a) $x^2 + y^2 + 4z^2 = 4$;
   (b) $x^2 + z^2 = y$;
   (c) $\frac{x^2}{4} - y^2 - z^2 = 1$;
   (d) $\frac{x^2}{4} - y^2 - z^2 = -1$.

2. A surface is generated by rotating the curve $F(y, z) = 0$ on the yz-plane about z-axis. If P is an arbitrary point on the surface, by considering the distance of P from the z-axis, show that the surface consists of points $X = (x, y, z)$ such that

$$F(\sqrt{x^2 + y^2}, z) = 0 \quad \text{or} \quad F(-\sqrt{x^2 + y^2}, z) = 0 \ .$$

3. Generalize the above result to the following cases:
   (a) Rotating $F(x, z) = 0$ about z-axis;
   (b) Rotating $F(x, y) = 0$ about x-axis.
   (c) Rotating $F(x, z) = 0$ about x-axis;
   (d) Rotating $F(x, y) = 0$ about y-axis;
   (e) Rotating $F(y, z) = 0$ about y-axis.

In Questions 4-9, find an equation of the surface obtained by revolving the given curve about the indicated axis.

4. $y = z^2$; about y-axis.
5. $x^2 + 2y^2 = 1$; about x-axis.
6. $4x^2 - z^2 = 4$; about x-axis.
7. $4x^2 - z^2 = 4$; about z-axis.
8. $x^2 - 4x + y^2 = 21$; about x-axis.
9. $x^2 + 2y + y^2 = 3$; about y-axis.

4.2 Cylinders and cones

Besides being a surface of revolution a circular cylinder is also generated by a straight line moving along a circle in such a way as to be always perpendicular to the plane of the circle. By a straightforward generalization of this procedure, an *elliptic cylinder* is generated by a moving line along an ellipse perpendicular to the plane of the curve. Thus we have a second method of generating quadratic surfaces by moving a straight line along a conic in a prescribed course. *Parabolic cylinders* and *hyperbolic cylinders* are obtained from parabolas and hyperbolas in this way. If the conics are placed on the xy-plane in the standard position the *quadric cylinders* (Figure 4-8) so obtained are then defined by the equations:

$$\frac{x^2}{a^2} + \frac{y^2}{b^2} = 1 \qquad \text{elliptic cylinder ;} \qquad (1)$$

$$x^2 = 4ay \qquad \text{parabolic cylinder ;} \qquad (2)$$

$$\frac{x^2}{a^2} - \frac{y^2}{b^2} = -1 \qquad \text{hyperbolic cylinder ;} \qquad (3)$$

which are all taken as equations in three variables x, y and z.

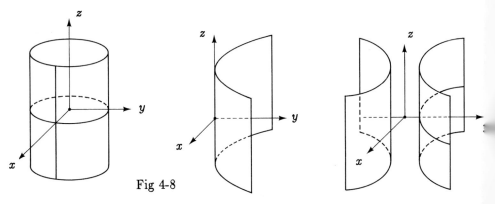

Fig 4-8

The conic on the xy-plane is called the *directrix* of the cylinder; every line that passes through the directrix and perpendicular to the plane of the directrix is called a *generator* or a *ruling* of the cylinder. More generally, if the directrix of a quadric cylinder is the quadratic curve

$$Ax^2 + Bxy + Cy^2 + Dx + Ey + F = 0$$

174

on the xy-plane, then the quadric cylinder itself is defined by the same quadratic equation, now regarded as an equation in three variables x, y and z.

Clearly a plane that passes through an axis of symmetry of the directrix and parallel to the rulings of a quadric cylinder is a *plane of symmetry* of the cylinder. For example both the xz-plane and yz-plane are planes of symmetry of the cylinders (1) and (3). Similarly every plane that is perpendicular to the rulings is also a plane of symmetry of the cylinder. If the directrix is a central conic, then the line which passes through the centre of the directrix and parallel the rulings is an *axis of symmetry* of the cylinder.

More generally we would expect that a plane section of an elliptic cylinder is an ellipse, and similar result would be obtained for the other types of quadric cylinders.

4.2.1 THEOREM *Every plane which is not parallel to the rulings of a quadric cylinder intersects it in a conic of the same type as the directrix. Moreover parallel plane sections of a quadric cylinder are congruent conics.*

PROOF Suppose that the directrix Γ on the xy-plane is a central conic with centre at O. Let a plane E intersect the quadric cylinder Q with directrix Γ in a curve Γ' and the xy-plane at an angle θ ($0 < \theta < \pi/2$) in a line L. By rotating the coordinate axes about the z-axis if necessary, we may assume that the x-axis is parallel to L and the y-axis is perpendicular to L. As the directrix Γ may be moved out of standard position, it is defined by an equation of the form

$$Ax^2 + Bxy + Cy^2 + F = 0 \tag{4}$$

on the xy-plane.

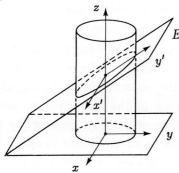

Fig 4-9

175

If we project the xy-plane onto the cutting plane E in the direction of the z-axis, then the plane section Γ' of Q is the projection of the directrix Γ. If we denote by x'-axis the projection of the x-axis and by y'-axis the projection of the y-axis, then the x'-axis and the y'-axis form a pair of coordinate axes of E. Now a point $P' = (x', y')$ of the $x'y'$-plane E is the projection of a point $P = (x, y)$ of the xy-plane if and only if

$$x = x' \quad \text{and} \quad y = y' \cos \theta .$$

Therefore the curve Γ' on E is the conic defined by

$$Ax'^2 + Bx'(y' \cos \theta) + C(y' \cos \theta)^2 + F = 0 . \tag{5}$$

The discriminant of Γ' is

$$B^2 \cos^2 \theta - 4AC \cos^2 \theta = (B^2 - 4AC) \cos^2 \theta .$$

Since $0 < \theta < \pi/2$ this has the same sign as $B^2 - 4AC$. Therefore (4) and (5) define the same type of conic. Hence the plane section Γ' and the directrix Γ are conics of the same type.

To prove the second statement of the theorem, we take a second plane F parallel to E. Then F intersects the xy-plane at the same angle θ in a line G parallel to L. Again we denote the projection of the x-axis and y-axis on F by x''-axis and y''-axis respectively. Then the section Γ'' by the $x''y''$-plane F is defined by an equation of the form

$$Ax'' + Bx''(y'' \cos \theta) + C(y'' \cos \theta)^2 + F = 0 .$$

Therefore the plane sections Γ' and Γ'' are congruent.

The proof for the parabolic case for which some minor adjustmnet will be necessary is left to the reader as an exercise.

A circular cone is also obtained by moving a straight line passing through a fixed point in space along a circle. The fixed point is the vertex of the cone and the moving line a ruling of the cone. A straightforward generalization of the circular cone produces the general quadric cone. It consists of the lines joining the points of a non-degenerate conic to a fixed point anywhere in space but not on the plane of the curve. Although there are three different types of conics, we do not get different types of quadric cones. This is born out by the investigation of Section 3.7 in which we have seen that a plane may intersect a circular cone in all three types of conics. A characteristic of the equation of a quadric cone is given in the following theorem.

4.2.2 THEOREM *A quadric surface is a cone with its vertex at the origin if and only if its equation is homogeneous in x, y, z.*

PROOF A cone with its vertex at the origin O is characterized by the fact that if $P\ (P \neq O)$ is a point on the cone, then every point on the line OP lies on the cone. In other words the equation of the cone is characterized by the property that if (a, b, c) satisfies the equation then (ra, rb, rc) also satisfies the equation for all values of r. But this is precisely the characteristic property of a homogeneous equation.

Surfaces that can be generated by straight lines are called *ruled surfaces*. Therefore cones and cylinders are *ruled quadric surfaces*, and they can be formed by bending thin metal sheets or cardboards. Later on we shall find two other types of ruled quadric surfaces: the hyperboloid of one sheet and the hyperbolic paraboloid. But these latter two types of surfaces cannot be formed by cardboard.

EXERCISES

1. Write an equation for each of the cylinders described below.
   (a) Axis of symmetry is the y-axis, the directrix is a circle on the xz-plane of radius 5.
   (b) Axis of symmetry is the x-axis, the directrix is $\frac{y^2}{4} + \frac{z^2}{25} = 1$, $x = 0$.
   (c) Axis of symmetry is the z-axis, the directrix is $9x^2 - 4y^2 = 36$, $z = 0$.

   What are the planes of symmetry in (a), (b) and (c)?
2. Find a cylinder that passes through the intersection of the surfaces $z = x^2 + y^2$ and $8 - z = x^2 + y^2$.
3. Identify the locus of a point such that the difference between the squares of its distances from the x-axis and y-axis is a constant.

4. Consider the surface Q whose equation is

$$\frac{x^2}{4} + \frac{y^2}{4} - \frac{z^2}{9} = 0 .$$

(a) Find the intersections of Q with xy-plane, yz-plane and xz-plane respectively.

(b) Find the intersections of Q with the planes $z = h$ for real h.

(c) What are the intersections of Q with the planes $x = k$ and $y = \ell$ for real k and ℓ?

By (a), (b) and (c), what can you conclude about the surface Q? Try to sketch the surface. Compare your result with Theorem 4.2.2.

5. Write an equation for each of the cones described below. Note that the vertices are the origin.

(a) Axis of symmetry is the x-axis, the intersection of the cone with the plane $x = 3$ is an ellipse $4y^2 + 9z^2 = 36$.

(b) Axis of symmetry is the z-axis, the intersection of the cone with the plane $z = 3$ is a circle of radius 3.

6. Identify the locus of a point such that the ratio between its distances to the xz-plane and the yz-plane is a constant.

7. Complete the proof of Theorem 4.2.1.

4.3 The ellipsoid

We have now found two large classes of different kinds of quadric surfaces: the quadric surfaces of revolution and the ruled quadric surfaces, corresponding to the two methods of generating surfaces. These are first identified geometrically, and their equations are then derived. For the next five types of quadric surfaces, we shall reverse the order of investigation: we take a particular quadratic equation and study the shape of the surface it defines.

The quadric surface defined by the equation

$$\frac{x^2}{a^2} + \frac{y^2}{b^2} + \frac{z^2}{c^2} = 1 \tag{1}$$

is called an *ellipsoid*. If two of the three positive numbers a, b, c are equal then it is a spheroid; if all three are equal then it is a sphere. The shape of a general ellipsoid where all three constants a, b, c are distinct is best revealed by its plane sections.

The ellipsoid (1) intersects the coordinate planes $x = 0$, $y = 0$, $z = 0$ respectively in the ellipses

$$\frac{y^2}{b^2} + \frac{z^2}{c^2} = 1, \quad \frac{x^2}{a^2} + \frac{z^2}{c^2} = 1, \quad \frac{x^2}{a^2} + \frac{y^2}{b^2} = 1 . \tag{2}$$

In Figure 4-10, these curves are drawn in the first quadrants of the coordinate planes to show the ellipsoid in the first octant.

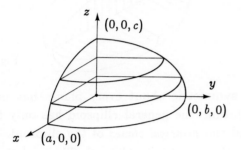

Fig 4-10

The section of the ellipsoid by a plane $z = k$ parallel to the xy-plane has the equation

$$\frac{x^2}{a^2} + \frac{y^2}{b^2} = 1 - \frac{k^2}{c^2} . \tag{3}$$

This is an ellipse if $k^2 < c^2$. The centre of this ellipse is at the point $(0, 0, k)$ on the z-axis; one axis lies on the xz-plane and the other on the yz-plane and have lengths

$$2a\sqrt{1 - (k/c)^2} \quad \text{and} \quad 2b\sqrt{1 - (k/c)^2}$$

respectively. From this we see that the sections of the ellipsoid by any two planes, both parallel to the xy-plane, are similar ellipses. As k increases from 0 towards c the ellipse (3) rises from the section $b^2 x^2 + a^2 y^2 = a^2 b^2$ on the xy-plane, becomes continuously smaller but remains similar; it shrinks finally to a point $(0, 0, c)$ when $k = c$ and disappears all together when $k > c$. The plane sections corresponding to negative values of k are entirely analogous. Therefore we may visualize the ellipsoid as a stack of similar ellipses perpendicular to the z-axis. Sections of the ellipsoid perpendicular to the y-axis and those perpendicular to the x-axis are related to one another in entirely the same way.

179

The ellipsoidal shape can often be recognized in stones found on beaches. A stone of any shape would increasingly resemble an ellipsoid as water and sand wear away at it.

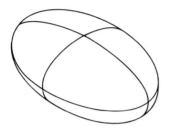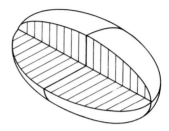

Fig 4-11

While a spheroid is symmetric about every plane that passes through the axis of revolution, the general ellipsoid has only three planes of symmetry called the *principal planes* of the ellipsoid. (For the ellipsoid (1) they are the coordinate planes.) These planes are mutually perpendicular and the three lines of intersection are called the *axes* of the ellipsoid. The axes meet at one point called the *centre* of the ellipsoid. The axes are cut by the ellipsoid into segments of unequal lengths. These segments are called the *major, mean* and *minor axes* of the ellipsoid in order of magnitude. Thus if $a > b > c$ in equation (1), then $2a$, $2b$, $2c$ will be the major, mean and minor axes of the ellipsoid (1).

It is not difficult to see that every plane section of an ellipsoid is an ellipse because it is a closed quadratic curve. The quadratic equation of the plane section will also show that parallel plane sections of an ellipsoid are similar ellipses. It is interesting to note that the general ellipsoid, like the elliptic cylinder, also has circular plane sections. In fact there is one circular plane section with a diameter falling on the mean axis of the ellipsoid. To see this, let $2a$, $2b$, $2c$ be the major, mean and minor axes of the ellipsoid. Take a plane that passes through the mean and the minor axes and rotate it about the mean axis towards the major axis. Throughout the rotation the plane cuts the ellipsoid in an ellipse one of whose axes is always the mean axis $2b$ of the ellipsoid. (See Figure 4-12). The second axis of the ellipse is $2c$ at the starting position. As the plane rotates about the mean axis of the ellipsoid, the second axis of the ellipse increases continuously to $2a$ at the final position. Consequently there must be a position in between at which the second axis of the ellipse takes on the intermediate value $2b$. At this position then the ellipse is a circle.

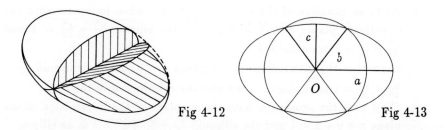

Fig 4-12   Fig 4-13

The above consideration establishes the existence of one circular plane section of the ellipsoid. By the symmetry with respect to the plane that passes through the mean and the minor axes of the ellipsoid, there is another circular section. Indeed the positions of these two circular section can be found by a construction illustrated in Figure 4-13. On the plane that contains the major and minor axes of the ellipsoid we have an ellipse with axes $2a$ and $2c$. The points of intersection of this ellipse and a circle of radius b with centre at the centre of the ellipse determine the positions of the planes of the two circular sections. But there are many more circular sections. In fact every plane parallel to a circular section intersects the ellipsoid in a circle. Thus there are two families of parallel circles on every ellipsoid. Based on this result we can make a model of the ellipsoid that consists of two sets of interlocking cardboard circular disks (see Figure 4-14). On the ellipsoid of revolution the two families of circles coincide.

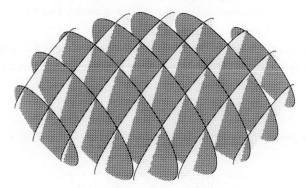

Fig 4-14

EXERCISES

1. Write an equation of the ellipsoid with $x-$, $y-$ and $z-$ intercepts ± 3, ± 5 and ± 7, respectively.

2. Write an equation of the ellipsoid such that when intersects with the coordinate planes $x = 0$ and $y = 0$, the ellipses $\frac{y^2}{9} + \frac{z^2}{16} = 1$ and $\frac{x^2}{25} + \frac{z^2}{16} = 1$ are obtained.

3. Show that the intersection of the plane $x = 3$ and the ellipsoid $\frac{x^2}{25} + \frac{y^2}{16} + \frac{z^2}{4} = 1$ is an ellipse. Find also the coordinates of the foci.

4. Show that the projection onto the xy-plane of the intersection of the plane $x + y + z = 1$ and the ellipsoid $\frac{x^2}{4} + \frac{y^2}{2} + z^2 = 1$ is an ellipse. [Hint: use Theorem 3.9.2].

5. A ray in the direction of the vector $[2, 3, 8]$ is drawn from the origin 0 to meet the ellipsoid $\frac{x^2}{4} + \frac{y^2}{9} + \frac{z^2}{16} = 1$ at a point P. Find $|OP|$.

6. Show that the equations

$$\begin{cases} x = a \cos \varphi \cos \theta \\ y = b \cos \varphi \sin \theta \\ z = c \sin \varphi \end{cases}$$

define an ellipsoid where a, b and c are constants, and φ and θ are variables.

7. (a) Show that the intersections of the planes $z = h$ and the ellipsoid $\frac{x^2}{16} + \frac{y^2}{9} + \frac{z^2}{4} = 1$ are ellipses for all values of h such that $-2 < h < 2$.
   (b) Show that the foci of the ellipses in (a) lie on the curve

$$\frac{x^2}{7} + \frac{z^2}{4} = 1 , \quad y = 0 .$$

8. Given an ellipsoid $\frac{x^2}{a^2} + \frac{y^2}{b^2} + \frac{z^2}{c^2} = 1$ $(0 < c < a < b)$. Suppose the plane $y = kz$ intersects the ellipsoid at a circle, find the values of k in terms of a, b and c.

9. As a generalization to Question 5, suppose a straight line with direction consines d_1, d_2 and d_3 is drawn from the origin O to meet the ellipsoid

$$\frac{x^2}{a^2} + \frac{y^2}{b^2} + \frac{z^2}{c^2} = 1$$

at the point P. Show that

$$\frac{1}{|OP|^2} = \frac{d_1{}^2}{a^2} + \frac{d_2{}^2}{b^2} + \frac{d_3{}^2}{c^2} .$$

10. Three pairwise perpendicular straight lines are drawn from the origin O to meet the ellipsoid $\frac{x^2}{a^2} + \frac{y^2}{b^2} + \frac{z^2}{c^2} = 1$ at the points P_1, P_2 and P_3 such that $|OP_1| = s_1$, $|OP_2| = s_2$ and $|OP_3| = s_3$. Prove that $\frac{1}{s_1{}^2} + \frac{1}{s_2{}^2} + \frac{1}{s_3{}^2} = \frac{1}{a^2} + \frac{1}{b^2} + \frac{1}{c^2}$.

11. Show that parallel plane sections of an ellipsoid are similar ellipses.

4.4 The hyperboloid of two sheets

The quadric surface defined by the quadratic equation

$$\frac{x^2}{a^2} + \frac{y^2}{b^2} - \frac{z^2}{c^2} = -1 \tag{1}$$

is called a *hyperboloid of two sheets*. If in particular $a = b$, it is a hyperboloid of revolution of two sheets. Let us examine the geometric shape of the surface in the general case where $a \neq b$ by its plane sections. The sections by the vertical coordinate planes $x = 0$ and $y = 0$ are the hyperbolas

$$\frac{y^2}{b^2} - \frac{z^2}{c^2} = -1 \quad \text{and} \quad \frac{x^2}{a^2} - \frac{z^2}{c^2} = -1 \tag{2}$$

respectively with their foci lying on the z-axis. The horizontal coordinate plane $z = 0$ does not intersect the surface. Therefore the surface is separated into two sheets by the horizontal coordinate plane. It is also true that the surface is separated by all horizontal planes $z = k$ with $k^2 < c^2$. The horizontal planes $z = \pm c$ are the tangent planes to the surface individually at the points $(0, 0, \pm c)$. The horizontal planes $z = k$, with $k^2 > c^2$, intersect the hyperboloid in similar ellipses, which increase in size as $|k|$ increases.

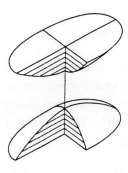

Fig 4-15

Similar to an ellipsoid, a hyperboloid of two sheets has three planes of symmetries which are perpendicular to each other. These are called the *principal planes* of the hyperboloid. (For the hyperboloid (1) they are the coordinate planes). The principal planes intersect each other at three lines, the *axes* of the hyperboloid and the axes intersect at one point, the *centre* of the hyperboloid. Like an ellipsoid, a hyperboloid of two sheets is a central quadric surface.

Recall that given a hyperbola $b^2x^2 - a^2y^2 = a^2b^2$ on the plane, we drop the constant term a^2b^2 to obtain the equation $b^2x^2 - a^2b^2 = (bx+ay)(bx-ay) = 0$ of the asymptotes of the hyperbola. Analogously we drop the constant term -1 of (1) to get

$$\frac{x^2}{a^2} + \frac{y^2}{b^2} - \frac{z^2}{c^2} = 0 \tag{3}$$

which by 4.2.2 defines a quadric cone with vertex at O. We call this cone the *asymptotic cone* of the hyperboloid. The relationship between the hyperboloid (1) and its asymptotic cone (3) is similar to the relationship between a hyperbola and its asymptotes. For example each sheet of the hyperboloid is within one nappe of the asymptotic cone, and points of the hyperboloid far from the centre O are arbitrarily close to the asymptotic cone.

The vertical coordinate planes $x = 0$ and $y = 0$ intersect the hyperboloid in the hyperbolas (2). They intersect the asymptotic cone (3) in two pairs of intersecting straight lines

$$\left(\frac{y}{b} - \frac{z}{c}\right)\left(\frac{y}{b} + \frac{z}{c}\right) = 0 \quad \text{and} \quad \left(\frac{x}{a} - \frac{z}{c}\right)\left(\frac{x}{a} + \frac{z}{c}\right) = 0$$

which are precisely the asymptotes of the hyperbolas (2). This is true also of any vertical plane

$$y = kx \tag{4}$$

which passes through the z-axis. In fact if we choose the line on the plane that passes the origin and perpendicular to the z-axis as t-axis of the plane (4), then from $y = kx$ and $t^2 = x^2 + y^2$ we obtain $(1 + k^2)x^2 = t^2$ and $(1 + k^2)y^2 = k^2t^2$. Therefore the plane section of the hyperboloid is the hyperbola

$$\frac{1}{1+k^2}\left(\frac{1}{a^2} + \frac{k^2}{b^2}\right)t^2 - \frac{z^2}{c^2} = -1$$

and the plane section of the asymptotic cone is the pair of asymptotes

$$\frac{1}{1+k^2}\left(\frac{1}{a^2} + \frac{k^2}{b^2}\right)t^2 - \frac{z^2}{c^2} = 0 \, ,$$

all on the tz-plane $y = kx$.

EXERCISES

1. Find equations of the hyperboloid of two sheets and its asymptotic cone such that when intersects with the coordinate planes $x = 0$ and $y = 0$, the hyperbolas $\frac{y^2}{4} - \frac{z^2}{16} = -1$ and $\frac{x^2}{36} - \frac{z^2}{16} = -1$ are obtained.

2. For the hyperboloid of two sheets

$$\frac{x^2}{a^2} + \frac{y^2}{b^2} - \frac{z^2}{c^2} = -1 \, , \quad a > b > 0$$

   find the foci of the intersection of the hyperboloid with
   (a) the plane $z = k$, where $k^2 > c^2$;
   (b) the plane $x = k$.
   What kinds of curves do we have in (a) and (b)?

4.5 The hyperboloid of one sheet

The quadric surface defined by the equation

$$\frac{x^2}{a^2} + \frac{y^2}{b^2} - \frac{z^2}{c^2} = 1 \tag{1}$$

is called a *hyperboloid of one sheet*. If $a = b$, it is a hyperboloid of revolution of one sheet. In general the vertical coordinate planes intersect the hyperboloid in the hyperbolas

$$\frac{y^2}{b^2} - \frac{z^2}{c^2} = 1 \quad \text{and} \quad \frac{x^2}{a^2} - \frac{z^2}{c^2} = 1 \, .$$

The sections by the planes $z = k$ perpendicular to the z-axis are all similar ellipses. The smallest one is the section by the coordinate plane $z = 0$; it is known as the *minimum ellipse*. Like a hyperboloid of two sheets, a hyperboloid of one sheet has three planes of symmetry, three axes and a centre. It is therefore also a central quadric surface. The quadric cone

$$\frac{x^2}{a^2} + \frac{y^2}{b^2} - \frac{z^2}{c^2} = 0$$

is also the asymptotic cone of the hyperboloid (1).

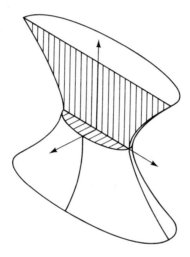

Fig 4-16

One of the most astonishing property of the hyperboloid of one sheet is that there are two families of straight lines lying on the surface. To show that this is true, we write the equation (1) of the hyperboloid as

$$\left(\frac{y}{b} + \frac{z}{c}\right)\left(\frac{y}{b} - \frac{z}{c}\right) = \left(1 + \frac{x}{a}\right)\left(1 - \frac{x}{a}\right) . \tag{2}$$

Then for every value of u, the linear equations

$$\left(\frac{y}{b} + \frac{z}{c}\right) = u\left(1 + \frac{x}{a}\right) \quad \text{and} \quad u\left(\frac{y}{b} - \frac{z}{c}\right) = \left(1 - \frac{x}{a}\right) \tag{3}$$

define two planes in space intersecting in a straight line $L(u)$ in space. Since (2) follows from (3) and (1) follows from (2), every point of $L(u)$ is a point of the hyperboloid (1). Therefore the line $L(u)$ lies entirely on the curved surface of the hyperboloid (1) for any arbitrary value of u. Dividing the equations of (3) by u and letting u approach infinity, we obtain one more straight line $L(\infty)$ defined by the linear equations

$$1 + \frac{x}{a} = 0 \quad \text{and} \quad \frac{y}{b} - \frac{z}{c} = 0 . \tag{4}$$

Clearly every point on the line $L(\infty)$ also lies on the hyperboloid (1). *Therefore the lines $L(u)$, one for each value of u, together with the limiting line $L(\infty)$ form a family of straight lines on the hyperboloid of one sheet. Similarly the lines $G(v)$, one for each value of v, defined by the equations*

$$\left(\frac{y}{b} - \frac{z}{c}\right) = v\left(1 + \frac{x}{a}\right) \quad \text{and} \quad v\left(\frac{y}{b} + \frac{z}{c}\right) = \left(1 - \frac{x}{a}\right)$$

together with the limiting line $G(\infty)$ defined by the equations

$$1 + \frac{x}{a} = 0 \quad \text{and} \quad \frac{y}{b} + \frac{z}{c} = 0$$

form another family of straight lines on the hyperboloid of one sheet. The lines of these families are called the *rulings* of the hyperboloid. As an exercise we shall leave the reader to show that *through each point of the hyperboloid (1) passes exactly one ruling of each family.* Thus a hyperboloid of one sheet is also a ruled quadric surface sharing a common feature with cones and cylinders. Based on this result string model of hyperboloid of one sheet can be made (See Figure 4-17).

Among conics, both ellipse and hyperbola have a centre and are called central conics, while a parabola is not a central conic. Among quadric surfaces, cones, elliptic cylinders, hyperbolic cylinders, ellipsoids, hyperboloids of two sheets and hyperboloids of one sheet are all central quadric surfaces. There are no more central quadric surfaces.

Fig 4-17

EXERCISES

1. Find the equation of the hyperboloid of revolution of one sheet which contains the points $(5, 0, 0)$ and $(5, 5, 2)$.

2. Use similar techniques as in equations (2) to (4), find equations of rulings of the hyperboloid $x^2 + y^2 - z^2 = 1$ which passes through the point $(1, 1, 1)$.

3. Find equations of rulings of the hyperboloid $\frac{x^2}{4} + \frac{y^2}{1} - \frac{z^2}{9} = 1$ which passes through the point $(2, 1, 3)$.

Answers to Questions 4 – 6 give a proof for the statement that through each point of the hyperboloid, $\frac{x^2}{a^2} + \frac{y^2}{b^2} - \frac{z^2}{c^2} = 1$, passes exactly one ruling of each family.

4. Show that if the point $P(x_0, y_0, z_0)$, with $x_0 \neq -a$, is on the hyperboloid of one sheet,

$$\frac{x^2}{a^2} + \frac{y^2}{b^2} - \frac{z^2}{c^2} = 1 \, ,$$

then there is a unique real value u for which

$$\frac{y_0}{b} + \frac{z_0}{c} = u\left(1 + \frac{x_0}{a}\right) \quad \text{and} \quad u\left(\frac{y_0}{b} - \frac{z_0}{c}\right) = 1 - \frac{x_0}{a} \, . \qquad (*)$$

5. Repeat Question 4 by replacing $(*)$ with

$$\frac{y_0}{b} - \frac{z_0}{c} = v\left(1 + \frac{x_0}{a}\right) \quad \text{and} \quad v\left(\frac{y_0}{b} + \frac{z_0}{c}\right) = 1 - \frac{x_0}{a} \, .$$

6. Show that with $x_0 = -a$, $P(x_0, y_0, z_0)$ lies on the line defined by

$$1 + \frac{x}{a} = 0 \quad \text{and} \quad \frac{y}{b} - \frac{z}{c} = 0$$

and on the line defined by

188

$$1 + \frac{x}{a} = 0 \quad \text{and} \quad \frac{y}{b} + \frac{z}{c} = 0 \, .$$

7. Show that the families of rulings

$$\left(\frac{y}{b} + \frac{z}{c}\right) = u\left(1 + \frac{x}{a}\right) \quad \text{and} \quad u\left(\frac{y}{b} - \frac{z}{c}\right) = \left(1 - \frac{x}{a}\right)$$

and

$$\left(\frac{y}{b} - \frac{z}{c}\right) = v\left(1 + \frac{x}{a}\right) \quad \text{and} \quad v\left(\frac{y}{b} + \frac{z}{c}\right) = \left(1 - \frac{x}{a}\right)$$

are distinct.

8. Let $Ax^2 + By^2 + Cz^2 = 1$ be a central quadric surface.

(a) A straight line with direction cosine d_1, d_2, d_3 is drawn from the origin O to the quadric surface at P. Show that

$$\frac{1}{|OP|^2} = Ad_1{}^2 + Bd_2{}^2 + Cd_3{}^2 \, .$$

(b) Three pairwise perpendicular straight lines are drawn from the origin O to the quadric surface at points P_1, P_2, P_3 such that $|OP_1| = s_1$, $|OP_2| = s_2$ and $|OP_3| = s_3$. Prove that

$$\frac{1}{s_1{}^2} + \frac{1}{s_2{}^2} + \frac{1}{s_3{}^2} = A + B + C \, .$$

4.6 The elliptic paraboloid

A paraboloid of revolution with vertical axis is defined by an equation of the form $x^2 + y^2 = 4az$. A general *elliptic paraboloid* is defined by an equation of the form

$$\frac{x^2}{a^2} + \frac{y^2}{b^2} = 4z \tag{1}$$

where $a \neq b$.

The horizontal coordinate plane $z = 0$ meets the hyperboloid only in the origin, the *vertex* of the paraboloid. It is therefore the tangent plane to the paraboloid at the vertex. Any horizontal plane $z = k$ $(k < 0)$ below it does not meet the surface, while a horizontal plane $z = k$ $(k > 0)$ above it intersects the surface in an ellipse

$$\frac{x^2}{a^2} + \frac{y^2}{b^2} = 4k$$

which increases in size as the height of plane increases. All horizontal sections are similar ellipses.

The vertical coordinate planes $x = 0$ and $y = 0$ cut the paraboloid (1) in parabolas

$$y^2 = 4b^2 z \quad \text{and} \quad x^2 = 4a^2 z ,$$

both opening upwards. In fact all planes passing the z-axis cut the paraboloid in parabolas.

The elliptic paraboloid has only two planes of symmetry (the vertical coordinate planes for the paraboloid (1)), one axis (the z-axis), one vertex (the origin) but no centre.

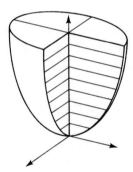

Fig 4-18

EXERCISES

1. Find an equation of the elliptic paraboloid whose intersection with the plane $z = 2$ is the ellipse $9x^2 + y^2 = 144$.

2. Show that the projection onto the xy-plane of the intersection of the plane $z = 2y$ and the elliptic paraboloid $z = x^2 + y^2$ is a circle. What is its radius and its centre?

3. Given that the intersection of the elliptic paraboloid $x^2 + \frac{y^2}{2} = 2z$ and the plane $x = kz$ $(k < 0)$ is a circle, find the value of k, the radius of the circle, and the coordinates of centre. [Hint: $(0, 0, 0)$ lies on the circle.]

4.7 The hyperbolic paraboloid

Probably the most interesting quadric surface is a *hyperbolic paraboloid* which is defined by

$$\frac{x^2}{a^2} - \frac{y^2}{b^2} = 4z \ . \tag{1}$$

It is never a surface of revolution, no matter what values are assigned to a and b. The vertical coordinate planes $x = 0$ and $y = 0$ cut the surface in two parabolas

$$y^2 = -4b^2 z \quad \text{and} \quad x^2 = 4a^2 z \ . \tag{2}$$

The first of these opens downwards on the yz-plane, and the second one opens upwards on the xz-plane. They touch each other at their common vertex O but are on separate planes perpendicular to each other.

The section by the horizontal coordinate plane $z = 0$ consists of two lines

$$bx + ax = 0 \quad \text{and} \quad bx - ax = 0 \tag{3}$$

meeting at the origin and askew (perpendicular if $a = b$) to each other. A horizontal plane $z = k$ $(k > 0)$ above the xy-plane cuts the surface in a hyperbola whose vertices are on the second upward parabola (2) of the xz-plane. A horizontal plane $z = k$ $(k < 0)$ below it cuts the surface in a hyperbola whose vertices are on the first downward parabola (2) of the yz-plane. The surface looks somewhat like a saddle. It rises along the parabola on the xz-plane and falls along the parabola on the yz-plane.

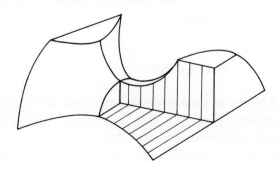

Fig 4-19

A hyperbolic paraboloid has two planes of symmetry. For the surface (1) they are the two vertical coordinate planes. The line of intersection of these two planes is the *axis* of the paraboloid which intersects the surface at its *vertex*. For the paraboloid (1), the axis is the z-axis and the vertex the origin.

Like a hyperboloid of one sheet, a hyperbolic paraboloid has two families of rulings. To find their equations, we write equation (1) into

$$\left(\frac{x}{a} + \frac{y}{b}\right)\left(\frac{x}{a} - \frac{y}{b}\right) = 4z$$

from which the families of rulings are obtained as:

$$L(u): \quad \frac{x}{a} + \frac{y}{b} = u, \quad u\left(\frac{x}{a} - \frac{y}{b}\right) = 4z \ ;$$

$$G(v): \quad \frac{x}{a} - \frac{y}{b} = v, \quad v\left(\frac{x}{a} + \frac{y}{b}\right) = 4z \ .$$

In particular the rulings $L(0)$ and $G(0)$ are the two lines (3) in which the horizontal coordinate plane $z = 0$ intersects the hyperbolic paraboloid. Moreover all rulings $L(u)$ of the first family lie on planes parallel to the plane $bx + ay = 0$. Similarly all rulings $G(v)$ of the second family lie on planes parallel to the plane $bx - ay = 0$. These two planes are known as the *directrix planes* of the hyperbolic paraboloid. It is readily verified that through every point on the hyperbolic paraboloid passes exactly one ruling of each family. Therefore the hyperolic paraboloid is also a ruled quadric surface. Moreover in contradistinction to the cone and the cylinder which have only one family of rulings, the hyperbolic paraboloid and the hyperboloid of one sheet are surfaces that contain two families of rulings. They are called *doubly ruled surfaces*. In fact the plane and these two quadrics are the only doubly ruled surfaces.

EXERCISES

1. Find equations of rulings of the hyperbolic paraboloid $4x^2 - y^2 = 16z$ which passes through the point $(3, 2, 2)$.
2. Given the hyperbolic paraboloid $x^2 - y^2 = 2z$.
   (a) Show that the two families of rulings of the hyperbolic paraboloid

have the directions of the vectors $[1, 1, v]$ and $[-1, 1, -u]$ respectively for real numbers v and u.

(b) Find equations of rulings which are parallel to the plane $x + y + 2z = 0$.

Answers to Question 3 – 6 give a proof for the statement that through every point on the hyperbolic paraboloid, $\frac{x^2}{a^2} - \frac{y^2}{b^2} = 4z$, passes exactly one ruling of each family.

3. Show that if the point $P(x_0, y_0, z_0)$ is on the hyperbolic paraboloid

$$\frac{x^2}{a^2} - \frac{y^2}{b^2} = 4z \ ,$$

then there is a unique real number u for which

$$\frac{x_0}{a} + \frac{y_0}{b} = u \quad \text{and} \quad u\left(\frac{x_0}{a} - \frac{y_0}{b}\right) = 4z_0 \ . \tag{$*$}$$

4. Show that, for each fixed value of u, if $P(x_0, y_0, z_0)$ is on the line of intersection of the planes with equations

$$\frac{x}{a} + \frac{y}{b} = u \quad \text{and} \quad u\left(\frac{x}{a} - \frac{y}{b}\right) = 4z \ , \tag{$**$}$$

then P is on the hyperbolic paraboloid $\frac{x^2}{a^2} - \frac{y^2}{b^2} = 4z$.

5. Repeat Question 3 by replacing $(*)$ with

$$\frac{x_0}{a} - \frac{y_0}{b} = v \quad \text{and} \quad v\left(\frac{x_0}{a} + \frac{y_0}{b}\right) = 4z_0 \ .$$

6. Repeat Question 4 by replacing $(**)$ with

$$\frac{x}{a} - \frac{y}{b} = v \quad \text{and} \quad v\left(\frac{x}{a} + \frac{y}{b}\right) = 4z \ .$$

4.8 Coordinate transformations

Now that we have found numerous types of quadric surfaces and their equations when in standard position, we would naturally wish to study equations of quadric surfaces that are moved out of standard position. Similar to our earlier study on conics, we shall consider parallel translations and rotations in space.

Every vector $[h, k, \ell]$ in space defines a *parallel translation* τ of the space that takes the point $X = (x, y, z)$ to the point $\tau(X) = (x + h, y + k, z + \ell)$ and takes the ray OX to the parallel ray $\tau(O)\tau(X)$. Therefore the images of the coordinate axes under τ form a right-hand coordinate system. If we denote the images of the x-axis, y-axis and z-axis by x'-axis, y'-axis and z'-axis respectively, then every point X in space have two sets of coordinates:

$$X = (x, y, z) \text{ relative to the } xyz\text{-coordinate system}$$
and $\quad X = ((x', y', z')) \text{ relative to the } x'y'z'\text{-coordinate system}$

where
$$\begin{cases} x = x' + h \\ y = y' + k \\ z = z' + \ell \end{cases} \quad \text{and} \quad \begin{cases} x' = x - h \\ y' = y - k \\ z' = z - \ell \,. \end{cases}$$

In particular for the origin O of the original xyz-system we have

$$O = (0, 0, 0) = ((-h, -k, -\ell))$$

and for the origin O' of the new $x'y'z'$-system

$$O' = ((0, 0, 0)) = (h, k, \ell)\,.$$

Similarly the locus of an equation

$$f(x, y, z) = 0$$

relative to the xyz-system is the locus of the equation

$$g(x', y', z') = f(x' + h, y' + k, z' + \ell) = 0$$

relative to the $x'y'z'$-system.

4.8.1 EXAMPLE   If we write the equation

$$2x^2 - 3y^2 - 4z^2 - 4x - 12y - 4z = 0$$

of a quadric surface as

$$2(x^2 - 2x + 1) - 3(y^2 + 4y + 4) - 4(z^2 + z + \frac{1}{4}) = 2 - 12 - 1$$

or $\quad 2(x - 1)^2 - 3(y + 2)^2 - 4(z + \frac{1}{2})^2 = -11$

or $\quad -2x'^2 + 3y'^2 + 4z'^2 = 11$

then the quadric surface is recognized as a hyperboloid of one sheet with its minimum ellipse parallel to the yz-plane.

Similarly a rotation of the coordinate axes affects a transformation of coordinates. Consider a rotation ρ of the space about the origin O. Obviously as a rigid motion, ρ takes a unit vector to a unit vector, an orthogonal pair to an orthogonal pair and a right-hand system to a right-hand system. If the vectors $e_1 = [1, 0, 0]$, $e_2 = [0, 1, 0]$, $e_3 = [0, 0, 1]$ are the unit coordinate vectors of the original coordinate axes, then their images $e_1' = \rho(e_1) = [\alpha_1, \alpha_2, \alpha_3]$, $e_2' = \rho(e_2) = [\beta_1, \beta_2, \beta_3]$, $e_3' = \rho(e_3) = [\gamma_1, \gamma_2, \gamma_3]$ under ρ are mutually orthogonal unit vectors. Clearly e_1', e_2', e_3' also form a right-hand coordinate system and determine three new coordinate axes, say x'-axis, y'-axis and z'-axis respectively. Consequently every vector x in space has two sets of components, $[x, y, z]$ and $[[x', y', z']]$, which are determined by

$$x = xe_1 + ye_2 + ze_3$$

and
$$x = x'e_1' + y'e_2' + z'e_3' \ .$$

Substituting $e_1' = \alpha_1 e_1 + \alpha_2 e_2 + \alpha_3 e_3$ etc. into the second equation, we obtain

$$x = (\alpha_1 x' + \beta_1 y' + \gamma_1 z')e_1 + (\alpha_2 x' + \beta_2 y' + \gamma_2 z')e_2 + (\alpha_3 x' + \beta_3 y' + \gamma_3 z')e_3.$$

Now it follows from the linear independence of the unit coordinate vectors e_1, e_2, e_3 that between the two sets of components of x we have

$$\begin{aligned}
x &= \alpha_1 x' + \beta_1 y' + \gamma_1 z' \\
y &= \alpha_2 x' + \beta_2 y' + \gamma_2 z' \\
z &= \alpha_3 x' + \beta_3 y' + \gamma_3 z'
\end{aligned} \tag{1}$$

expressing x, y, z in terms of x', y', z'. The equations (1) are therefore the *equations of transformation*. Before we proceed to write down the equations of transformation in reverse, a review of the 2-dimensional case would be very helpful.

4.8.2 REMARKS. If we carry out the above calculation for a rotation ρ of the plane about O through an angle θ, then for $\rho(e_1) = e_1' = [\alpha_1, \alpha_2]$ and $\rho(e_2) = e_2' = [\beta_1, \beta_2]$ we would get

$$\alpha_1 = \cos\theta, \alpha_2 = \sin\theta; \ \beta_1 = -\sin\theta, \beta_2 = \cos\theta.$$

Therefore the equations of transformation expressing x, y in terms of x', y' have the form

$$x = \alpha_1 x' + \beta_1 y' \qquad\qquad x = x' \cos\theta - y' \sin\theta$$
$$\text{or}$$
$$y = \alpha_2 x' + \beta_2 y' \qquad\qquad y = x' \sin\theta + y' \cos\theta \ .$$

By the result of Section 3.8, the equations of transformation in reverse are

$$x' = x\cos\theta + y\sin\theta \qquad\qquad x' = \alpha_1 x + \alpha_2 y$$
$$\text{or}$$
$$y' = -x\sin\theta + y\cos\theta \qquad\qquad y' = \beta_1 x + \beta_2 y \ .$$

A comparison of these sets of equations of transformation with (1) leads us to the "intelligent guess" that

$$x' = \alpha_1 x + \alpha_2 y + \alpha_3 z$$
$$y' = \beta_1 x + \beta_2 y + \beta_3 z \qquad\qquad (2)$$
$$z' = \gamma_1 x + \gamma_2 y + \gamma_3 z$$

would be the equations of transformation in reverse.

In order to verify (2) we examine the constants α_1, α_2 etc. that appear in equations (1) and (2). Firstly it follows from the fact that e_1', e_2' and e_3' are unit vectors that

$$\alpha_1{}^2 + \alpha_2{}^2 + \alpha_3{}^2 = 1$$
$$\beta_1{}^2 + \beta_2{}^2 + \beta_3{}^2 = 1 \qquad\qquad (3)$$
$$\gamma_1{}^2 + \gamma_2{}^2 + \gamma_3{}^2 = 1 \ .$$

In fact the α's, β's, γ's are just the direction cosines of the x'-, y'-, z'-axes with respect to the original axes. Secondly the vectors e_1', e_2', e_3' are mutually orthogonal. Therefore

$$\alpha_1\beta_1 + \alpha_2\beta_2 + \alpha_3\beta_3 = 0$$
$$\beta_1\gamma_1 + \beta_2\gamma_2 + \beta_3\gamma_3 = 0 \qquad\qquad (4)$$
$$\gamma_1\alpha_1 + \gamma_2\alpha_2 + \gamma_3\alpha_3 = 0 \ .$$

Finally the validity of (2) can be directly verified by direct substitution. For example

$$\alpha_1 x + \alpha_2 y + \alpha_3 z$$
$$= \alpha_1(\alpha_1 x' + \beta_1 y' + \gamma_1 z') + \alpha_2(\alpha_2 x' + \beta_2 y' + \gamma_2 z') + \alpha_3(\alpha_3 x' + \beta_3 y' + \gamma_3 z')$$
$$= (\alpha_1{}^2 + \alpha_2{}^2 + \alpha_3{}^2)x' + (\alpha_1\beta_1 + \alpha_2\beta_2 + \alpha_3\beta_3)y' + (\alpha_1\gamma_1 + \alpha_2\gamma_2 + \alpha_3\gamma_3)z'$$
$$= x' \ .$$

What is carried out on the position vectors can be done to the points in space. Therefore by rotating the coordinate axes to a new position about the origin, every point X in space have two sets of coordinates, (x, y, z) relative to the original axes and $((x', y', z'))$ relative to the new axes, which are related to one another by the following equations of transformation

$$
\begin{aligned}
x &= \alpha_1 x' + \beta_1 x' + \gamma_1 z' & x' &= \alpha_1 x + \alpha_2 y + \alpha_3 z \\
y &= \alpha_2 x' + \beta_2 y' + \gamma_2 z' \quad \text{and} & y' &= \beta_1 x + \beta_2 y + \beta_3 z \\
z &= \alpha_3 x' + \beta_3 y' + \gamma_3 z' & z' &= \gamma_1 x + \gamma_2 y + \gamma_3 z .
\end{aligned} \tag{5}
$$

In particular for the origin O and the unit points E_1, E_2, E_3 and E_1', E_2', E_3' on the coordinate axes, we have

$$O = (0,0,0) = ((0,0,0))$$

$$
\begin{aligned}
E_1 &= (1,0,0) = ((\alpha_1,\beta_1,\gamma_1)) & E_1' &= ((1,0,0)) = (\alpha_1,\alpha_2,\alpha_3) \\
E_2 &= (0,1,0) = ((\alpha_2,\beta_2,\gamma_2)) & E_2' &= ((0,1,0)) = (\beta_1,\beta_2,\beta_3) \\
E_3 &= (0,0,1) = ((\alpha_3,\beta_3,\gamma_3)) & E_3' &= ((0,0,1)) = (\gamma_1,\gamma_2,\gamma_3) .
\end{aligned}
$$

The equations (5) of transformation are conveniently arranged into the following scheme

	x'	y'	z'
x	α_1	β_1	γ_1
y	α_2	β_2	γ_2
z	α_3	β_3	γ_3

Reading across we obtain the first set of equations of (5) and reading down we get the second set. Also the rows give the direction consines of the old axes with respect to the new axes, and the columns those of the new axes with respect to the old axes.

4.8.3 REMARKS   Recall that any triple h, k, ℓ of real numbers determine a shift of the axes to the new origin (h, k, ℓ) resulting in a transformation of coordinates

$$
\begin{cases}
x = x' + h \\
y = y' + k \quad \text{and} \\
z = z' + \ell
\end{cases}
\begin{cases}
x' = x - h \\
y' = y - k \\
z' = z - \ell .
\end{cases}
$$

It is far from being true that given nine arbitrary real numbers α_1, α_2, α_3, β_1, β_2, β_3, γ_1, γ_2, γ_3 the equations

$$\begin{cases} x = \alpha_1 x' + \beta_1 y' + \gamma_1 z' \\ y = \alpha_2 x' + \beta_2 y' + \gamma_2 z' \\ z = \alpha_3 x' + \beta_3 y' + \gamma_3 z' \end{cases} \text{ and } \begin{cases} x' = \alpha_1 x + \alpha_2 y + \alpha_3 z \\ y' = \beta_1 x + \beta_2 y + \beta_3 z \\ z' = \gamma_1 x + \gamma_2 y + \gamma_3 z \end{cases}$$

are the equations of a coordinate transformation affected by a rotation of axes. It is not difficult to see that this is true if and only if the vectors $\mathbf{a} = [\alpha_1, \alpha_2, \alpha_3]$, $\mathbf{b} = [\beta_1, \beta_2, \beta_3]$, $\mathbf{c} = [\gamma_1, \gamma_2, \gamma_3]$ are mutually orthogonal unit vectors that form a right-hand system, i.e. the following conditions are satisfied

(i) $\mathbf{a} \cdot \mathbf{a} = \mathbf{b} \cdot \mathbf{b} = \mathbf{c} \cdot \mathbf{c} = 1$
(ii) $\mathbf{a} \cdot \mathbf{b} = \mathbf{b} \cdot \mathbf{c} = \mathbf{c} \cdot \mathbf{a} = 0$
(iii) $\mathbf{a} \times \mathbf{b} = \mathbf{c}$, $\mathbf{b} \times \mathbf{c} = \mathbf{a}$, $\mathbf{c} \times \mathbf{a} = \mathbf{b}$.

4.8.4 EXAMPLE   Transform the equation of the surface

$$25x^2 + 16y^2 + 22z^2 - 4xy - 16yz + 20xz - 36 = 0$$

by a transformation of coordinates so that the new axes go through the origin O and have direction cosines $-1/3$, $2/3$, $2/3$; $2/3$, $-1/3$, $2/3$; $2/3$, $2/3$, $-1/3$ respectively.

SOLUTION   It is obvious that the three conditions of Remarks 4.8.3 are satisfied. Therefore the equations of transformation are

$$x = \frac{1}{3}(-x' + 2y' + 2z')$$

$$y = \frac{1}{3}(2x' - y' + 2z')$$

$$z = \frac{1}{3}(2x' + 2y' - z') .$$

Substituting these values in the equation of surface and simplifying the result, we obtain

$$x'^2 + 4y'^2 + 2z'^2 = 4 .$$

Hence the surface is an ellipsoid whose centre is at O and whose principal planes are $x - 2y - 2z = 0$, $2x - y + 2z = 0$, $2x + 2y - z = 0$.

4.8.5 EXAMPLES   What surface is represented by the equation

$$25x^2 + 16y^2 + 22z^2 - 4xy - 16yz + 20xz - 94x + 68y - 124z + 169 = 0 \ ?$$

SOLUTION We try to get rid of the linear terms of the equation by a transformation of the form

$$x = x' + h, \quad y = y' + k, \quad z = z' + \ell \ .$$

Substituting the values of x, y, z given in terms of x', y', z', h, k, ℓ in the equation of surface, collecting terms and then putting the coefficients of x', y', z' equal to zero, we obtain the equations:

$$50h - 4k + 20\ell - 94 = 0$$
$$-4h + 32k - 16\ell + 68 = 0$$
$$20h - 16k + 44\ell - 124 = 0 \ .$$

These equations have a unique solution, namely $h = 1$, $k = -1$, $\ell = 2$. If the equation of the given surface is transformed by shifting the origin to $(1, -1, 2)$, it becomes

$$25x'^2 + 16y'^2 + 22z'^2 - 4x'y' - 16y'z' + 20x'z' - 36 = 0 \ .$$

But this is the same equation in x', y', z' as the equation of the surface of the last example in x, y, z. Therefore the surface of the present example is the same as the one of last example but with its centre at $(1, -1, 2)$.

4.8.6 EXAMPLE Use appropriate coordinate transformations to simplify the quadratic equation

$$4x^2 - y^2 - z^2 - 2yz - 8x + (10/\sqrt{2})y + (6/\sqrt{2})z - 6 = 0 \ .$$

SOLUTION Since the cross product terms xy and xz are missing in the equation, to eliminate the only non-vanishing cross product term $2yz$, we need only to rotate the y-axis and the z-axis about the x-axis. This means that we apply a transformation of coordinates

$$x = x', \quad y = y' \cos\theta - z' \sin\theta, \quad z = y' \sin\theta + z' \cos\theta$$

to the given equation. Putting the term in $x'y'$ equal to zero, we obtain the desired angle of rotation as $\theta = \pi/4$. With this value of θ, the given equation in x and y is now transformed into

$$2x'^2 - y'^2 - 4x' + 4y' - z' - 3 = 0$$

which can be also written as

$$2(x' - 1)^2 - (y' - 2)^2 = z' + 1 .$$

Therefore by a shift of the origin

$$x'' = x' - 1, \quad y'' = y' - 2, \quad z'' = x + 1$$

the equation is further simplified as

$$2x''^2 - y''^2 = z''$$

which represents a hyperbolic paraboloid.

EXERCISES

1. A parallel translation is defined by the equations $x = x' + 3$, $y = y' - 2$ and $z = z' + 1$. Find the new coordinates of the points $(0, 0, 0)$, $(5, -3, 7)$ and $(8, 7, -6)$.

2. After a parallel translation, the point $(3, -6, 1)$ becomes $((0, 1, 4))$. Write down the equations for the translation.

3. After a parallel translation, the point $(2, 1, -3)$ lies on the $x'y'$-plane, the point $(3, 8, -4)$ lies on the $y'z'$-plane, and the point $(5, -1, 0)$ lies on the $x'z'$-plane. Write down the corresponding equations for the translation.

4. A parallel translation takes the origin to the point $(1, -4, -3)$. Find the equation of each of the following after the translation.
   (a) $2x - y + 3z = 5$;
   (b) $\frac{x-1}{4} = \frac{y+2}{3} = \frac{z-3}{2}$;
   (c) $x^2 + 3z^2 - 6xy - 26x + 6y + 18z + 63 = 0$.

5. Find the point to which the origin should be shifted by a parallel translation such that the equation of surface

$$2x^2 + 3y^2 + 4z^2 - 4x - 12y - 24z + 49 = 0$$

does not contain any term of the first degree.

6. Prove that under the parallel translation $x = x' + h$, $y = y' + k$ and $z = z' + \ell$, the distance between any two points remains unchanged.

7. Find the parallel translation such that the equation of surface

$$Ax^2 + By^2 + Cz^2 + 2Dx + 2Ey + 2Fz + G = 0$$

where $A \cdot B \cdot C \neq 0$ does not contain any term of the first degree.

8. Under two parallel tanslations, the origin O is brought to $O_1 = (a_1, b_1, c_1)$ and $O_2 = (a_2, b_2, c_2)$ respectively. Find the coordinates of the middle point of O_1 and O_2 under the two new systems of coordinates.

9. After a rotation, the new axes through the origin O have direction cosines $-\frac{1}{3}, \frac{2}{3}, \frac{2}{3}; \frac{2}{3}, -\frac{1}{3}, \frac{2}{3}; \frac{2}{3}, \frac{2}{3}, -\frac{1}{3}$. Write down the equations of transformation and find the new coordinates of the points $(0, 0, 0)$, $(5, -3, 7)$ and $(8, 7, -6)$.

10. After a rotation, the new axes through the origin O have unit vectors e_1', e_2' and e_3' such that $e_1' = [\frac{2}{3}, \frac{2}{3}, -\frac{1}{3}]$ and $e_2' = [-\frac{1}{3}, \frac{2}{3}, \frac{2}{3}]$. Find e_3' and hence write down the equations of transformation.

11. Use a similar method as that employed in 4.8.6, eliminate the xy term in the equation $x^2 + xy + y^2 - z = 0$ and identify the surface.

12. Transform the equation of the surface

$$10x^2 + 13y^2 + 13z^2 - 4xy - 4xz + 8yz - 4 = 0$$

to new axes through the origin O whose direction cosines are respectively $\frac{2}{3}, \frac{2}{3}, -\frac{1}{3}; -\frac{1}{3}, \frac{2}{3}, \frac{2}{3}; \frac{2}{3}, -\frac{1}{3}, \frac{2}{3}$.

13. Prove that a rotation of the coordinate axes does not affect the distance between any two points in space.

14. Consider the plane $2x - y + 2z + 7 = 0$.

   (a) Find a parallel translation such that the equation of the plane becomes $2x' - y' + 2z' = 0$.

   (b) Then find a suitable rotation so that the plane becomes $x'' = 0$.
   [Hint: Find suitable coordinate unit vectors first.]

4.9 The general quadratic equation

By definition a quadric surface is a surface in space whose points $X = (x, y, z)$ satisfy a quadratic equation

$$Ax^2 + By^2 + Cz^2 + Hxy + Jyz + Kxz + Dx + Ey + Fz + G = 0 . \quad (1)$$

So far we have identified nine different types of quadric surfaces.

They are namely
- (i) the cones,
- (ii) the elliptic cylinders,
- (iii) the hyperbolic cylinders,
- (iv) the parabolic cylinders,
- (v) the ellipsoids,
- (vi) the hyperboloids of two sheets,
- (vii) the hyperboloids of one sheet,
- (viii) the elliptic paraboloids,
- (ix) the hyperbolic paraboloids.

Among them, surfaces of types (i), (ii), (iii), (v), (vi) and (vii) all have a centre of symmetry; they are the *central quadric surfaces*. The *ruled quadric surfaces* are those of types (i), (ii), (iii), (iv), (vii) and (ix) because they can be generated by one or two families of moving lines. All surfaces of the above nine types are *non-degenerate quadric surfaces*. In addition there is a tenth type:

- (x) the degenerate quadric surfaces,

which comprise of the empty set, single points, straight lines, single planes, pairs of parallel planes and pairs of intersecting planes. For example each of the following quadratic equations defines a degenerate quadric surfaces:

$$x^2 + y^2 + z^2 + 1 = 0 , \qquad \text{(the empty set)} ;$$
$$x^2 + y^2 + z^2 = 0 , \qquad \text{(a single point)} ;$$
$$y^2 = 0 \qquad \text{(a plane)} ;$$
$$z^2 - 1 = 0 \qquad \text{(a pair of parallel planes)} ;$$
$$x^2 - y^2 = 0 \qquad \text{(a pair of intersecting planes)} .$$

We now wish to know if there are still other types of quadric surfaces. It turns out that the above mentioned ten types constitute a complete classification of all possible quadric surfaces in space. In other words every surface in space which is defined by a quadratic equation is of one of these types. To see this we proceed in two steps. We shall first show that every quadratic equation of the special form

$$Ax^2 + By^2 + Cz^2 + Dx + Ey + Fz + G = 0 \qquad (2)$$

with vanishing cross product terms can be transformed into a quadratic equation which defines a surface of one of the known types. Then finally we shall have to see that every quadratic equation of the general form (1) can transformed into an equation of the special form (2).

Let us consider the quadratic equation (2). As it is of degree two, at least one of the quadratic terms must be non-zero. Then by an appropriate rotation of axes and by multiplying the equation by -1 if necessary, we may assume that the three quadratic coefficients A, B, C have the following pattern of signs.

(a) $A > 0$, $B = C = 0$ *if exactly one of them is non-zero.*
(b) $A > 0$, $B \neq 0$, $C = 0$ *if exactly two of them are non-zero.*
(c) $A > 0$, $B > 0$, $C \neq 0$ *if all three of them are non-zero.*

For example if given in (2) $A = B = 0$ and $C \neq 0$, then the following substitution

$$x' = z, \quad y' = y, \quad z' = -x$$

which corresponds to a rotation about the y-axis will transform equation (2) into an equivalent equation in which $A' = C \neq 0$, $B' = B = 0$, $C' = A = 0$. Should the leading coefficient A' be negative, we need only multiply the equation through by -1. Therefore if exactly one of the quadratic coefficents is non-zero we can assume that we have pattern (a). Cases in which more than one quadratic coefficients are non-zero can be treated similarly.

Next by an appropriate shift of the origin, we can assume that the linear coefficients D, E, F have the following pattern.

(d) *If any one of the quadratic coefficents A, B or C is non-zero (e.g. $B \neq 0$) then the corresponding linear coefficient D, E or F is zero (i.e. $E = 0$).*

For example we can write

$$x^2 + 2y^2 + 4x - 12y - 2z + 12$$
$$= (x + 2)^2 + 2(y - 3)^2 - 2z - 4 - 18 + 12$$
$$= x'^2 + 2y'^2 - 2z' - 10$$

eliminating two linear coefficients.

Finally the constant term G can be assumed to fit into the following pattern.

(e) *If any one of the linear coefficients D, E or F is non-zero (e.g. F ≠ 0) while the corresponding quadratic coefficient is zero (i.e. C = 0), then the constant term G is zero.*

An example of this would be

$$x^2 - y^2 - 6z + 12$$
$$= x^2 - y^2 - 6(z - 2)$$
$$= x'^2 - y'^2 - 6z' .$$

With the seven coefficients $A, B, \cdots, G$ subject to the patterns (a) – (e) by appropriate coordinate transformations, it is now a simple exercise to make an exhaustive list of the different possibilities for the coefficients $A, B, \cdots, G$. There are 17 possibilities falling into five groups which are tabulated below.

		$A\ B\ C$	$D\ E\ F$	G	Surface
I	1	+ 0 0	0 0 0	+	Empty set
	2	+ 0 0	0 0 0	0	plane
	3	+ 0 0	0 0 0	−	Two parallel planes
	4	+ 0 0	0 ± *	0	Parabolic cylinder
II	1	+ + 0	0 0 0	+	Empty set
	2	+ + 0	0 0 0	0	Line
	3	+ + 0	0 0 0	−	Elliptic cylinder
	4	+ + 0	0 0 ±	0	Elliptic paraboloid
III	1	+ − 0	0 0 0	0	Two intersecting planes
	2	+ − 0	0 0 0	±	Hyperbolic cylinder
	3	+ − 0	0 0 ±	0	Hyperbolic paraboloid
IV	1	+ + +	0 0 0	+	Empty set
	2	+ + +	0 0 0	0	Single point
	3	+ + +	0 0 0	−	Ellipsoid
V	1	+ + −	0 0 0	+	Hyperboloid of two sheets
	2	+ + −	0 0 0	0	Cone
	3	+ + −	0 0 0	−	Hyperboloid of one sheet

Here the signs $+$, $-$, 0, $\pm$ indicate that the coefficient is positive, negative, zero, non-zero respectively while an asterisk $*$ means that the coefficient can have any arbitrary value. For example in group III we have one quadratic coefficient equal to zero while the other two quadratic coefficients having opposite signs. Then by assumption (b) A, B, C have the configuration $+,-$, 0. Upon application of (d) we obtain 0, 0 for the linear coefficients D, E. In the particular case III.3 the remaining linear coefficent F happens to be non-zero; hence the sign $\pm$. Therefore by (e) we get 0 for the constant term G. Therefore the quadratic equation of III.3 has the form

$$a^2 x^2 - b^2 y^2 = Fz$$

with a, b, F non-zero and represents a hyperbolic paraboloid.

By the above analysis, we can now conclude that every quadratic equation of special form (2) in three variables x, y, z represents either a non-degenerate quadric surface (i.e. cases I.4; II.3,4; III.2,3; IV.3; V.1,2,3) which has been studied in the previous sections or a degenerate quadric surface (i.e. Cases I.1,2,3; II.1,2; III.1; IV.1,2). This completes the first step of our investigation.

For the second step, we would have to show that there exists a rotation of coordinate axes such that the cross product terms Hxy, Jyz, Kxz in (1) can be eliminated. We have seen in Example 4.8.6 of the last section that if $H = K = 0$, we can use a rotation about the x-axis to eliminate the remaining coefficient J. Similarly if any two of the coefficients H, J, K are zero, the third one can be made to vanish by a rotation about a coordinate axis. It might seem possible, at first sight, that given a quadratic equation (1) with arbitrary H, J, K we could use rotations about a coordinate axis to eliminate these coefficients one by one. Unfortunately this turns out to be wrong. Indeed we may eliminate J in (1) and obtain an equation

$$A'x'^2 + B'y'^2 + C'z'^2 + H'x'y' + K'x'z' + D'x' + E'y' + F'z' + G' = 0 \quad (3)$$

by some substitution

$$x = x' \; ; \; y = \beta_2 y' + \gamma_2 z' \; ; \; z = \beta_3 y' + \gamma_3 z'$$

as in Example 4.8.6. However in trying to eliminate the coefficient K' by substituting

$$x' = \alpha'_1 x'' + \gamma'_1 z'' \; ; \quad y' = y'' \; ; \quad z' = \alpha'_3 x'' + \gamma'_3 z''$$

into (3) we shall find the cross product terms to be:

$$H'' x'' y'' = (\alpha'_1 H') x'' y'' \; ;$$
$$J'' y'' z'' = (\gamma'_1 H') y'' z'' \; ;$$
$$K'' x'' z'' = (2\alpha'_1 \gamma'_1 A' + 2\alpha'_3 \gamma'_3 C' + \alpha'_1 \gamma'_3 K' + \gamma'_1 \alpha'_3 K') x'' z'' \; .$$

In general it may not be possible to find suitable values of α'_1, α'_3, γ'_1, γ'_3 so that $K'' = 0$ and at the same time keeping $J'' = 0$! Therefore the process of successive eliminations of mixed product terms breaks down. However methods are available in more advanced linear algebra that provide suitable rotation of coordinate axes to bring an equation of the general form (1) into an equation of special form (2). These methods which involve solving cubic equations and simultaneous linear equations will not be given here. In accepting this without a proof, we now conclude that there are only ten types of quadric surfaces in space.

4.9.1 REMARKS For a quadratic equation in two variables x, y we have found an invariant $B^2 - 4AC$ in Section 3.9 by which the type of conic can be determined. For the general quadratic equation (1) in three variables x, y, z the expression

$$\Delta = \begin{vmatrix} A & H & K \\ H & B & J \\ K & J & C \end{vmatrix} = ABC + 2HJK - AJ^2 - BK^2 - CH^2$$

is an invariant under rotations of the coordinate axes. A rather crude classification of the quadric surface according to the value of this invariant is as follows.

$$\Delta \neq 0 : \quad \text{central quadric, cone, point or empty set}$$
$$\Delta = 0 : \quad \text{paraboloid, cylinder, or empty set .}$$

CHAPTER FIVE

HIGHER DIMENSIONAL VECTOR SPACES

In the first two chapters, we have learnt the language and techniques of linear algebra of vectors in $\mathbf{R}^2$ and $\mathbf{R}^3$. Instead of moving up one dimension from $\mathbf{R}^3$ to $\mathbf{R}^4$, we shall study the general n-dimensional vector space $\mathbf{R}^n$ for any positive integer n. However the nature of the present course only allows us a restricted scope of study. We shall therefore concentrate on the notions of linear independence and of subspace of $\mathbf{R}^n$. In order to study the notion of dimension properly, we find it necessary to introduce matrices and elementary transformations on matrices.

5.1 The vector space $\mathbf{R}^n$

Let n be any positive integer. If $n = 2$, then an n-dimensional vector is an order pair $[x_1, x_2]$ of real numbers. If $n = 3$, then an n-dimensional vector is an order triple $[x_1, x_2, x_3]$ of real numbers. For an arbitrary positive integer n, an *n-dimensional vector* is an ordered n-tuple $\mathbf{x} = [x_1, x_2, \cdots, x_n]$ of real numbers x_i. The real number x_i, $i = 1, 2, \cdots, n$ is called the *i-th component* of the vector $\mathbf{x} = [x_1, x_2, \cdots, x_n]$. Thus two n-dimensional vectors $\mathbf{x} = [x_1, x_2, \cdots, x_n]$ and $\mathbf{y} = [y_1, y_2, \cdots, y_n]$ are equal if and only if they have identical components: $\mathbf{x} = \mathbf{y}$ if and only if $x_i = y_i$ for $i = 1, 2, \cdots, n$. The set of all n-dimensional vectors shall be denoted by $\mathbf{R}^n$:

$$\mathbf{R}^n = \{[x_1, x_2, \cdots, x_n] \mid x_i \in \mathbf{R},\ i = 1, 2, \cdots, n\}.$$

We take note that here we continue to follow the convention of the earlier chapters by using lower case bold-faced types to denote vectors and brackets to enclose the components of a vector. For example $[1, -1, 0, 1]$ is a 4-dimensional vector and $[1, -1, 0, 1] \in \mathbf{R}^4$ while $[1, -1, 0, 1, 0, 0]$ is a 6-dimensional vector and $[1, -1, 0, 1, 0, 0] \in \mathbf{R}^6$.

In $\mathbf{R}^n$ the n vectors

$$\mathbf{e}_1 = [1, 0, \cdots, 0], \ \mathbf{e}_2 = [0, 1, 0, \cdots, 0], \cdots, \ \mathbf{e}_n = [0, \cdots, 0, 1]$$

are called the *unit coordinate vectors* of $\mathbf{R}^n$. We note that the i-th unit coordinate vector $\mathbf{e}_i$ has the i-th component equal to 1 and all other components equal to zero. For example the 3-rd unit vector of $\mathbf{R}^4$ is $[0, 0, 1, 0]$ while the 2-nd unit vector of $\mathbf{R}^6$ is the 6-tuple $[0, 1, 0, 0, 0, 0]$.

To express the components of the unit coordinate vectors more conveniently we can make use of the *Kronecker symbols* δ_{ij} which is defined by

$$\delta_{ij} = \begin{cases} 0 & \text{if } i \neq j \\ 1 & \text{if } i = j \end{cases}$$

for all $i = 1, 2, \cdots, n$ and $j = 1, 2, \cdots, n$. For example $\delta_{23} = \delta_{16} = 0$ and $\delta_{11} = \delta_{33} = 1$. Now the i-th unit coordinate vector of $\mathbf{R}^n$ is simply

$$\mathbf{e}_i = [\delta_{i1}, \delta_{i2}, \cdots, \delta_{in}] = [\delta_{1i}, \delta_{2i}, \cdots, \delta_{ni}], \ i = 1, 2, \cdots, n \ .$$

Next we specify the *sum* of two vectors $\mathbf{x} = [x_1, x_2, \cdots, x_n]$ and $\mathbf{y} = [y_1, y_2, \cdots, y_n]$ of $\mathbf{R}^n$ to be the vector

$$\mathbf{x} + \mathbf{y} = [x_1 + y_1, x_2 + y_2, \cdots, x_n + y_n] \ .$$

With respect to the addition of vectors, the *zero vector* $\mathbf{0} = [0, 0, \cdots, 0]$, whose n components are all zero, plays the role of the *additive identity* in the sense that $\mathbf{x} + \mathbf{0} = \mathbf{0} + \mathbf{x} = \mathbf{x}$ for all vectors of $\mathbf{R}^n$.

Finally the *scalar multiple* $r\mathbf{x}$ of a vector $\mathbf{x}$ of $\mathbf{R}^n$ by a real number r is the vector

$$r\mathbf{x} = [rx_1, rx_2, \cdots, rx_n] \ .$$

If $r = 1$, then $1\mathbf{x} = \mathbf{x}$. If $r = -1$, we denote the scalar multiple $(-1)\mathbf{x} = [-x_1, -x_2, \cdots, -x_n]$ by $-\mathbf{x}$. The vector $-\mathbf{x}$ is then the *additive inverse* of $\mathbf{x}$ in the sense that $\mathbf{x} + (-\mathbf{x}) = (-\mathbf{x}) + \mathbf{x} = \mathbf{0}$.

Now the set $\mathbf{R}^n$ of all n-dimensional vectors together with the addition and the scalar multiplication will be called the *n-dimensional vector space $\mathbf{R}^n$*. For $n = 1$, $\mathbf{R}^1$ is essentially the same as the system $\mathbf{R}$ of real numbers. For $n = 2$ and $n = 3$, $\mathbf{R}^2$ and $\mathbf{R}^3$ are just the vector space of plane vectors and the vector space of space vectors respectively. Furthermore, like $\mathbf{R}^2$ and $\mathbf{R}^3$, many properties of the

n-dimensional vector space $\mathbf{R}^n$ can be directly derived from the fundamental properties contained in the follow theorem.

5.1.1 THEOREM *Let* $\mathbf{a}, \mathbf{b}, \mathbf{c}$ *be vectors of the vector space* $\mathbf{R}^n$, *and* r, s *be scalars. Then the following statements hold.*

(1) $\mathbf{a} + \mathbf{b} = \mathbf{b} + \mathbf{a}$.

(2) $(\mathbf{a} + \mathbf{b}) + \mathbf{c} = \mathbf{a} + (\mathbf{b} + \mathbf{c})$.

(3) *There exists a unique vector* $\mathbf{0}$ *such that* $\mathbf{a} + \mathbf{0} = \mathbf{a}$.

(4) *For* $\mathbf{a}$ *there is a unique vector* $-\mathbf{a}$ *such that* $\mathbf{a} + (-\mathbf{a}) = \mathbf{0}$.

(5) $(rs)\mathbf{a} = r(s\mathbf{a})$.

(6) $(r + s)\mathbf{a} = r\mathbf{a} + s\mathbf{a}$.

(7) $r(\mathbf{a} + \mathbf{b}) = r\mathbf{a} + r\mathbf{b}$.

(8) $1\mathbf{a} = \mathbf{a}$.

For example the second proof of Theorem 1.2.4 can be taken verbatim as a proof of the following theorem.

5.1.2 THEOREM *Let* $\mathbf{a}$ *be a vector of* $\mathbf{R}^n$ *and* r *a scalar. Then* $r\mathbf{a} = \mathbf{0}$ *if and only if* $r = 0$ *or* $\mathbf{a} = \mathbf{0}$.

EXERCISES

1. Show that $\mathbf{x} = \mathbf{0}$ is the only vector in $\mathbf{R}^n$ such that

$$\mathbf{x} + \mathbf{a} = \mathbf{a} = \mathbf{a} + \mathbf{x}$$

for all vectors $\mathbf{a}$ in $\mathbf{R}^n$.

2. Let $\mathbf{a}$ be a vector in $\mathbf{R}^n$. Show that $\mathbf{w} = -\mathbf{a}$ is the only vector in $\mathbf{R}^n$ such that

$$\mathbf{a} + \mathbf{w} = \mathbf{0} = \mathbf{w} + \mathbf{a} .$$

3. Prove Theorem 5.1.2.

4. By Theorem 5.1.1 and Theorem 5.1.2, show that for any real number r and any vector $\mathbf{a}$ in $\mathbf{R}^n$,

   (a) $(-r)\mathbf{a} = -(r\mathbf{a})$;   (b) $-(-\mathbf{a}) = \mathbf{a}$;

   (c) $r(-\mathbf{a}) = -(r\mathbf{a})$;   (d) $(-r)(-\mathbf{a}) = r\mathbf{a}$.

5. Show that the commutative law can be deduced from other properties in Theorem 5.1.1. [Hint: Consider the vector $-(\mathbf{b} + \mathbf{a}) = (-\mathbf{b}) + (-\mathbf{a})$.]

5.2 Linear combinations and subspaces

Let q be any positive integer. For any q vectors $x_1, x_2, \cdots, x_q$ of $\mathbf{R}^n$ and any q real numbers (scalars) $r_1, r_2, \cdots, r_q$ the vector $r_1 x_1 + r_2 x_2 + \cdots + r_q x_q$ of $\mathbf{R}^n$ is called a *linear combination* of the vectors $x_1, x_2, \cdots, x_q$. Using the familiar summation sign Σ we may write this vector as

$$\sum_{i=1}^{q} r_i x_i \quad \text{or} \quad \sum r_i x_i \; .$$

Here the subscript i is the *summation index* which can be replaced by any other letter, say j, without affecting the linear combination. For example

$$\sum_{i=1}^{q} r_i x_i = \sum_{j=1}^{q} r_j x_j \quad \text{or} \quad \sum r_i x_i = \sum r_j x_j$$

as long as the range from 1 to q of the summation remains unchanged. Obviously different ranges of summation will produce different linear combinations. For example

$$\sum_{i=1}^{1} r_i x_i = r_1 x_1 \; ;$$

$$\sum_{i=1}^{2} r_i x_i = r_1 x_1 + r_2 x_2 \; ;$$

$$\cdots\cdots\cdots\cdots$$

$$\sum_{i=1}^{q-1} r_i x_i = r_1 x_1 + r_2 x_2 + \cdots + r_{q-1} x_{q-1} \; ;$$

$$\sum_{i=1}^{q} r_i x_i = r_1 x_1 + r_2 x_2 + \cdots + r_{q-1} x_{q-1} + r_q x_q \; .$$

Suppose that we have p sets of q scalars,

$$
\begin{array}{cccc}
r_{11}, & r_{12}, & \cdots, & r_{1q} \\
r_{21}, & r_{21}, & \cdots, & r_{2q} \\
\multicolumn{4}{c}{\cdots\cdots\cdots\cdots\cdots} \\
r_{p1}, & r_{p2}, & \cdots, & r_{pq}
\end{array} \quad .
$$

Then together with q fixed vectors $\mathbf{x}_1, \mathbf{x}_2, \cdots, \mathbf{x}_q$ of $\mathbf{R}^n$, each set of q scalars produce a linear combination:

$$\mathbf{y}_1 = r_{11}\mathbf{x}_1 + r_{12}\mathbf{x}_2 + \cdots + r_{1q}\mathbf{x}_q \; ;$$
$$\mathbf{y}_2 = r_{21}\mathbf{x}_1 + r_{22}\mathbf{x}_2 + \cdots + r_{2q}\mathbf{x}_q \; ;$$
$$\cdots\cdots\cdots\cdots\cdots\cdots\cdots\cdots$$
$$\mathbf{y}_p = r_{p1}\mathbf{x}_1 + r_{p2}\mathbf{x}_2 + \cdots + r_{pq}\mathbf{x}_q \; .$$

In other words we have p linear combinations

$$\mathbf{y}_i = \sum_{j=1}^{q} r_{ij}\mathbf{x}_j = r_{i1}\mathbf{x}_1 + r_{i2}\mathbf{x}_2 + \cdots + r_{iq}\mathbf{x}_q, \quad \text{for} \quad i = 1, 2, \cdots, p \; .$$

Let $\mathbf{z} = s_1\mathbf{y}_1 + s_2\mathbf{y}_2 + \cdots + s_p\mathbf{y}_p$ be a linear combination of the p vectors $\mathbf{y}_1, \mathbf{y}_2, \cdots, \mathbf{y}_p$. Then

$$
\begin{aligned}
\mathbf{z} ={} & s_1\mathbf{y}_1 + s_2\mathbf{y}_2 + \cdots + s_p\mathbf{y}_p \\
={} & s_1(r_{11}\mathbf{x}_1 + r_{12}\mathbf{x}_2 + \cdots + r_{1q}\mathbf{x}_q) \\
& + s_2(r_{21}\mathbf{x}_1 + r_{22}\mathbf{x}_2 + \cdots + r_{2q}\mathbf{x}_1) \\
& + \quad \cdots \qquad \cdots \qquad \cdots \\
& + s_p(r_{p1}\mathbf{x}_1 + r_{p2}\mathbf{x}_2 + \cdots + r_{pq}\mathbf{x}_q) \\
={} & (s_1 r_{11} + s_2 r_{21} + \cdots + s_p r_{p1})\mathbf{x}_1 \\
& + (s_1 r_{12} + s_2 r_{22} + \cdots + s_p r_{p2})\mathbf{x}_2 \\
& + \quad \cdots \qquad \cdots \qquad \cdots \\
& + (s_1 r_{1q} + s_2 r_{2q} + \cdots + s_p r_{pq})\mathbf{x}_q \\
={} & t_1\mathbf{x}_1 + t_2\mathbf{x}_2 + \cdots + t_q\mathbf{x}_q
\end{aligned}
$$

where $t_j = s_1 r_{1j} + s_2 r_{2j} + \cdots + s_p r_{pj}$ is a scalar for $j = 1, 2, \cdots, q$. This can be written as

$$\mathbf{z} = \sum_{i=1}^{p} s_i\mathbf{y}_i = \sum_{i=1}^{p} s_i\left(\sum_{j=1}^{q} r_{ij}\mathbf{x}_j\right) = \sum_{j=1}^{q}\left(\sum_{i=1}^{p} s_i r_{ij}\right)\mathbf{x}_j = \sum_{j=1}^{q} t_j\mathbf{x}_j \; .$$

Clearly a correct interpretation of the result is that if each $\mathbf{y}_i$ is a linear combination of $\mathbf{x}_1, \mathbf{x}_2, \cdots, \mathbf{x}_q$, then every linear combination of $\mathbf{y}_1, \mathbf{y}_2, \cdots, \mathbf{y}_p$ is a linear combination of $\mathbf{x}_1, \mathbf{x}_2, \cdots, \mathbf{x}_p$. This can also be rephrased as follows: *every linear combination of linear combinations of* $x_1, x_2, \cdots, x_q$ *is a linear combination of* $x_1, x_2, \cdots, x_q$.

5.2.1 EXAMPLE   Let $y_1 = 2x_1 + 3x_2, y_2 = -x_1 + x_2$ and $z = 3y_1 - 2y_2$. Then $z = 3(2x_1 + 3x_2) - 2(-x_1 + x_2) = 8x_1 + 7x_2$. If $x_1 = [0, 1, 2, 0]$, $x_2 = [-1, 3, 0, 1]$, then $y_1 = [-3, 11, 4, 3]$, $y_2 = [-1, 2, -2, 1]$, $z = [-7, 29, 16, 7]$.

The above result leads us to consider the set X of all linear combinations of the q vectors $x_1, x_2, \cdots, x_q$:

$$X = \{r_1 x_1 + r_2 x_2 + \cdots + r_q x_q \mid r_i \in \mathbf{R}, \ i = 1, 2, \cdots, q\} \ .$$

By the above discussion, we know that all linear combinations of vectors of X belong to X. Because of this said property, the subset X of $\mathbf{R}^n$ is called *the subspace of the vector space* $\mathbf{R}^n$ *generated by the vectors* $x_1, x_2, \cdots, x_q$.

For example the set of vectors of the form

$$[-s, r + 3s, 2r, s]$$

for arbitrary scalars r and s is the subspace of $\mathbf{R}^4$ generated by the two vectors $[0, 1, 2, 0]$ and $[-1, 3, 0, 1]$.

In general a subspace of the vector space $\mathbf{R}^n$ is defined as follows.

5.2.2 DEFINITION   *A non-empty subset A of the vector space $\mathbf{R}^n$ is called a subspace of $\mathbf{R}^n$ if all linear combinations of vectors of A belong to A.*

It follows from the definition that a non-empty set A of the set $\mathbf{R}^n$ is a subspace of the vector space $\mathbf{R}^n$ if and only for any positive integer q, any q vectors $x_1, x_2, \cdots, x_q$ of A and any q scalars $r_1, r_2, \cdots, r_q$ of $\mathbf{R}$ the linear combination $r_1 x_1 + r_2 x_2 + \cdots + r_q x_q$ belongs to the subset A.

We note that the zero vector is always a linear combination of vectors of a subspace A of $\mathbf{R}^n$. Therefore every subspace of $\mathbf{R}^n$ contains the zero vector of $\mathbf{R}^n$. We also observe that the subspace generated by a finite number of vectors $x_1, x_2, \cdots, x_q$ of $\mathbf{R}^n$ is a subspace of $\mathbf{R}^n$ in the sense of the above definition. This subspace is denoted by $\langle x_1, x_2, \cdots, x_q \rangle$. The singleton subset that consists of the zero vector 0 alone is trivially a subspace of $\mathbf{R}^n$, which will be called the *zero subspace* or the *null subspace* of $\mathbf{R}^n$ and is denoted by 0. Though the same notation is used to denote a vector and a subspace, confusion is not likely to occur. If a subspace contains a non-zero vector $\mathbf{x}$, then it will contain all its scalar multiples $r\mathbf{x}$. Therefore the null space is the only subspace of $\mathbf{R}^n$ that consists of a single vector. Since the empty set ϕ is by defintion not a subspace of $\mathbf{R}^n$, the null subspace is the smallest subspace of $\mathbf{R}^n$. On the other hand since all linear combinations of vectors of $\mathbf{R}^n$ are vectors of $\mathbf{R}^n$, the vector space $\mathbf{R}^n$ itself is trivially a subspace of $\mathbf{R}^n$. Infact $\mathbf{R}^n$ is the largest subspace of $\mathbf{R}^n$. The smallest subspace $\mathbf{0}$ and the largest subspace $\mathbf{R}^n$ of $\mathbf{R}^n$ are called the *trivial subspaces* of $\mathbf{R}^n$.

In order to verify that a given non-empty subset of $\mathbf{R}^n$ is a subspace, we can make use of the simple conditions of the following theorem.

5.2.3 THEOREM *A non-empty subset X of the vector space $\mathbf{R}^n$ is a subspace of $\mathbf{R}^n$ if and only if either of the following equivalent conditions is satisfied.*

(S1) *If $\mathbf{x}$ and $\mathbf{y}$ are any two vectors of X, then all linear combinations $r\mathbf{x} + s\mathbf{y}$ belong to X.*

(S2) *If $\mathbf{x}$ and $\mathbf{y}$ are any two vectors of X and r is any scalar, then the multiple $r\mathbf{x}$ and the sum $\mathbf{x} + \mathbf{y}$ belong to X.*

PROOF It is clear that conditions $(S1)$ and $(S2)$ are equivalent and that both are necessary conditions for X to be a subspace. Suppose now that $(S1)$ is satisfied. To show that X is a subspace $\mathbf{R}^n$, we need to prove that for any positive integer q, and any q vectors $\mathbf{x}_1, \mathbf{x}_2, \cdots, \mathbf{x}_q$ of X the linear combination $r_1\mathbf{x}_1 + r_2\mathbf{x}_2 + \cdots + r_q\mathbf{x}_q$ belongs to X for all choices of scalars $r_1, r_2, \cdots, r_q$. We propose to prove this by induction on q. For $q = 1$, this is obviously true. Assume that the said statement is true for an integer $q \geq 1$. Let $\mathbf{x}_1, \mathbf{x}_2, \cdots, \mathbf{x}_q, \mathbf{x}_{q+1}$ be any $q + 1$ vectors of X and $r_1, r_2, \cdots, r_q, r_{q+1}$ be any $q + 1$ scalars. Then by the induction assumption the vectors $r_1\mathbf{x}_1 + r_2\mathbf{x}_2 + \cdots + r_q\mathbf{x}_q$ and $r_{q+1}\mathbf{x}_{q+1}$ are both vectors of X. Now by $(S1)$ the vector

$$\left(r_1\mathbf{x}_1 + r_2\mathbf{x}_2 + \cdots + r_q\mathbf{x}_q\right) + r_{q+1}\mathbf{x}_{q+1}$$
$$= r_1\mathbf{x}_1 + r_2\mathbf{x}_2 + \cdots + r_q\mathbf{x}_q + r_{q+1}\mathbf{x}_{q+1}$$

is a vector of X. Hence X is a subspace of $\mathbf{R}^n$ if (S1) is satisfied. The proof is complete.

Let us consider some examples of non-trivial subspace of $\mathbf{R}^n (n \neq 1)$.

5.2.4 EXAMPLE Let $\mathbf{x} = [x_1, x_2, \cdots, x_n]$ be a non-zero vector. Then the subspace $\langle \mathbf{x} \rangle = \{r\mathbf{x} = [rx_1, rx_2, \cdots, rx_n] \mid r \in \mathbf{R}\}$ generated by $\mathbf{x}$ is a non-trivial subspace of $\mathbf{R}^n$. To see this, we need only observe that $0 \neq \mathbf{x} \in \langle \mathbf{x} \rangle$ and that if $x_1 = a \neq 0$ then $[0, a, \cdots, a] \notin \langle \mathbf{x} \rangle$.

5.2.5 EXAMPLE Let $r_1, r_2, \cdots, r_n$ be $n (n \neq 1)$ scalars not all equal to zero. Then for every $\mathbf{x} = [x_1, x_2, \cdots, x_n]$ of $\mathbf{R}^n$ we put $f(\mathbf{x}) = r_1 x_1 + r_2 x_2 + \cdots + r_n x_n$ to define a mapping $f : \mathbf{R}^n \to \mathbf{R}$. Consider the set F of all vectors of $\mathbf{R}^n$ which are mapped into 0 by f:

$$F = \{\mathbf{x} \in \mathbf{R}^n \mid f(\mathbf{x}) = 0\} .$$

It is easy to verify that F is non-empty (e.g. $0 \in F$) and satisfies the conditions of Theorem 5.2.3. Therefore F is a subspace of $\mathbf{R}^n$. We wish to show that F is a non-trivial subspace. Consider the vector $\mathbf{r} = [r_1, r_2, \cdots, r_n]$. It follows from the initial assumption on the components r_i that

$$f(\mathbf{r}) = r_1{}^2 + r_2{}^2 + \cdots + r_n{}^2 \neq 0 .$$

Hence $\mathbf{r} \notin F$ and $F \neq \mathbf{R}^n$. To see that F is not the null subspace we have to find a non-zero vector $\mathbf{x}$ of $\mathbf{R}^n$ such that $f(\mathbf{x}) = 0$. Assume without loss of generality that $r_1 \neq 0$. Put $\mathbf{x} = [-r_2, r_1, 0, \cdots, 0]$. Then $\mathbf{x} \neq 0$ and

$$f(\mathbf{x}) = -r_1 r_2 + r_2 r_1 = 0 .$$

Therefore $\mathbf{x} \in F$ and F is a non-trivial subspace. Finally we note that if the n given scalars $r_1, r_2, \cdots, r_n$ are all zero then $F = \mathbf{R}^n$.

Let X and Y be two subspaces of $\mathbf{R}^n$. As subsets of $\mathbf{R}^n$, X and Y have an intersection subset $X \cap Y$ and a union subset $X \cup Y$. We want to know if these two subsets are subspaces of $\mathbf{R}^n$. It is easy to verify that the intersection $X \cap Y$ satisfies the conditions of Theorem 5.2.3 and constitutes a subspace.

As for the union, consider the unit coordinate vectors $e_1 = [1, 0]$ and $e_2 = [0, 1]$ of the 2-dimensional vector space $\mathbf{R}^2$. If $A = \langle e_1 \rangle$ and $B = \langle e_2 \rangle$ then

$$A \cup B = \{[r, s] \mid r = 0 \text{ or } s = 0\} .$$

Now the vector $e_1 + e_2 = [1, 1]$ does not belong to $A \cup B$. Therefore the condition $(S2)$ of Theorem 5.2.3 is not satisfied by the subset $A \cup B$. Hence $A \cup B$ is not a subspace of $\mathbf{R}^2$ though both A and B are subspaces. On the other hand if two spaces C and D of $\mathbf{R}^n$ are such that $C \supset D$, then $C \cup D = C$ is a subspace of $\mathbf{R}^n$. Therefore for subspaces X and Y of $\mathbf{R}^n$, their union $X \cup Y$ is *not always* a subspace of $\mathbf{R}^n$.

We leave to the reader as an exercise to show that the complement $\mathbf{R}^n \backslash X = \{\mathbf{x} \in \mathbf{R}^n | \mathbf{x} \notin X\}$ of a subspace X of $\mathbf{R}^n$ is never a subspace of $\mathbf{R}^n$.

5.2.6 THEOREM *Let X and Y be two subspaces of the vector space $\mathbf{R}^n$. Then their intersection $X \cap Y$ is a subspace of $\mathbf{R}^n$ while their union $X \cup Y$ is not necessarily a subspace of $\mathbf{R}^n$.*

An immediate consequence of 5.2.5 and 5.2.6 is the following example.

5.2.7 EXAMPLE Given any $m \times n$ real numbers $a_{ij}, i = 1, 2, \cdots, m$ and $j = 1, 2, \cdots, n$, the set of all vectors $\mathbf{x} = [x_1, x_2, \cdots, x_n]$ whose components satisfy the following m homogeneous linear equations

$$
\begin{aligned}
a_{11}x_1 + a_{12}x_2 + \cdots + a_{1n}x_n &= 0 \\
a_{21}x_1 + a_{22}x_2 + \cdots + a_{2n}x_n &= 0 \\
&\cdots\cdots\cdots\cdots \\
a_{m1}x_1 + a_{m2}x_2 + \cdots + a_{mn}x_n &= 0
\end{aligned}
$$

form a subspace S of $\mathbf{R}^n$. S is sometimes called the *solution space* of the m homogenous linear equations. This follows from the fact that

$$S = F_1 \cap F_2 \cap \cdots \cap F_m$$

where $F_i = \{[x_1, x_2, \cdots, x_n] \mid a_{i1}x_1 + a_{i2}x_2 + \cdots + a_{in}x_n = 0\}$. Since F_i for each $i = 1, 2, \cdots, m$ is a subspace, S is a subspace of $\mathbf{R}^n$ by repeated applications of 5.2.6.

Though the union $X \cup Y$ of two subspaces X and Y may fail to be a subspace, the set of all sums of one vector of X and one vector of Y is always a subspace. More precisely we define the *sum* $X + Y$ of the subspaces X and Y to be

$$X + Y = \{\mathbf{x} + \mathbf{y} \mid \mathbf{x} \in X \text{ and } \mathbf{y} \in Y\}.$$

Then for any two vectors $\mathbf{x}, \mathbf{x}'$ of X and any two vectors $\mathbf{y}, \mathbf{y}'$ of Y, we have

$$(\mathbf{x} + \mathbf{y}) + (\mathbf{x}' + \mathbf{y}') = (\mathbf{x} + \mathbf{x}') + (\mathbf{y} + \mathbf{y}')$$

and for any scalar r we have

$$r(\mathbf{x} + \mathbf{y}) = r\mathbf{x} + r\mathbf{y} \ .$$

Since $(\mathbf{x} + \mathbf{x}')$, $r\mathbf{x}$ are vectors of X and $(\mathbf{y} + \mathbf{y}')$, $r\mathbf{y}$ are vectors of Y, condition $(S2)$ of Theorem 5.2.3 is satisfied by the subset $X + Y$. Therefore *the sum $X + Y$ of two subspaces X and Y of $\mathbf{R}^n$ is a subspace of $\mathbf{R}^n$.*

EXERCISES

1. Given $\mathbf{u} = [1, -1, 1, 0]$, $\mathbf{v} = [2, 1, -1, 3]$, and $\mathbf{w} = [-1, -4, 1, 1]$. Find
   (a) $3\mathbf{u} - \mathbf{v} - \mathbf{w}$; (b) $5\mathbf{u} + 3\mathbf{v} - \mathbf{w}$.

2. Which of the following are linear combinations of $\mathbf{u} = [1, 0, 0, 0]$ and $\mathbf{v} = [2, -1, 3, 0]$?
   (a) $[6, -2, 6, 0]$; (b) $[0, 0, 0, 0]$; (c) $[1, 3, 5, 7]$.

3. Find the values of m and n such that $\mathbf{u} = [2, -1, m, n]$ is a linear combination of $\mathbf{v} = [5, 0, -2, 0]$ and $\mathbf{w} = [3, -1, -5, 0]$.

4. Show that every vector in $\mathbf{R}^3$ is a linear combination of $[1, 0, 0]$, $[0, 1, 0]$ and $[0, 0, 2]$. This means that $\langle [1, 0, 0], [0, 1, 0], [0, 0, 2] \rangle = \mathbf{R}^3$.

5. Show that $\langle [2, 3], [1, -1] \rangle = \mathbf{R}^2$.

6. In $\mathbf{R}^2$, show that $[4, 9]$ belongs to $\langle [1, 2], [0, 1] \rangle$, but does not belong to $\langle [1, 2], [3, 6] \rangle$.

7. In each of the following, show that X is a subspace of $\mathbf{R}^3$.
   (a) $X = \{[r, s, 0] | r \text{ and } s \text{ are real numbers}\}$.
   (b) $X = \{[r, s, t] | r + s + t = 0\}$.

8. By considering suitable counter example, show that in each of the following, X is not a subspace of $\mathbf{R}^3$.
   (a) $X = \{[r, s, t] | r \geq 0\}$;
   (b) $X = \{[r, s, t] | r, s, t \in \mathbf{Q}\}$;
   (c) $X = \{[r, s, t] | r \cdot s = 0\}$.

9. Let $X = \{\cdots, -2, -1, 0, 1, 2, \cdots\}$. Is X a subspace of $\mathbf{R}$?

10. Let $X = \{[r, s] | r^2 + s^2 \leq c\}$. For what values of c is X a subspace of $\mathbf{R}^2$?

11. Show that the set $F = \{\mathbf{x} \in \mathbf{R}^n | f(\mathbf{x}) = 0\}$ in Example 5.2.5 is a subspace of $\mathbf{R}^n$.

12. For any two non-zero vectors $\mathbf{u}$ and $\mathbf{v}$ of $\mathbf{R}^n$, show that if $\mathbf{u} \in \langle \mathbf{v} \rangle$, then $\mathbf{v} \in \langle \mathbf{u} \rangle$ and hence deduce that $\langle \mathbf{u} \rangle = \langle \mathbf{v} \rangle$.

13. For any two vectors $\mathbf{u}$ and $\mathbf{v}$ of $\mathbf{R}^n$, show that $\langle \mathbf{u}, \mathbf{v} \rangle = \langle \mathbf{u} + \mathbf{v}, \mathbf{u} - \mathbf{v} \rangle$.

In Exercises 14 and 15, X is a non-empty subset of $\mathbf{R}^n$ and $\langle X \rangle$ the subspace generated by the elements of X.

14. For any two non-zero vectors $\mathbf{u}$ and $\mathbf{v}$ of $\mathbf{R}^n$, if $\mathbf{v} \in \langle X \cup \{\mathbf{u}\} \rangle$ but $\mathbf{v} \notin \langle X \rangle$, show that $\mathbf{u} \in \langle X \cup \{\mathbf{v}\} \rangle$.

15. Show that
    (a) $\langle X \rangle = X$ if and only if X is a subspace of $\mathbf{R}^n$.
    (b) $\langle \langle X \rangle \rangle = \langle X \rangle$.

In Exercises 16 – 19, X and Y are subspaces of $\mathbf{R}^n$.

16. Show that $\mathbf{R}^n \setminus X$ is never a subspace of $\mathbf{R}^n$.

17. Show that $X + Y = X$ if and only if $Y \subset X$.

18. (a) Show that $X \cap Y$ is a subspace of $\mathbf{R}^n$.
    (b) Show that the intersection of any finite number of subspaces of $\mathbf{R}^n$ is a subspace of $\mathbf{R}^n$.

19. Show that $X \cup Y$ is a subspace of $\mathbf{R}^n$ if and only if $X \subset Y$ or $Y \subset X$. (Hint: For "only if" part, use proof by contradiction.)

20. Show that the only non-trivial subspace of $\mathbf{R}^2$ are the "lines through the origin", that is subsets L of the type

$$L = \{[x, y] \in \mathbf{R}^2 | ax + by = 0\}$$

for some fixed real numbers a and b, not both zero.

In Exercises 21 – 23, $\mathbf{u} = [a, b]$, $\mathbf{v} = [c, d]$, and $\mathbf{w} = [e, f]$ are vectors in $\mathbf{R}^2$.

21. If $ad - bc \neq 0$, find real numbers ℓ_1, ℓ_2, m_1 and m_2 such that

$$\ell_1[a, b] + m_1[c, d] = [1, 0],$$

and $\qquad \ell_2[a, b] + m_2[c, d] = [0, 1].$

Hence deduce that $\langle [a, b], [c, d] \rangle = \mathbf{R}^2$.

22. If $ad - bc = 0$, we want to show that $\langle [a, b], [c, d] \rangle \neq \mathbf{R}^2$.
    (a) Why is this obvious if either $[a, b]$ or $[c, d] = [0, 0]$?
    (b) If neither $[a, b]$ nor $[c, d]$ is the zero vector, show that there are

non-zero real numbers ℓ and m such that $\ell[a, b] + m[c, d] = [0, 0]$. Hence deduce that $\langle[a, b], [c, d]\rangle = \langle[a, b]\rangle \neq \mathbf{R}^2$.

23. If $\langle \mathbf{u}, \mathbf{v}, \mathbf{w} \rangle = \mathbf{R}^2$, use answers to Questions 21 and 22 to prove that either $\langle \mathbf{u}, \mathbf{v} \rangle$, $\langle \mathbf{v}, \mathbf{w} \rangle$ or $\langle \mathbf{u}, \mathbf{w} \rangle$ equals $\mathbf{R}^2$.

24. Let X, Y, A, B be subspaces of $\mathbf{R}^n$.

   (a) Show that if $A \subset X$ and $A \subset Y$, then $A \subset X \cap Y$.

   (b) Show that if $X \subset B$ and $Y \subset B$, then $X + Y \subset B$.

5.3 Linear independence

The notions of linear dependence and linear independence which have played an important role in the linear algebra of $\mathbf{R}^2$ and $\mathbf{R}^3$ can be easily extended to vectors of the n-dimensional vector space $\mathbf{R}^n$. Let $\mathbf{x}_1$, $\mathbf{x}_2, \cdots, \mathbf{x}_m$ ($m \geq 2$) be vectors of $\mathbf{R}^n$. We say that the m vectors $\mathbf{x}_1$, $\mathbf{x}_2, \cdots, \mathbf{x}_m$ are *linearly dependent* if one of them is a linear combination of the other $m-1$ vectors, otherwise they are said to be *linearly independent*. In particular $\mathbf{x}_1$, $\mathbf{x}_2, \cdots, \mathbf{x}_m$ are linearly dependent if any two of them are identical or if any one of them is the zero vector. This leads us to accept that for $m = 1$, a single vector $\mathbf{x}_1$ by itself is said to be linearly dependent if $\mathbf{x}_1 = \mathbf{0}$. A necessary and sufficient condition for linear dependence expressed in terms of linear combination is given in the following theorem.

5.3.1 THEOREM *Let m be any positive integer and $\mathbf{x}_1$, $\mathbf{x}_2, \cdots, \mathbf{x}_m$ any m vectors of the vector space $\mathbf{R}^n$. Then $\mathbf{x}_1$, $\mathbf{x}_2, \cdots, \mathbf{x}_m$ are linearly dependent if and only if there exist m scalars $r_1, r_2, \cdots, r_m$ not all equal to zero such that $r_1\mathbf{x}_1 + r_2\mathbf{x}_2 + \cdots + r_m\mathbf{x}_m = \mathbf{0}$.*

PROOF Suppose that the given vectors are linearly dependent. Then one of them, say $\mathbf{x}_1$, is a linear combination of the other $m - 1$ vectors $\mathbf{x}_2$, $\mathbf{x}_3, \cdots, \mathbf{x}_m$. This means that for some $m - 1$ scalars $r_2, r_3, \cdots, r_m$

$$\mathbf{x}_1 = r_2\mathbf{x}_2 + r_3\mathbf{x}_3 + \cdots + r_m\mathbf{x}_m \ .$$

Put $r_1 = -1$. Then

$$r_1\mathbf{x}_1 + r_2\mathbf{x}_2 + \cdots + r_m\mathbf{x}_m$$
$$= -\mathbf{x}_1 + r_2\mathbf{x}_2 + \cdots + r_m\mathbf{x}_m = \mathbf{0}$$

where the m scalars $r_1, r_2, \cdots, r_m$ are not all zero. The same argument applies to any other x_i being a linear combination of the other $m-1$ vectors.

Conversely assume that there are m scalars $r_1, r_2, \cdots, r_m$ not all equal to zero such that

$$r_1\mathbf{x}_1 + r_2\mathbf{x}_2 + \cdots + r_m\mathbf{x}_m = \mathbf{0} \ .$$

Assume without loss of generality that $r_1 \neq 0$ (because the same argument applies to any other $x_i \neq 0$). Then

$$\mathbf{x}_1 = \frac{1}{r_1}(r_1\mathbf{x}_1) = \frac{1}{r_1}(-r_2\mathbf{x}_2 - r_3\mathbf{x}_3 - \cdots - r_m\mathbf{x}_m)$$
$$= (-\frac{r_2}{r_1})\mathbf{x}_2 + (-\frac{r_3}{r_1})\mathbf{x}_3 + \cdots + (-\frac{r_m}{r_1})\mathbf{x}_m$$

is a linear combination of the other $m-1$ vectors $\mathbf{x}_2, \mathbf{x}_3, \cdots, \mathbf{x}_m$. Therefore the m vectors are linearly dependent.

5.3.2 COROLLARY *If $\mathbf{x}_1, \mathbf{x}_2, \cdots, \mathbf{x}_m$ are m linearly dependent vectors of $\mathbf{R}^n$, then for any vector $\mathbf{y}$ of $\mathbf{R}^n$, the $m+1$ vectors $\mathbf{x}_1, \mathbf{x}_2, \cdots, \mathbf{x}_m, \mathbf{y}$ are linearly dependent.*

5.3.3 COROLLARY *Let $\mathbf{x}_1, \mathbf{x}_2, \cdots, \mathbf{x}_m$ be m vectors of $\mathbf{R}^n$. Then these m vectors are linearly independent if and only if $r_1\mathbf{x}_1 + r_2\mathbf{x}_2 + \cdots + r_m\mathbf{x}_m = 0$ always implies that $r_1 = r_2 = \cdots = r_m = 0$.*

5.3.4 COROLLARY *If $\mathbf{x}_1, \mathbf{x}_2, \cdots, \mathbf{x}_m$ are m linearly independent vectors of $\mathbf{R}^n$, then any p $(p \leq m)$ vectors among them are linear independent.*

5.3.5 COROLLARY *Let $\mathbf{x}_1, \mathbf{x}_2, \cdots, \mathbf{x}_m$ be m linearly independent vectors and $\mathbf{y}$ any vector of $\mathbf{R}^n$. Then the $m+1$ vectors $\mathbf{x}_1, \mathbf{x}_2, \cdots, \mathbf{x}_m, \mathbf{y}$ are linearly dependent if and only if $\mathbf{y}$ is a linear combination of $\mathbf{x}_1, \mathbf{x}_2, \cdots, \mathbf{x}_m$.*

PROOF Clearly if $\mathbf{y}$ is a linear combination of $\mathbf{x}_1, \mathbf{x}_2, \cdots, \mathbf{x}_m$, then the $m+1$ vectors are linearly dependent. Conversely suppose that the $m+1$ vectors are linearly dependent. Then by Theorem 5.3.1 there are $m+1$ scalars r_1, $r_2, \cdots, r_m$, s not all equal to zero such that $r_1\mathbf{x}_1 + r_2\mathbf{x}_2 + \cdots + r_m\mathbf{x}_m + s\mathbf{y} = 0$. Therefore it suffices to show that $s \neq 0$ among the $m+1$ scalars. Suppose to the contrary that $s = 0$. Then among the m scalars $r_1, r_2, \cdots, r_m$ at least one is non-zero. On the other hand

$$r_1\mathbf{x}_+ r_2\mathbf{x}_2 + \cdots + r_m\mathbf{x}_m$$
$$= r_1\mathbf{x} + r_2\mathbf{x}_2 + \cdots + r_m\mathbf{x}_m + s\mathbf{y} = 0$$

where the m scalars $r_1, r_2, \cdots, r_m$ are not all zero. But this is absurd since the m vectors $x_1, x_2, \cdots, x_m$ are linearly independent by hypothesis.

5.3.6 COROLLARY *Let* $x_1, x_2, \cdots, x_m$ *be* m *linearly independent vectors and* y *any vector of* $\mathbf{R}^n$. *Then the* $m + 1$ *vectors* $x_1, x_2, \cdots, x_m, y$ *are linearly independent if and only if* y *is not a linear combination of* $x_1, x_2, \cdots, x_m$.

EXERCISES

1. Determine whether the following are linearly dependent or independent.
   (a) $[1, 2, -1, 1], [1, 2, 1, 3], [0, 0, -3, -3]$.
   (b) $[1, 0, -1, 1, 2], [0, 1, 1, 2, 3], [1, 1, 0, 1, 0]$.
2. Show that for any vector u in $\mathbf{R}^n$ and for any real number r, u and ru are linearly dependent.
3. Show that if u and v are linearly independent vectors of $\mathbf{R}^n$, then $u+v$ and $u - v$ are also linearly independent.
4. For any u, v and w in $\mathbf{R}^n$, show that the vectors $u - v$, $v - w$, and $w - u$ are linearly dependent.
5. Given u, v, w are linearly independent vectors of $\mathbf{R}^n$. Show that the vectors $u + v$, $v + w$, $w + u$ are still linearly independent.
6. Let u and v be linearly independent vectors in $\mathbf{R}^n$.
   (a) Show that u and $u + v$ are linearly independent, and so are v and $u + v$.
   (b) Show that u, v and $u + v$ are not linearly independent.
7. Prove that for any linearly independent vectors $x_1, x_2, \cdots, x_m$, and any non-zero real numbers $r_1, r_2, \cdots, r_m$ (not necessarily distinct), the vectors $r_1 x_1, r_2 x_2, \cdots, r_m x_m$ are also linearly independent.
8. Let $x_1, x_2, \cdots, x_m$ be linearly independent vectors such that $y \notin \langle x_1, x_2, \cdots, x_m \rangle$. Show that $x_1 + y, x_2 + y, \cdots, x_m + y$ are also linearly independent.
9. Prove Corollary 5.3.2.
10. Prove Corollary 5.3.3.
11. Prove Corollary 5.3.4.
12. Prove Corollary 5.3.6.
13. Show that if U and W are subspaces of $\mathbf{R}^n$, then $U \cap W = \{0\}$ if and only if u and w are linearly independent for any non-zero vectors u in U and w in W.

14. As an extension of Exercise 3, given $\mathbf{u}$ and $\mathbf{v}$ be linearly independent vectors of $\mathbf{R}^n$. Let $\mathbf{u}_1 = a\mathbf{u} + b\mathbf{v}$ and $\mathbf{v}_1 = c\mathbf{u} + d\mathbf{v}$ for some real numbers a, b, c and d. Show that $\mathbf{u}_1$ and $\mathbf{v}_1$ are linearly independent if and only if $ad - bc \neq 0$.

15. Let X be a finite subset of $\mathbf{R}^n$, and let X' be a non-empty subset of X. Prove that
    (a) If X' is linearly dependent, then X is linearly dependent.
    (b) If X is linearly independent, then X' is linearly independent.

5.4 Elementary transformations of matrices

Given m vectors $\mathbf{a}_1, \mathbf{a}_2, \cdots, \mathbf{a}_m$ of $\mathbf{R}^n$. How do we find out if they are linearly independent? Unfortunately with few exceptions, it is seldom that this question can be answered by a direct application of the theorem and its corollaries of the last section. However there is an effective computational procedure which will always lead to a definite answer though it may involve a certain amount of calculation.

Let m be any positive integer, and for each $i = 1, 2, \cdots, m$ let $\mathbf{a}_i = [a_{i1}, a_{i2}, \cdots, a_{in}]$ be a vector of the vector space $\mathbf{R}^n$. We wish to determine if the m vectors $\mathbf{a}_i$ with components a_{ij} are linearly independent. For $n = 1$ the proposed problem is trivial. For $n = 2, 3$ the various results of Chapters 1 and 2 could be applied. In the general case we first arrange the $m \times n$ components a_{ij} of the given vectors into a rectangular array as follows

$$\begin{pmatrix} a_{11} & a_{12} & \cdots & a_{1n} \\ a_{21} & a_{22} & \cdots & a_{2n} \\ \cdots\cdots\cdots\cdots\cdots\cdots\cdots\cdots \\ a_{m1} & a_{m2} & \cdots & a_{mn} \end{pmatrix}$$

to facilitate the ensuing calculation. We shall call such array of $m \times n$ real numbers an $m \times n - matrix$. Here the integer m indicates the number of vectors $\mathbf{a}_i$ and the integer n the number of components in each vector.

If we denote the matrix by A, then each of the $m \times n$ numbers a_{ij} is called an *element* of the matrix A. Each horizontal sequence of n real numbers is called a *row* and each vertical sequence of m numbers is called a *column* of the matrix A. Under this arrangement the element a_{ij} belongs to the i-th row and the j-th column of the matrix A. Consequently the first subscript i of a_{ij} is called the *row index* and the second subscript j the *column index* of the element a_{ij}. Later in Chapter Six we shall study matrices as a new kind of algebraic entities. Here they are only used as a systematic way of presenting the components of m vectors in $\mathbf{R}^n$.

For the present, since it is the components of the given vectors $\mathbf{a}_i$ that give rise to the matrix A, we first look at the horizontal rows of A. The i-th row of the $m \times n$-matrix A is a vector of the vector space $\mathbf{R}^n$ and is called the *i-th row vector* of the matrix A, denoted by $\mathbf{r}_i(A)$. Thus $\mathbf{r}_i(A) = [a_{i1}, a_{i2}, \cdots, a_{in}]$ is identical to the given vector $\mathbf{a}_i$.

Instead of simply answering the question whether the vectors $\mathbf{a}_1$, $\mathbf{a}_2, \cdots, \mathbf{a}_m$ are linearly independent, we shall try to find the maximum number p of linearly independent vectors among the m row vectors of the matrix A. More precisely the number p is such that

(i) there are p vectors among $\mathbf{r}_1(A)$, $\mathbf{r}_2(A), \cdots, \mathbf{r}_m(A)$ which are linearly independent, and

(ii) any $p + 1$ vectors among $\mathbf{r}_1(A)$, $\mathbf{r}_2(A), \cdots, \mathbf{r}_m(A)$ are linearly dependent.

This number p ($p \leq m$) is called the *row rank* of the matrix A. Since $\mathbf{a}_i = \mathbf{r}_i(A)$ for $i = 1, 2, \cdots, m$, the vectors $\mathbf{a}_1$, $\mathbf{a}_2, \cdots, \mathbf{a}_m$ are linearly independent if and only if $p = m$.

Similarly the j-th column of A gives the *j-th column vector* $\mathbf{c}_j(A) = [a_{1j}, a_{2j}, \cdots, a_{mj}]$ of A which is a vector of the vector space $\mathbf{R}^m$, consisting of the j-th components of the given vectors $\mathbf{a}_1, \mathbf{a}_2, \cdots, \mathbf{a}_m$. The maximum number q of linearly independent vectors among the n column vectors $\mathbf{c}_1(A)$, $\mathbf{c}_2(A), \cdots, \mathbf{c}_n(A)$ is called the *column rank* of the matrix A. We shall see in the next section (Theorem 5.5.3) that the row rank and the column rank of A are actually identical, i.e. $p = q$; they are simply called the *rank* of A.

Take a very simple example. Here $\mathbf{a}_1 = \mathbf{e}_1$, $\mathbf{a}_2 = \mathbf{e}_2$, $\mathbf{a}_3 = \mathbf{e}_3$ are the first three unit coordinate vectors of $\mathbf{R}^6$ and $\mathbf{a}_4 = \mathbf{0}$ is the zero vector. Then

$$E = \begin{pmatrix} 1 & 0 & 0 & 0 & 0 & 0 \\ 0 & 1 & 0 & 0 & 0 & 0 \\ 0 & 0 & 1 & 0 & 0 & 0 \\ 0 & 0 & 0 & 0 & 0 & 0 \end{pmatrix}$$

is a 4×6 – matrix of row rank 3. The four given vectors $\mathbf{a}_1$, $\mathbf{a}_2$, $\mathbf{a}_3$, $\mathbf{a}_4$ are linearly dependent.

We shall now describe a procedure by which any given $m \times n$ – matrix is gradually changed by a series of *elementary row transformations* into a matrix not much more complicated than the above matrix E from which the row rank of A can be easily determined. There are three types of elementary row transformations that operate on the row vectors of a matrix and all of them leave the row rank of the matrix invariant (see Theorem 5.5.1). We remark that for our present purpose it is not necessary to use elementary row transformations of the third type. In fact elementary row transformations of the second type alone would suffice, but it is more convenient to have also those of the first type at our disposal. The three types of the elementary row transformation are as follows.

FIRST TYPE   For a fixed row index i, the i-th row vector $\mathbf{a}_i$ is replaced by the scalar multiple $s\mathbf{a}_i$ with a non-zero scalar s. This transformation may be denoted by $\mathbf{a}_i \to s\mathbf{a}_i$, $\mathbf{r}_i(A) \to s\mathbf{r}_i(A)$ or $\mathbf{r}_i \to s\mathbf{r}_i$.

SECOND TYPE   For fixed row indices i and j, the i-th row vector $\mathbf{a}_i$ is replaced by $\mathbf{a}_i + s\mathbf{a}_j$ for any scalar s; notation: $\mathbf{a}_i \to \mathbf{a}_i + s\mathbf{a}_j$; $\mathbf{r}_i(A) \to \mathbf{r}_i(A) + s\mathbf{r}_j(A)$ or $\mathbf{r}_i \to \mathbf{r}_i + s\mathbf{r}_j$.

THIRD TYPE   For fixed row indices i and j, the i-th row vector $\mathbf{a}_i$ is replaced by $\mathbf{a}_j$ and the j-th row vector $\mathbf{a}_j$ is replaced by $\mathbf{a}_i$; notation: $\mathbf{a}_i \leftrightarrow \mathbf{a}_j$, $\mathbf{r}_i(A) \leftrightarrow \mathbf{r}_j(A)$ or $\mathbf{r}_i \leftrightarrow \mathbf{r}_j$.

Let A' denote the resulting matrix after an elementary row transformation is carried out on A. Then after a transformation of the first type we find $\mathbf{r}_i(A') = s\mathbf{r}_i(A)$ and $\mathbf{r}_k(A') = \mathbf{r}_k(A)$ for $k \neq i$. Similarly after a transformation of the second type: $\mathbf{r}_i(A') = \mathbf{r}_i(A) + s\mathbf{r}_j(A)$ and $\mathbf{r}_k(A') = \mathbf{r}_k(A)$ for $k \neq i$, and after a transformation of the third type: $\mathbf{r}_i(A') = \mathbf{r}_j(A)$, $\mathbf{r}_j(A') = \mathbf{r}_i(A)$ and $\mathbf{r}_k(A') = \mathbf{r}_k(A)$ for $k \neq i, j$. For example the matrix

$$A = \begin{pmatrix} 2 & 7 & -2 & -5 \\ 1 & -1 & 3 & 4 \\ 1 & 0 & 6 & 2 \\ 0 & 8 & -11 & -11 \end{pmatrix}$$

is transformed into

$$\begin{pmatrix} 2 & 7 & -2 & -5 \\ 3 & -3 & 9 & 12 \\ 1 & 0 & 6 & 2 \\ 0 & 8 & -11 & -11 \end{pmatrix} \quad \text{by } \mathbf{a}_2 \to 3\mathbf{a}_2, \text{ or into}$$

$$\begin{pmatrix} 0 & 7 & -14 & -9 \\ 1 & -1 & 3 & 4 \\ 1 & 0 & 6 & 2 \\ 0 & 8 & -11 & -11 \end{pmatrix} \quad \text{by } \mathbf{a}_1 \to \mathbf{a}_1 - 2\mathbf{a}_3, \text{ or into}$$

$$\begin{pmatrix} 2 & 7 & -2 & -5 \\ 1 & 0 & 6 & 2 \\ 1 & -1 & 3 & 4 \\ 0 & 8 & -11 & -11 \end{pmatrix} \quad \text{by } \mathbf{a}_2 \leftrightarrow \mathbf{a}_3 \ .$$

5.4.1 EXAMPLE Simplify the above 4×4-matrix A by a series of appropriate elementary row transformations so as to change as many column vectors into unit coordinate vectors as possible.

SOLUTION

$$\begin{pmatrix} 2 & 7 & -2 & -5 \\ 1 & -1 & 3 & 4 \\ 1 & 0 & 6 & 2 \\ 0 & 8 & -11 & -11 \end{pmatrix} \xrightarrow[\mathbf{r}_2 \to \mathbf{r}_2 - \mathbf{r}_3]{\mathbf{r}_1 \to \mathbf{r}_1 - 2\mathbf{r}_3} \begin{pmatrix} 0 & 7 & -14 & -9 \\ 0 & -1 & -3 & 2 \\ 1 & 0 & 6 & 2 \\ 0 & 8 & -11 & -11 \end{pmatrix}$$

$$\xrightarrow[]{\mathbf{r}_2 \to -\mathbf{r}_2} \begin{pmatrix} 0 & 7 & -14 & -9 \\ 0 & 1 & 3 & -2 \\ 1 & 0 & 6 & 2 \\ 0 & 8 & -11 & -11 \end{pmatrix} \xrightarrow[\mathbf{r}_4 \to \mathbf{r}_4 - 8\mathbf{r}_2]{\mathbf{r}_1 \to \mathbf{r}_1 - 7\mathbf{r}_2} \begin{pmatrix} 0 & 0 & -35 & 5 \\ 0 & 1 & 3 & -2 \\ 1 & 0 & 6 & 2 \\ 0 & 0 & -35 & 5 \end{pmatrix}$$

$$\xrightarrow[]{\mathbf{r}_1 \to \frac{1}{5}\mathbf{r}_1} \begin{pmatrix} 0 & 0 & -7 & 1 \\ 0 & 1 & 3 & -2 \\ 1 & 0 & 6 & 2 \\ 0 & 0 & -35 & 5 \end{pmatrix} \xrightarrow[\substack{\mathbf{r}_3 \to \mathbf{r}_3 - 2\mathbf{r}_1 \\ \mathbf{r}_4 \to \mathbf{r}_4 - 5\mathbf{r}_1}]{\mathbf{r}_2 \to \mathbf{r}_2 + 2\mathbf{r}_1} \begin{pmatrix} 0 & 0 & -7 & 1 \\ 0 & 1 & -11 & 0 \\ 1 & 0 & 20 & 0 \\ 0 & 0 & 0 & 0 \end{pmatrix} .$$

The procedure of the solution consists of three stages corresponding roughly to the three lines above. At the first stage we pay special attention to the first column and use its non-zero third element as a *pivot* to eliminate all other elements of the first column by appropriate row transformations of the second type. After the first stage the first column of the matrix becomes the unit coordinate vector $\mathbf{e}_3$ of

224

$\mathbf{R}^4$. We observe that both the third and the fourth column vectors undergo some changes because they have a non-zero third component, while the second column vector remains unchanged because its third component equals zero. The second stage changes the second column into another unit coordinate vector e_2. Here the non-zero second element on the second column serves as the pivot of the operations. Therefore the first column vector which is already in the desired form undergoes no alternation because its second component is zero, while the last columns undergo changes. Finally the non-zero first element on the fourth column is chosen as the pivot at the last stage. Operations of this stage do not change any element on the first and second columns that are already in the desired form but turn the fourth column into yet another unit coordinate vector e_1. Thus at different stages, non-zero elements at different positions on different columns are used as pivots to ensure that columns already in the desired form are not disturbed.

Let us now examine the matrix D at the end of the procedure. It has three non-zero row vectors $\mathbf{d}_1 = [0, 0, -7, 1]$, $\mathbf{d}_2 = [0, 1, -11, 0]$ and $\mathbf{d}_3 = [1, 0, 20, 0]$. A linear combination $r\mathbf{d}_1 + s\mathbf{d}_2 = [0, s, -7r - 11s, r]$ of $\mathbf{d}_1$ and $\mathbf{d}_2$ will have a zero first component; therefore $\mathbf{d}_3$ cannot be a linear combination of $\mathbf{d}_1$ and $\mathbf{d}_2$. For similar reason, $\mathbf{d}_2$ is not a linear combination of $\mathbf{d}_1$ and $\mathbf{d}_3$, nor is $\mathbf{d}_1$ a linear combination of $\mathbf{d}_2$ and $\mathbf{d}_3$. Hence the row vectors $\mathbf{d}_1$, $\mathbf{d}_2$, $\mathbf{d}_3$ are linearly independent. Since the fourth row vector $\mathbf{d}_4$ is the zero vector, the row rank of this matrix D is 3. Taking for granted that elementary row transformations do not alter the row rank of the matrix, we conclude that the given matrix A also has a row rank equal to 3.

We observe firstly that the third column of D cannot be further simplified by elementary transformations operating on the rows of the matrix without changing the elements on the other columns which we do not wish to disturb. Secondly even though the third column vector of D is not a unit coordinate vector of $\mathbf{R}^4$, it presents no difficulty in the evaluation of the row rank of the matrix D.

In general the different stages of the procedure for a given $m \times n$-matrix A are as follows. If A is the zero matrix, then obviously A has rank 0 and there is no need to proceed. Otherwise we choose a non-zero element, say the element $a_{i_1 j_1}$ on the i_1-th row and the j_1-th column, as the pivotal element of the first stage. Then by a transformation of the first type we make the pivot $a_{i_1 j_1}$ into 1, and by a series of transformations of the second type we eliminate all other elements $a_{k j_1}$ on the j_1-th column. Consequently some other columns of A will undergo certain changes. Thus at the conclusion of the first stage the given matrix A is transformed into a matrix B whose j_1-th column is the unit coordinate vector $\mathbf{e}_{i_1}$ of $\mathbf{R}^m$. Moreover B has at least one non-zero row, namely the i_1-th row which has a non-zero j_1-th component 1.

If all other rows of B are zero, then obviously B has row rank 1 and there is no need to proceed any further. Otherwise we choose a non-zero element, say $b_{i_2 j_2}$ on any other non-zero row as the pivot of the second stage. Then the indices of $b_{i_2 j_2}$ are such that $i_2 \neq i_1$ and $j_2 \neq j_1$. As in the first stage we make the pivot $b_{i_2 j_2}$ into 1 and eliminate all other elements of the j_2-th column by transformations of the first and the second types. Other columns of B undergo consequential changes. As the j_1-th column of B is the vector $\mathbf{e}_{i_1}$ whose i_2-th component is zero, all its elements remain unchanged throughout the second stage. Therefore at the conclusion of the second stage we have a matrix C whose j_1-th column and j_2-th column are the distinct coordinate vectors $\mathbf{e}_{i_1}$ and $\mathbf{e}_{i_2}$ respectively. Moreover C has at least two (non-zero) linearly independent row vectors, namely $\mathbf{r}_{i_1}(C)$ and $\mathbf{r}_{i_2}(C)$.

If all other rows of C are zero then C has row rank 2 and a third stage is unnecessary. Otherwise we proceed with a non-zero element $c_{i_3 j_3}$ on any other non-zero row (i.e. with $i_3 \neq i_2, i_2; j_3 \neq j_1, j_2$) as the pivot of the third stage and carry out similar operations. At the end of the third stage we obtain a matrix D with $\mathbf{e}_{i_1}, \mathbf{e}_{i_2}, \mathbf{e}_{i_3}$ on its j_1-th, j_2-th, j_3-th columns respectively and three linearly independent row vectors $\mathbf{r}_{i_1}(D), \mathbf{r}_{i_2}(D), \mathbf{r}_{i_3}(D)$.

It is now clear that after say r (with $r \leq \min(m, n)$) stages, we would have transformed the given matrix A into a much simplified $m \times n$-matrix H which has

(i) r distinct unit coordinate vectors $\mathbf{e}_{i_1}, \mathbf{e}_{i_2}, \cdots, \mathbf{e}_{i_r}$ of $\mathbf{R}^m$ on its

j_1-th, j_2-th, $\cdots$, j_r-th columns respectively, i.e. $c_{j_1}(H) = e_{i_1}$, $c_{j_2}(H) = e_{i_2}, \cdots, c_{j_r}(H) = e_{i_r}$,

(ii) r linearly independent row vectors $r_{i_1}(H), r_{i_2}(H), \cdots, r_{i_r}(H)$, and

(iii) all remaining rows equal to zero.

This final matrix H has row rank r. Taking for granted (for the time being until it is proved to be true in Theorem 5.5.1) that elementary row transformations do not alter row ranks, we conclude that the given matrix A also has row rank r.

5.4.2 EXAMPLE   Find the row rank of the matrix

$$A = \begin{pmatrix} 1 & 2 & 2 & -1 & 0 & 1 \\ 5 & 5 & 10 & -3 & 1 & 1 \\ 4 & 3 & 8 & -2 & 1 & 0 \\ -1 & -1 & -2 & 0 & 2 & 0 \end{pmatrix}.$$

SOLUTION   In this solution we shall only write down the results of the consecutive stages of the procedure. The reader shall have no difficulty in finding out the elementary transformations involved.

$$A \rightarrow \begin{pmatrix} 1 & 2 & 2 & -1 & 0 & 1 \\ 0 & -5 & 0 & 2 & 1 & -4 \\ 0 & -5 & 0 & 2 & 1 & -4 \\ 0 & 1 & 0 & -1 & 2 & 1 \end{pmatrix} \rightarrow \begin{pmatrix} 1 & 0 & 2 & 1 & -4 & -1 \\ 0 & 0 & 0 & -3 & 11 & 1 \\ 0 & 0 & 0 & -3 & 11 & 1 \\ 0 & 1 & 0 & -1 & 2 & 1 \end{pmatrix} \rightarrow$$

$$\begin{pmatrix} 1 & 0 & 2 & 0 & -1/3 & -2/3 \\ 0 & 0 & 0 & 0 & 0 & 0 \\ 0 & 0 & 0 & 1 & -11/3 & -1/3 \\ 0 & 1 & 0 & 0 & -5/3 & 2/3 \end{pmatrix}.$$

A linear combination of the first and the last row vectors of the final matrix has a zero fourth component; therefore the third row vector is not a linear combination of the first and the fourth row vectors. Using similar argument, we see that the three non-zero row vectors of the last matrix are linearly independent. Therefore the row rank is 3. Taking for granted that elementary transformations leave the row rank invariant, we conclude that the row rank of A is 3.

5.4.3 EXAMPLE Evaluate the row rank of the matrix

$$A = \begin{pmatrix} -1 & 2 & -2 & 6 \\ 8 & -9 & 16 & -27 \\ 4 & 5 & 8 & 15 \\ 7 & 1 & 14 & 3 \\ 3 & -4 & 6 & -12 \end{pmatrix}.$$

SOLUTION

$$A \rightarrow \begin{pmatrix} 1 & -2 & 2 & -6 \\ 0 & 7 & 0 & 21 \\ 0 & 13 & 0 & 39 \\ 0 & 15 & 0 & 45 \\ 0 & 2 & 0 & 6 \end{pmatrix} \rightarrow \begin{pmatrix} 1 & 0 & 2 & 0 \\ 0 & 0 & 0 & 0 \\ 0 & 0 & 0 & 0 \\ 0 & 0 & 0 & 0 \\ 0 & 1 & 0 & 3 \end{pmatrix}.$$

The row rank of A is 2.

EXERCISES

1. In (a) to (c), write down the matrices after elementary transformations (i) to (iv) are performed consecutively:
   (i) $r_2 \rightarrow 2r_2$ (ii) $r_3 \rightarrow r_3 - 3r_1$ (iii) $r_1 \rightarrow r_1 + r_2$ (iv) $r_2 \leftrightarrow r_3$.

   (a) $\begin{pmatrix} 1 & 4 & 7 \\ 2 & 5 & 8 \\ 3 & 6 & 9 \end{pmatrix}$; (b) $\begin{pmatrix} 8 & -11 & 0 \\ -9 & 4 & 3 \\ 10 & 7 & -5 \end{pmatrix}$; (c) $\begin{pmatrix} 1 & 9 & -6 \\ -6 & -1 & 9 \\ 9 & -6 & 1 \end{pmatrix}$.

2. Find the row rank of each of the following matrices:

   (a) $\begin{pmatrix} 1 & 1 \\ 0 & 1 \end{pmatrix}$; (b) $\begin{pmatrix} 1 & 1 \\ 1 & 0 \\ 1 & 1 \end{pmatrix}$; (c) $\begin{pmatrix} 1 & 1 & 1 \\ 1 & 0 & 1 \\ 1 & 1 & 1 \\ 1 & 0 & -1 \end{pmatrix}$.

3. Find the row rank of each of the following matrices:

   (a) $\begin{pmatrix} 1 & 2 & 1 \\ 2 & 1 & -4 \\ -1 & 1 & 5 \end{pmatrix}$; (b) $\begin{pmatrix} 1 & 2 & 4 & 1 \\ 2 & 1 & -1 & 1 \\ 1 & 3 & 7 & 1 \end{pmatrix}$; (c) $\begin{pmatrix} 0 & 6 & 6 & 1 \\ -8 & 7 & 2 & 3 \\ -3 & 2 & 1 & 1 \\ 1 & 1 & -1 & 0 \end{pmatrix}$.

4. Let x_1, x_2 and x_3 be three distinct real numbers.
   (a) Show that rank of $\begin{pmatrix} 1 & x_1 \\ 1 & x_2 \end{pmatrix} = 2$.
   (b) Show that rank of $\begin{pmatrix} 1 & x_1 & x_1{}^2 \\ 1 & x_2 & x_2{}^2 \\ 1 & x_3 & x_3{}^2 \end{pmatrix} = 3$.

5.5 The rank of a matrix

In the present section which is an appendix to the last one the relationship between the row rank, the column rank and elementary transformations is investigated. Here are the main results of the section.

5.5.1 THEOREM *Elementary row transformations do not alter the row rank of a matrix.*

5.5.2 THEOREM *Elementary row transformations do not alter the column rank of a matrix.*

5.5.3 THEOREM *The row rank and the column rank of a matrix are equal.*

Readers who find the notations and arguments in the following proofs too complicated may simply accept the validity of these theorems.

Before we proceed to the proofs, it is important to observe that an elementary row transformation T on a matrix A is a reversible process. More precisely if $A' = T(A)$, then there is an elementary row transformation T^{-1}, called the *inverse* of T, such that $T^{-1}(A') = A$; in other words T^{-1} reverses everything that T does. Indeed if T is of the first type, say $\mathbf{a}_i \to s\mathbf{a}_i = \mathbf{a}_i{}'$ then T^{-1} is just the elementary transformation $\mathbf{a}_i{}' \to \frac{1}{s}\mathbf{a}_i{}' = \mathbf{a}_i$. Similarly it is easy to see that every elementary transformation of the second or third type has an inverse.

PROOF OF 5.5.1 By the above remark on the reversibility of elementary transformations it is sufficient to show that the row rank of a matrix A will not be increased by any elementary transformation T:

row rank of $T(A) \le$ row rank of A.

Indeed when this is proved then also:

row rank of $A =$ row rank of $T^{-1}(T(A)) \le$ row rank of $T(A)$,

and the equality of the row ranks of A and $T(A)$ follows. Therefore we have to show that if T is an elementary transformation T and A is a matrix of row rank p, then every $p + 1$ row vectors of the transformed matrix $T(A) = A'$ are linearly dependent. For simplicity of notation, we shall carry out the proof of this for the first $p + 1$ row vectors $\mathbf{a}'_1, \mathbf{a}'_2, \cdots, \mathbf{a}'_{p+1}$ of A', other cases being proved similarly.

Let T be of the first type such that $\mathbf{a}_i \to s\mathbf{a}_i = \mathbf{a}_i{}'$ for a fixed index

229

i and a scalar $s \neq 0$. As the row rank of A is p, it follows from the linear dependence of any $p+1$ row vectors of A that

$$t_1\mathbf{a}_1 + t_2\mathbf{a}_2 + \cdots + t_p\mathbf{a}_p + t_{p+1}\mathbf{a}_{p+1} = \mathbf{0}$$

for some $p+1$ scalars t_k not all equal to zero. If the fixed row index i is not among $1, 2, \cdots, p+1$, then $\mathbf{a}_k = \mathbf{a}'_k$ for $k = 1, 2, \cdots, p+1$. Therefore $\mathbf{a}'_1$, $\mathbf{a}'_2, \cdots, \mathbf{a}'_{p+1}$ are linearly dependent. If the index i is among $1, 2, \cdots, p+1$, then we assume without loss of generality that $i = p+1$. Then $\mathbf{a}_k = \mathbf{a}'_k$ for $k = 1, 2, \cdots, p$ and $\mathbf{a}_{p+1} = \frac{1}{s}\mathbf{a}'_{p+1}$. Therefore

$$t_1\mathbf{a}'_1 + t_2\mathbf{a}'_2 + \cdots + t_p\mathbf{a}'_p + (t_{p+1}/s)\mathbf{a}'_{p+1} = \mathbf{0}$$

where the $p+1$ scalars $t_1, t_2, \cdots, t_p, t_{p+1}/s$ are not all equal to zero. Hence $\mathbf{a}'_1, \mathbf{a}'_2, \cdots, \mathbf{a}'_{p+1}$ are linearly dependent.

Let T be of the second type such that $\mathbf{a}_i \rightarrow \mathbf{a}_i + s\mathbf{a}_j = \mathbf{a}_i'$. If i is not among $1, 2, \cdots, p+1$, then $\mathbf{a}'_1, \mathbf{a}'_2, \cdots, \mathbf{a}'_{p+1}$ are identical to $\mathbf{a}_1$, $\mathbf{a}_2, \cdots, \mathbf{a}_{p+1}$ and hence linearly dependent. Otherwise assume again that $i = p+1$, i.e. $\mathbf{a}'_{p+1} = \mathbf{a}_{p+1} + s\mathbf{a}_j$. Consider the p row vectors $\mathbf{a}_1, \mathbf{a}_2, \cdots, \mathbf{a}_p$. Either they are linearly dependent or they are linearly independent. Then by 5.3.2 the linear dependence of $\mathbf{a}'_1, \mathbf{a}'_2, \cdots, \mathbf{a}'_{p+1}$ would follow if the first p vectors $\mathbf{a}'_1 = \mathbf{a}_1, \mathbf{a}'_2 = \mathbf{a}_2, \cdots, \mathbf{a}'_p = \mathbf{a}_p$ are linearly dependent. For the remaining case, assume that the p vectors $\mathbf{a}_1, \mathbf{a}_2, \cdots, \mathbf{a}_p$ are linearly independent. Then it follows from 5.3.5 that both $\mathbf{a}_{p+1}$ and $\mathbf{a}_j$ are linear combinations of these p vectors:

$$\mathbf{a}_{p+1} = u_1\mathbf{a}_1 + u_2\mathbf{a}_2 + \cdots + u_p\mathbf{a}_p$$

$$\mathbf{a}_j = v_1\mathbf{a}_1 + v_2\mathbf{a}_2 + \cdots + r_p\mathbf{a}_p \ .$$

Therefore

$$\begin{aligned}
\mathbf{a}'_{p+1} &= \mathbf{a}_{p+1} + s\mathbf{a}_j \\
&= (u_1\mathbf{a}_1 + \cdots + u_p\mathbf{a}_p) + s(v_1\mathbf{a}_1 + \cdots + v_p\mathbf{a}_p) \\
&= (u_1 + sv_1)\mathbf{a}'_1 + (u_2 + sv_2)\mathbf{a}'_2 + \cdots + (u_p + sv_p)\mathbf{a}'_p
\end{aligned}$$

is a linear combination of $\mathbf{a}'_1, \mathbf{a}'_2, \cdots, \mathbf{a}'_p$. Hence the $p+1$ row vectors $\mathbf{a}'_1$, $\mathbf{a}'_2, \cdots, \mathbf{a}'_{p+1}$ are linearly dependent.

Finally let T be of the third type. Then row vectors of A' are the same as the row vectors of A. In this case the two matrices clearly have the same row rank. The proof is now complete.

PROOF OF 5.5.2 Let the column rank of the matrix A be q. Then using the same argument as in the last proof it is sufficient to prove that any $q+1$ column vectors of A' are linearly dependent. Again for simplicity of notation we only show that the first $q+1$ column vectors $c_1(A')$, $c_2(A')$, $\cdots$, $c_{q+1}(A')$ of the transformed matrix $A' = T(A)$ by an elementary row transformation T remain linearly dependent.

Let T be of the first type such that $a_i \rightarrow sa_i = a'_i$. Without loss of generality we may further assume that $i = 1$. Then for all $k = 1, 2, \cdots, q+1$, the k-th column vectors of the matrices A and A' are respectively

$$c_k(A) = [a_{1k}, a_{2k}, \cdots, a_{mk}] \text{ and } c_k(A') = [sa_{1k}, a_{2k}, \cdots, a_{mk}] .$$

In other words, except for their first components, they have identical components: $a'_{1k} = sa_{1k}$ and $a'_{jk} = a_{jk}$ for all $j = 2, 3, \cdots, m$. The column rank of A being q,

$$t_1 c_1(A) + t_2 c_2(A) + \cdots + t_{q+1} c_{q+1}(A) = 0$$

for $q + 1$ scalars t_1, t_2, $\cdots$, t_{q+1} not all equal to 0. Therefore in terms of their components

$$t_1 a_{11} + t_2 a_{12} + \cdots + t_{q+1} a_{1\,q+1} = 0$$
$$t_1 a_{21} + t_2 a_{22} + \cdots + t_{q+1} a_{2\,q+1} = 0$$
$$\cdots \quad \cdots \quad \cdots$$
$$t_1 a_{m1} + t_2 a_{m2} + \cdots + t_{q+1} a_{m\,q+1} = 0 .$$

Hence for the components of the column vectors of A', we get

$$t_1 a'_{11} + \cdots + t_{q+1} a'_{1\,q+1} = s(t_1 a_{11} + \cdots + t_{q+1} a_{1\,q+1}) = 0$$
$$t_1 a'_{21} + \cdots + t_{q+1} a'_{2\,q+1} = t_1 a_{21} + \cdots + t_{q+1} a_{2\,q+1} = 0$$
$$\cdots \quad \cdots \quad \cdots \quad \cdots \quad \cdots$$
$$t_1 a'_{m1} + \cdots + t_{q+1} a'_{m\,q+1} = t_1 a_{m1} + \cdots + t_{q+1} a_{m\,q+1} = 0$$

which means that

$$t_1 c_1(A') + t_2 c_2(A') + \cdots + t_{q+1} c_{q+1}(A') = 0 .$$

Therefore the first $q + 1$ column vectors of A' are linearly dependent. Using the same method we can show that the first $q + 1$ column vectors of $T(A)$

are linearly dependent under an elementary transformation of the second or the third type. The detail of this will be left to the interested reader as an exercise.

Corresponding to the three types of elementary row transformations that operate on the row vectors of a matrix A, we have three types of *elementary column transformations* that operate on the column vectors of A. The procedure of these is as follows.

FIRST TYPE   For a fixed column index i, the i-th column vector $c_i(A)$ is replaced by the scalar multiple $sc_i(A)$ with a non-zero scalar s; notation: $c_i(A) \to sc_i(A)$ or $c_i \to sc_i$.

SECOND TYPE   For fixed column indices i and j, the i-th column vector $c_i(A)$ is replaced by $c_i(A) + sc_j(A)$ for any scalar s; notation: $c_i(A) \to c_i(A) + sc_j(A)$ or $c_i \to c_i + sc_j$.

THIRD TYPE   For fixed column indices i and j, the i-th column vector $c_i(A)$ is replaced by $c_j(A)$ and the j-th column vector $c_j(A)$ by $c_i(A)$; notation: $c_i(A) \leftrightarrow c_j(A)$ or $c_i \leftrightarrow c_j$.

Analogously given any $m \times n$-matrix A, we can transform it by a series of elementary column transformations into a much simplified $m \times n$-matrix G which has

(i) s distinct unit coordinate vectors $e_{\ell_1}, e_{\ell_2}, \cdots, e_{\ell_s}$ of R^n on its k_1-th, k_2-th, $\cdots, k_s$-th rows respectively, i.e. $r_{k_1}(G) = e_{\ell_1}, r_{k_2}(G) = e_{\ell_2}, \cdots, r_{k_s}(G) = e_{\ell_s}$,

(ii) s linearly independent column vectors $c_{\ell_1}(G), c_{\ell_2}(G), \cdots, c_{\ell_s}(G)$, and

(iii) all remaining columns equal to zero.

This matrix G has column rank s. Rewriting the proofs of 5.5.1 and 5.5.2 for column transformations in the obvious way, we obtain the validity of the following corresponding theorems.

5.5.4 THEOREM   *Elementary column transformations do not alter the column rank of a matrix.*

5.5.5 THEOREM   *Elementary column transformations do not alter the row rank of a matrix.*

Alternatively we can consider for any given $m \times n$-matrix

$$A = \begin{pmatrix} a_{11} & a_{12} & \cdots & a_{1n} \\ a_{21} & a_{22} & \cdots & a_{2n} \\ \cdots\cdots\cdots\cdots\cdots\cdots \\ a_{m1} & a_{m2} & & a_{mn} \end{pmatrix}$$

its *transpose* A^t which is obtained by converting every row of A into a column and hence every column into a row. Therefore the transpose A^t of A is the $n \times m$-matrix

$$A^t = \begin{pmatrix} a_{11} & a_{21} & \vdots & a_{m1} \\ a_{12} & a_{22} & \vdots & a_{m2} \\ \vdots & \vdots & \vdots & \vdots \\ a_{1m} & a_{2m} & \vdots & a_{mn} \end{pmatrix} .$$

For example

$$\begin{pmatrix} 1 & 2 & 2 & -1 & 0 & 1 \\ 5 & 5 & 10 & -3 & 1 & 0 \\ 4 & 3 & 8 & -2 & 1 & 0 \\ -1 & -1 & -2 & 0 & 2 & 0 \end{pmatrix}^t = \begin{pmatrix} 1 & 5 & 4 & -1 \\ 2 & 5 & 3 & -1 \\ 2 & 10 & 8 & -2 \\ -1 & -3 & -2 & 0 \\ 0 & 1 & 1 & 2 \\ 1 & 0 & 0 & 0 \end{pmatrix} .$$

Formally the transpose A^t of an $m \times n$-matrix A is the $n \times m$-matrix

$$A^t = B = \begin{pmatrix} b_{11} & b_{12} & \cdots & b_{1m} \\ b_{21} & b_{22} & \cdots & b_{2m} \\ \cdots\cdots\cdots\cdots\cdots\cdots \\ b_{n1} & b_{n2} & \cdots & b_{nm} \end{pmatrix}$$

where $b_{ij} = a_{ji}$ for all $i = 1, 2, \cdots, n$ and $j = 1, 2, \cdots, m$. Therefore among row vectors and column vectors of A and A^t we have

$$\mathbf{r}_i(A^t) = \mathbf{c}_i(A) \text{ for } i = 1, 2, \cdots, n \text{ ;}$$
$$\mathbf{c}_j(A^t) = \mathbf{r}_j(A) \text{ for } j = 1, 2, \cdots, m \text{ ;}$$
$$\text{number of rows of } A = \text{number of columns of } A^t \text{ ;}$$
$$\text{number of column of } A = \text{number of row of } A^t \text{ .}$$

In particular every row transformation on A^t is a column transformation A, every column transformation on A^t is a row transformation on A, and vice versa. Furthermore the row rank of A^t is the column rank of A and the column rank of A^t is the row rank of A. Since it has been seen that elementary row transformations on A do not change the row rank and the column rank of A, we conclude that elementary row transformations on A^t do not change the row rank and the column rank of A^t. Therefore elementary column transformations on A do not change the column rank and the row rank of A.

Let us now use the invariance of row rank and column rank under elementary transformations to prove Theorem 5.5.3 that the two ranks of a matrix are actually equal.

PROOF OF 5.5.3 Let A be an $m \times n$-matrix of row rank r. Then using elementary row transformations alone, we transform A into H such that there are exactly r (non-zero) linearly independent row vectors $\mathbf{r}_{i_1}(H)$, $\mathbf{r}_{i_2}(H), \cdots, \mathbf{r}_{i_r}(H)$ in H, the remaining ones being all zero. Moreover among the column vectors of H we find the unit coordinate vectors $\mathbf{c}_{j_1}(H) = \mathbf{e}_{i_1}, \mathbf{c}_{j_2}(H) = \mathbf{e}_{i_2}, \cdots, \mathbf{c}_{j_r}(H) = \mathbf{e}_{i_r}$ of $\mathbf{R}^m$.

We now apply to H a series of elementary column transformations. We use the non-zero $h_{i_1 j_1} = 1$ as pivot to eliminate all other elements the i_1-th row by column transformations of the second type, turning the i_1-th row of H into the unit coordinate vector $\mathbf{e}_{j_1}$ of $\mathbf{R}^n$. We observe that since $h_{i_1 j_1} = 1$ is the only non-zero element of the j_1-th column of H, throughout this stage of transformation all other rows of H remain unchanged. Clearly we carry out similar column transformations to turn each non-zero row vector $\mathbf{r}_{i_k}(H)$ of H into a unit coordinate vector $\mathbf{e}_{j_k}$ of $\mathbf{R}^n$.

At the end of the procedure, H is transformed into a matrix G by a series of column transformations of the second type such that

(i) G has exactly r non-zero rows vectors which are the unit coordinate vectors $\mathbf{e}_{j_1}, \mathbf{e}_{j_2}, \cdots, \mathbf{e}_{j_r}$ of $\mathbf{R}^n$, and

(ii) G has exactly r non-zero column vectors (the same as those of H) which are the coordinate unit vectors $\mathbf{e}_{i_1}, \mathbf{e}_{i_2}, \cdots, \mathbf{e}_{i_r}$ of $\mathbf{R}^m$.

With G in this simple form, we see that row rank of G and column rank of G are both equal to r. On the other hand, by Theorems 5.5.1, 5.5.2, 5.5.4, 5.5.5, the three matrices A, H, G have the same row rank and the same column rank. Therefore row rank and column rank of A are indeed equal. The proof of 5.5.3 is now complete.

One consequence of Theorem 5.5.3 is that no distinction need to be made between the row rank and the column rank of a matrix. Hence they are referred to as the *rank* of a matrix.

EXERCISES

1. Write down the inverse row transformations for each of the following:

(a) $\begin{pmatrix} 1 & 3 \\ 2 & 4 \end{pmatrix} \to \begin{pmatrix} 1 & 0 \\ 0 & 1 \end{pmatrix}$;

(b) $\begin{pmatrix} 2 & 3 & -1 \\ 4 & 0 & 1 \\ 0 & 0 & 1 \end{pmatrix} \to \begin{pmatrix} 1 & 0 & 0 \\ 0 & 1 & 0 \\ 0 & 0 & 1 \end{pmatrix}$;

(c) $\begin{pmatrix} 3 & 2 & 3 & 1 \\ 4 & 3 & 5 & 2 \\ 2 & 1 & 1 & 0 \end{pmatrix} \to \begin{pmatrix} 1 & 0 & -1 & -1 \\ 0 & 1 & 3 & 2 \\ 0 & 0 & 0 & 0 \end{pmatrix}$.

2. Write down the transpose of each of the following matrix:

(a) $\begin{pmatrix} 5 & -1 \\ 3 & 4 \end{pmatrix}$;

(b) $\begin{pmatrix} 3 & 4 & 5 & 6 & 7 \\ -2 & 5 & -8 & 13 & 1 \end{pmatrix}$;

(c) $\begin{pmatrix} 0 & 1 & 0 & 0 \\ 0 & 0 & 2 & 0 \\ 1 & 2 & 0 & 3 \\ 0 & 0 & 3 & 4 \end{pmatrix}$.

3. (a) Can a 3×4 matrix have
     (i) all linearly independent column vectors?
     (ii) all linearly independent row vectors?
   (b) If A is 5×4 matrix and rank $A = 3$, can A have
     (i) all linearly independent column vectors?
     (ii) all linearly independent row vectors?
   (c) Can an $m \times n$ matrix with $m \neq n$ have all its row vectors linearly independent and all its column vectors linearly independent?

4. As an extension to Question 4 of Section 5.4, show that for n pairwise distinct real numbers $x_1, x_2, \cdots, x_n$,

$$\begin{pmatrix} 1 & x_1 & x_1{}^2 & \cdots & x_1{}^{n-1} \\ 1 & x_2 & x_2{}^2 & \cdots & x_2{}^{n-1} \\ \cdots\cdots\cdots\cdots\cdots\cdots\cdots \\ 1 & x_n & x_n{}^2 & \cdots & x_n{}^{n-1} \end{pmatrix}$$

has rank n. (Hint: Consider elementary column transformations and elementary row transformations.)

5.6 Dimension and base

In this last section of the chapter, we return to the notion of subspace which is introduced in Section 5.2. We shall see that equipped with the results of the last three sections, we shall be able to produce more precise results on the subspaces of $\mathbf{R}^n$. Indeed we shall assign to each subspace of $\mathbf{R}^n$ a definite non-negative integer, called the dimension, as an accurate measurement of its size. For this purpose, we first introduce the notion of dimension in the following definition, and then show step by step that every subspace of $\mathbf{R}^n$ has a definite dimension.

5.6.1 DEFINITION *Let X be a subspace of the vector space $\mathbf{R}^n$. The maximum number of linearly independent vectors among the vectors of X is called the* dimension *of X and is denoted by* dim X.

By this definition, a non-negative integer p is the dimension of a subspace X of $\mathbf{R}^n$ (i.e. $p = \dim X$) if and only if

(a) there are p linearly independent vectors in X, and

(b) any $p + 1$ vectors of X are linearly dependent.

Clearly it follows that for each subspace X there can be at most one such integer p. In other words, X cannot have two different dimensions. In the following discussion we shall show that every subspace X of $\mathbf{R}^n$ always has a dimension which is an integer that satisfies (a) and (b).

Let us first evaluate the dimensions of the trivial subspaces 0 and $\mathbf{R}^n$ of the vector space $\mathbf{R}^n$. The only vector of the zero subspace 0 of $\mathbf{R}^n$ is the zero vector $\mathbf{0}$ which by itself is linear dependent. This means that the zero subspace has no linearly independent vector. Therefore $\dim 0 = 0$. Next we claim that $\dim \mathbf{R}^n = n$. The unit coordinate vectors $e_1, e_2, \cdots, e_n$ of $\mathbf{R}^n$ are linearly independent; therefore n satisfies the condition (a). Let $a_1, a_2, \cdots, a_{n+1}$ be any $n + 1$ vectors of $\mathbf{R}^n$. If $a_i = [a_{i1}, a_{i2}, \cdots, a_{in}]$, we consider the $(n + 1) \times n$-matrix

$$A = \begin{pmatrix} a_{11} & a_{12} & \cdots & a_{1n} \\ a_{21} & a_{22} & \cdots & a_{2n} \\ \cdots & \cdots & \cdots & \cdots \\ a_{n1} & a_{n2} & \cdots & a_{nn} \\ a_{n+1\,1} & a_{n+1\,2} & \cdots & a_{n+1\,n} \end{pmatrix}.$$

Then we get row rank of A = column rank $A \leq n$ since A has only n column vectors. Therefore the $n + 1$ vectors $a_1, a_2, \cdots, a_{p+1}$ are linearly dependent. Hence $\dim \mathbf{R}^n = n$.

5.6.2 THEOREM $\dim 0 = 0$ and $\dim \mathbf{R}^n = n$. If $m > n$ then any m vectors of $\mathbf{R}^n$ are linearly dependent. Therefore no subspace X of $\mathbf{R}^n$ can have a dimension greater than n.

We take note that the above theorem affords a retrospective justification to calling $\mathbf{R}^n$ the n-dimensional vector space.

After we have seen that the trivial subspaces of $\mathbf{R}^n$ have dimensions 0 and n, we now examine subspaces of $\mathbf{R}^n$ which are generated by a finite number of vectors. Let x be a non-zero vector of $\mathbf{R}^n$. Consider the subspace

$$\langle x \rangle = \{ rx \mid r \in \mathbf{R} \}$$

generated by x. By itself, x is linearly independent; there is one linearly independent vector in $\langle x \rangle$. Take any two vectors rx and sx of $\langle x \rangle$. If r or s is zero, then $rx = 0$ or $sx = 0$, and the two vectors rx and sx are linearly dependent. Otherwise $-s(rx) + r(sx) = 0$. Therefore in all cases rx and sx are linearly dependent. Hence $\dim \langle x \rangle = 1$, i.e. every subspace generated by one non-zero vector has dimension 1. More generally we have the following theorem on the dimension of a subspace generated by a finite number of vectors.

5.6.3 THEOREM Let $x_1, x_2, \cdots, x_m$ be m vectors of $\mathbf{R}^n$ and p the maximum number of linearly independent vectors among them. Then the subspace $X = \langle x_1, x_2, \cdots, x_m \rangle$ generated by the m vectors has dimension $p : \dim X = p$.

PROOF Clearly it only remains to prove that any $p+1$ vectors $\mathbf{y}_1, \mathbf{y}_2, \cdots,$ $\mathbf{y}_{p+1}$ of X are linear dependent or equivalently that the $(p+1) \times n$-matrix A whose row vectors are $\mathbf{y}_1, \mathbf{y}_2, \cdots, \mathbf{y}_{p+1}$ has a rank less than $p+1$. Assume without loss of generality that the first p vectors $\mathbf{x}_1, \mathbf{x}_2, \cdots, \mathbf{x}_p$ among the m given vectors are linearly independent. Then the last $m - p$ vectors $\mathbf{x}_{p+1}, \mathbf{x}_{p+2}, \cdots, \mathbf{x}_m$ are all linear combinations of $\mathbf{x}_1, \mathbf{x}_2, \cdots, \mathbf{x}_p$ by 5.3.5. On the other hand as a vector of X, each $\mathbf{y}_i$ is a linear combination of the m vectors $\mathbf{x}_1, \mathbf{x}_2, \cdots, \mathbf{x}_m$ and hence a linear combination of the p vectors $\mathbf{x}_1, \mathbf{x}_2, \cdots, \mathbf{x}_p$. Consider the $(2p + 1) \times n$-matrix B whose row vectors are $\mathbf{x}_1, \mathbf{x}_2, \cdots, \mathbf{x}_p, \mathbf{y}_1, \cdots, \mathbf{y}_{p+1}$. Then rank of $B \geq$ rank of A. Therefore it is sufficient to show that rank $B \leq p$. But this is obvious since being linear combinations of the first p row vectors, the last $p+1$ row vectors of B can be made into zero by appropriate elementary row transformations of the second type. The proof is complete.

By virtue of Theorem 5.6.3 the only thing that we need for the conclusion that every subspace has a dimension is now the following theorem.

5.6.4 THEOREM *Every subspace X of $\mathbf{R}^n$ is generated by a finite number of vectors.*

PROOF If X is the zero subspace, then $X = \langle 0 \rangle$. Otherwise X has a non-zero vector $\mathbf{x}_1$. Then either $X = \langle \mathbf{x}_1 \rangle$ or $X \neq \langle \mathbf{x}_1 \rangle$. In the former case X is generated by $\mathbf{x}_1$. Otherwise there is a vector $\mathbf{x}_2$ of X which is not a linear combination of $\mathbf{x}_1$. Then $\mathbf{x}_1$ and $\mathbf{x}_2$ are linearly independent. Now either $X = \langle \mathbf{x}_1, \mathbf{x}_2 \rangle$ or $X \neq \langle \mathbf{x}_1, \mathbf{x}_2 \rangle$. In the former case X is generated by $\mathbf{x}_1$ and $\mathbf{x}_2$. Otherwise there is a vector $\mathbf{x}_3$ of X which is not a linear combination of $\mathbf{x}_1$ and $\mathbf{x}_2$. Then $X = \langle \mathbf{x}_1, \mathbf{x}_2, \mathbf{x}_3 \rangle$ or $X \neq \langle \mathbf{x}_1, \mathbf{x}_2, \mathbf{x}_3 \rangle$. In the former case X is generated by three vectors. Otherwise we can find a vector $\mathbf{x}_4$ and so forth. Since $\dim \mathbf{R}^n = n$, every $n + 1$ vectors of $\mathbf{R}^n$ must be linearly dependent; hence the process must terminate in no more than n steps. Therefore X must contain p $(p \leq n)$ linearly independent vectors which generate X. The proof is complete.

Combining 5.6.3 and 5.6.4, we have the main result of this section in the following statements.

5.6.5 THEOREM *Every subspace of the vector space $\mathbf{R}^n$ has a dimension p such that $0 \leq p \leq n$.*

5.6.6 COROLLARY AND DEFINITION *Every p-dimensional subspace X of $\mathbf{R}^n$ contains p linearly independent vectors which generate it. Any p such vectors of X are said to form a* base *of X.*

In particular the unit coordinate vectors $e_1, e_2, \cdots, e_n$ form a base of the vector space $\mathbf{R}^n$, and any p vectors among them form a base of a subspace of dimension p. To justify that the dimension of a subspace is a measurement of its size, we have the following theorem.

5.6.7 THEOREM *Let X and Y be subspaces of the vector space of $\mathbf{R}^n$ such that $X \subset Y$. Then $\dim X \leq \dim Y$ and the equality sign holds if and only if $X = Y$.*

PROOF Let $p = \dim X$ and $q = \dim Y$. Since every p linearly independent vectors of X are linearly independent vectors of Y, it follows that $p \leq q$. Obviously if $X = Y$ then $p = q$. Conversely suppose that $p = q$. Let $x_1, x_2, \cdots, x_p$ be a base of X and let y be any vector of Y. It suffices to show that $y \in X$. Since $\dim Y = p$, the $p + 1$ vectors $x_1, x_2, \cdots, x_p$, y of Y are linearly dependent and by 5.3.5 y is a linear combination of $x_1, x_2, \cdots, x_p$. Since the latter p vectors form a base of X, they generate X. Hence y is a vector of X. Therefore $X = Y$.

We observe that for the conclusions of the above theorem to be valid, the assumption that $X \subset Y$ is essential. Take, for example, $X_1 = \langle e_1 \rangle$, $X_2 = \langle e_2, e_3 \rangle$ and $X_3 = \langle e_1 + e_2 \rangle$ of $\mathbf{R}^3$. Then $\dim X_1 = 1$, $\dim X_2 = 2$, $\dim X_3 = 1$ and yet $X_1 \neq X_3$, $X_1 \not\subset X_2$ and $X_3 \not\subset X_2$.

In the following example we sketch a method of finding a base of a subspace, the justification of which is left to the interested reader as an exercise.

5.6.8 EXAMPLE Find a base of the subspace X of $\mathbf{R}^4$ geneated by $a_1 = [1, -1, 2, 3]$, $a_2 = [4, 5, 8, -9]$, $a_3 = [2, 1, 4, -1]$, $a_4 = [2, -5, 4, 13]$.

SOLUTION Use the given vectors as row vectors to get a 4×4-matrix

$$A = \begin{pmatrix} 1 & -1 & 2 & 3 \\ 4 & 5 & 8 & -9 \\ 2 & 1 & 4 & -1 \\ 2 & -5 & 4 & 13 \end{pmatrix}.$$

Use elementary row transformations of the first and the second type to transform A into

$$A \rightarrow \begin{pmatrix} 1 & -1 & 2 & 3 \\ 0 & 9 & 0 & -21 \\ 0 & 3 & 0 & -7 \\ 0 & -3 & 0 & 7 \end{pmatrix} \rightarrow \begin{pmatrix} 1 & -1 & 2 & 3 \\ 0 & 0 & 0 & 0 \\ 0 & 3 & 0 & -7 \\ 0 & 0 & 0 & 0 \end{pmatrix}.$$

The first and the third row vectors of the last matrix are linearly independent. We conclude $\dim X = 2$ and that a_1 and a_3 form a base of X. In fact $a_2 = -2a_1 + 3a_3$ and $a_4 = 4a_1 - a_3$. But the set $\{a_1, a_3\}$ is not the only base of X; for example, the first and the third row vectors of the last matrix also form a base of X. In fact for this particular case any two vectors among a_1, a_2, a_3, a_4 form a base of X.

EXERCISES

1. Without doing any calculation, give reasons why:
   (a) $[3, 4, 5, 6]$, $[1, 3, 5, 7]$, $[2, -1, 14, 0]$ do not generate $\mathbf{R}^4$.
   (b) $[2, 5, 1, 2]$, $[1, 3, 4, 4]$, $[4, 3, 2, -1]$, $[3, 1, -1, -1]$, $[2, 4, -6, -8]$ are not linearly independent.
   (c) $[1, 3, 1]$, $[1, 0, 3]$, $[2, 5, 6]$, $[7, 3, 4]$ do not form a base for $\mathbf{R}^3$.

2. The result of this question is another version of Corollary 5.6.6 which is easier to be used in practice. For a p-dimensional subspace X of $\mathbf{R}^n$, take any p linearly independent vectors $x_1, x_2, \cdots, x_p$ in X.
   (a) By Corollary 5.3.5 and the definition of dimension, show that any vector in X is a linear combination of x_i's.
   (b) Hence deduce that any p linearly independent vectors of X form a base of X.

3. Determine whether or not each of the following form a base of $\mathbf{R}^2$:
   (a) $[1, 2]$, $[1, -1]$;
   (b) $[0, 2]$, $[0, -3]$.

4. Determine whether or not each of the following form a base of $\mathbf{R}^3$:
   (a) $[1, 1, 1]$, $[1, -1, 1]$, $[0, 1, 1]$;
   (b) $[1, 1, 2]$, $[1, 2, 5]$, $[5, 3, 4]$.

5. Using the technique introduced in Example 5.6.8, find a base for $\langle [2, 6, 0, 1, 3], [1, 1, 0, -1, 4], [0, 0, 0, 1, 1] \rangle$.

6. By following the arguments used in the proof of Theorem 5.6.7, prove that if X is a subspace of $\mathbf{R}^n$ such that $\dim X = n$, then $X = \mathbf{R}^n$. (Try not to apply the theorem directly!)

7. Let $\{u_1, u_2, u_3\}$ be a base for $\mathbf{R}^3$. Show that $\{v_1, v_2, v_3\}$ is also a base where $v_1 = u_1$, $v_2 = u_1 + u_2$, and $v_3 = u_1 + u_2 + u_3$.

8. Let $X = \{[x_1, x_2, x_3] | x_1 + x_2 = x_3,\ 2x_1 + x_2 = 0\} \subset \mathbf{R}^3$.

   (a) Show that every vector in X can be written as a scalar multiple of $[-1, 2, 1]$. What is then the dimension of X?

   (b) Find two vectors u and v such that $[-1, 2, 1]$, u, v form a base of $\mathbf{R}^3$.

9. This question generalizes 8(b). Let u be any non-zero vector in $\mathbf{R}^3$.

   (a) Show that there are vectors v_1 and v_2 such that $\{u, v_1, v_2\}$ is a base of $\mathbf{R}^3$.

   (b) Show that any linearly independent subset X of $\mathbf{R}^3$ can be extended to a base of $\mathbf{R}^3$.

10. Let X be a subspace of $\mathbf{R}^n$. Consider any non-zero vector u in $\mathbf{R}^n$ not in X. Let

    $$Y = \{au + v | v \text{ is in } X \text{ and } a \text{ is a real number}\}$$

    (a) Prove that Y is a subspace of $\mathbf{R}^n$.

    (b) Prove that $\dim Y = \dim X + 1$.

11. Let $\{x_1, x_2, \cdots, x_p\}$ be a set that generates $\mathbf{R}^n$, prove that the following are equivalent:

    (a) $\{x_1, x_2, \cdots, x_p\}$ is a linearly independent set

    (b) If $x \in \mathbf{R}^n$, the expression $x = a_1 x_1 + \cdots + a_p x_p$ is unique, for real numbers a_i's.

    (c) $n = p$.

    Thus any element in $\mathbf{R}^n$ can be expressed as a unique linear combination of elements in any base.

12. Let A be the $m \times n$-matrix whose row vectors are m given vectors $a_1, a_2, \cdots, a_m$ of $\mathbf{R}^n$ and $X = \langle a_1, a_2, \cdots, a_m \rangle$. Suppose by using elementary row transformations of the first and the second types, we transform A into an $m \times n$-matrix H such that (i) there are only p non-zero row vectors $r_{i_1}(H),\ r_{i_2}(H), \cdots, r_{i_p}(H)$ in H, and (ii) among the column vectors of H there are p distinct unit coordinate vectors $c_{j_1}(H) = e_{i_1},\ c_{j_2}(H) = e_{i_2}, \cdots, c_{j_p}(H) = e_{i_p}$ of $\mathbf{R}^m$.

    (a) Show that $\dim X = p$.

    (b) Also show that both $a_{i_1}, a_{i_2}, \cdots, a_{i_p}$ and $r_{i_1}(H), r_{i_2}(H), \cdots, r_{i_p}(H)$ form bases of X.

13. Let X and Y be two subspaces of $\mathbf{R}^n$. Suppose $\dim(X \cap Y) = r$, $\dim X = r + p$ and $\dim Y = r + q$. Use the argument in the Proof of Theorem 5.6.4 to prove the following statements.

   (a) If $z_1, z_2, \cdots, z_r$ form a base of $X \cap Y$, then there are p vectors $x_1, x_2, \cdots, x_p$ of X and q vectors $y_1, y_2, \cdots, y_q$ of Y such that $z_1, z_2, \cdots, z_r, x_1, x_2, \cdots, x_p$ form a base of X and $z_1, z_2, \cdots, z_r$, $y_1, y_2, \cdots, y_q$ form a base of Y.

   (b) The $r + p + q$ vectors $z_1, z_2, \cdots, z_r, x_1, x_2, \cdots, x_p, y_1, y_2, \cdots, y_q$ form a base of $X + Y$.

   (c) $\dim X + \dim Y = \dim(X \cap Y) + \dim(X + Y)$.

CHAPTER SIX

MATRIX AND DETERMINANT

Matrices are introduced in the last chapter as a systematic way of presenting the components of m vectors of $\mathbf{R}^n$ so that we can keep track of certain calculations being carried out on them. The chief concern of such calculations is to evaluate the rank of a matrix and to select linearly independent row vectors.

In this chapter matrices are treated as individual algebraic entities on their own right. Sums and products of matrices as well as their properties are studied in the first part of this chapter. A particularly interesting and useful result is the interpretation of elementary transformations on matrices as multiplications by elementary matrices.

Determinants of order 2 and 3 are mentioned briefly in the earlier chapters; they will be studied in some detail in the second part of this chapter. In conclusion we also make some useful observations on determinants of higher order so that readers will be able to see how they can be defined and what properties they have in common with determinants of lower order.

6.1 Terminology

We recall that for any two positive integers m and n, an $m \times n$-matrix A is a rectangular array of mn real numbers:

$$A = \begin{pmatrix} a_{11} & a_{12} & \cdots & a_{1n} \\ a_{21} & a_{22} & \cdots & a_{2n} \\ \hdotsfor{4} \\ a_{m1} & a_{m2} & \cdots & a_{mn} \end{pmatrix}$$

in m rows and n columns. For easy reference, we shall introduce the following specific terminology and notations. The ordered pair (m, n) is called the *order* of the matrix A. Each of the mn real numbers

a_{ij} $(i = 1, 2, \cdots, m; \; j = 1, 2, \cdots, n)$ is an *element* of A. The integer i is the *row index* and the integer j is the *column index* of the element a_{ij}; they indicate the position of the element a_{ij} in the matrix A. A matrix is usually denoted by an italic capital letter and its elements by the lower case italics of the same letter. The elements of a matrix are usually enclosed by a pair of elongated parentheses.

Two matrices A and B are equal if and only if they have the same order and the same elements. Thus for an $m \times n$-matrix A and a $p \times q$-matrix B, $A = B$ if and only if (i) $m = p$ and $n = q$, and (ii) $a_{ij} = b_{ij}$ for all i and j.

Each horizontal sequence of n numbers of an $m \times n$-matrix of A is called a *row* of A. The i-th row of A is a vector $\mathbf{r}_i(A) = [a_{i1}, a_{i2}, \cdots, a_{in}]$ of the vector space $\mathbf{R}^n$. As its elements are arranged horizontally in A, it is also a $1 \times n$-matrix or a *row matrix*:

$$\mathbf{r}_i(A) = \begin{pmatrix} a_{i1} & a_{i2} & \cdots & a_{in} \end{pmatrix} .$$

Similarly the j-th column of A is a vector $\mathbf{c}_j(A) = [a_{1j}, a_{2j}, \cdots, a_{mj}]$ of $\mathbf{R}^m$ and is also an $m \times 1$-matrix or a *column matrix*:

$$\mathbf{c}_j(A) = \begin{pmatrix} a_{1j} \\ a_{2j} \\ \vdots \\ a_{mj} \end{pmatrix} .$$

Finally by converting rows of A into columns, we obtain the *transpose* A^t of A:

$$A^t = \begin{pmatrix} a_{11} & a_{21} & \vdots & a_{m1} \\ a_{12} & a_{22} & \vdots & a_{m2} \\ \vdots & \vdots & \vdots & \vdots \\ a_{1n} & a_{2n} & \vdots & a_{mn} \end{pmatrix}$$

which is an $n \times m$-matrix. Therefore $(A^t)^t = A$. If we call the elements $a_{11}, a_{22}, \cdots, a_{ii}, \cdots$ of A the *diagonal elements* of A, then A has $\min(m, n)$ diagonal elements which are identical to the diagonal elements of A^t.

EXERCISES

1. Construct the 3×3 matrix defined by

$$a_{ij} = 0 \qquad \text{for } i \neq j$$
$$a_{ii} = 3i \, .$$

2. Construct the 4×4 matrix such that

$$a_{ij} = \text{least common multiple of } i \text{ and } j \, .$$

3. Construct the 5×5 matrix such that

$$a_{ij} = 1 \qquad \text{if } |i - j| \text{ is even or zero}$$
$$a_{ij} = 0 \qquad \text{otherwise} \, .$$

4. Find real numbers x, y, z such that

$$\begin{pmatrix} 6 & 4 \\ 5 & 3 \end{pmatrix} = \begin{pmatrix} x + y + z & x + z \\ y + z & x + y \end{pmatrix} \, .$$

5. Find real numbers a, b, c and d if

$$\begin{pmatrix} a & c \\ b & d \end{pmatrix} = \begin{pmatrix} c - 3d & 2a + d \\ -d & a + b \end{pmatrix} \, .$$

6.2 Scalar multiple and sum

We now begin treating the set of all $m \times n$-matrices as a set of algebraic entities for any fixed integers m and n. Infact we shall carry out certain algebraic operations on them in a rather similar way as we have done with vectors. The *multiple* of an $m \times n$-matrix A by a real number (scalar) r is the $m \times n$-matrix rA whose elements are ra_{ij} $(i = 1, 2, \cdots, m; \ j = 1, 2, \cdots, n)$:

$$rA = \begin{pmatrix} ra_{11} & ra_{12} & \cdots & ra_{1n} \\ ra_{21} & ra_{22} & \cdots & ra_{2n} \\ \multicolumn{4}{c}{\dotfill} \\ ra_{m1} & ra_{m2} & \cdots & ra_{mn} \end{pmatrix} \, .$$

Therefore if $r = 1$, then $1A = A$. If $r = 0$, then all elements of $0A$ are zero. This particular matrix is called the *zero matrix of order* (m, n) and denoted by 0. Thus $0A = 0$, where the 0 on the left-hand side is the real number 0 and the one on the right-hand side is the zero matrix. Similarly if $r = -1$, we shall denote the matrix $(-1)A$ by $-A$.

The *sum* of two $m \times n$-matrices A and B is the $m \times n$-matrix

$$A + B = \begin{pmatrix} a_{11} + b_{11} & a_{12} + b_{12} & \cdots & a_{1n} + b_{1n} \\ a_{21} + b_{21} & a_{22} + b_{22} & \cdots & a_{2n} + b_{2n} \\ \cdots\cdots\cdots\cdots\cdots\cdots\cdots\cdots\cdots\cdots\cdots \\ a_{m1} + b_{m1} & a_{m2} + b_{m2} & \cdots & a_{mn} + b_{mn} \end{pmatrix}.$$

For example

$$3 \begin{pmatrix} 2 & 3 & 1 & -2 \\ -1 & 2 & 3 & 5 \\ 0 & 1 & -1 & 4 \end{pmatrix} = \begin{pmatrix} 6 & 9 & 3 & -6 \\ -3 & 6 & 9 & 15 \\ 0 & 3 & -3 & 12 \end{pmatrix}$$

$$\begin{pmatrix} 2 & 3 & 1 & -2 \\ -1 & 2 & 3 & 5 \\ 0 & 1 & -1 & 4 \end{pmatrix} + \begin{pmatrix} -1 & -7 & 2 & 4 \\ 3 & 4 & -5 & 6 \\ 1 & -2 & 1 & -3 \end{pmatrix} = \begin{pmatrix} 1 & -4 & 3 & 2 \\ 2 & 6 & -2 & 11 \\ 1 & -1 & 0 & 1 \end{pmatrix}.$$

It is easy to verify the following formal properties of addition and scalar multiplication which are entirely similar to the corresponding properties of $\mathbf{R}^n$.

(1) $A + B = B + A$;

(2) $(A + B) + C = A + (B + C)$;

(3) $A + 0 = A$;

(4) $A + (-A) = 0$;

(5) $r(sA) = (rs)A$;

(6) $r(A + B) = rA + rB$;

(7) $(r + s)A = rA + sA$;

(8) $1A = A$.

Finally we remark that two matrices of different orders do not have a sum. This is again similar to the fact that we do not add two vectors of different vector spaces. Furthermore it is possible to extend the notions of linear combination, linear dependence and dimension for matrices. But we shall not need them in the work of this course.

EXERCISES

1. Let $A = \begin{pmatrix} 2 & -1 & 7 \\ 3 & 0 & 4 \end{pmatrix}$ and $B = \begin{pmatrix} -3 & 4 & -2 \\ -1 & -2 & 5 \end{pmatrix}$. Compute the following matrices:

   (a) $2A$; (b) $A - B$; (c) $3A + 5B$.

2. Let $A = \begin{pmatrix} 1 & 7 & 2 \\ 2 & -6 & 5 \\ 3 & 0 & 8 \end{pmatrix}$ and $B = \begin{pmatrix} 3 & 6 & 9 \\ 5 & 7 & 8 \\ 2 & 2 & 1 \end{pmatrix}$. Find

   (a) $3B^t$; (b) $(A + B)^t$; (c) $A^t + B^t$.

3. Let A and B be two $m \times n$ matrices. Show that $(A + B)^t = A^t + B^t$.

4. Find matrix A if:

   (a) $7A - \begin{pmatrix} 5 & 3 \\ 2 & 1 \end{pmatrix} = 3A + \begin{pmatrix} 3 & 1 \\ 2 & 7 \end{pmatrix}$;

   (b) $3A + 2\begin{pmatrix} 2 \\ 1 \end{pmatrix} = 5A - 7\begin{pmatrix} 0 \\ 4 \end{pmatrix}$.

5. Find x, y, z and w if $5\begin{pmatrix} x & y \\ z & w \end{pmatrix} = \begin{pmatrix} x & 1 \\ -2 & 3w \end{pmatrix} + \begin{pmatrix} 4 & x+y \\ z+w & 8 \end{pmatrix}$.

6. Find matrices A and B such that

$$A + B = \begin{pmatrix} 2 & 1 & 3 \\ 0 & -7 & 4 \\ 5 & 8 & 6 \end{pmatrix} \quad \text{and} \quad A - B = \begin{pmatrix} 4 & 3 & 1 \\ 2 & 5 & 0 \\ 1 & 2 & 2 \end{pmatrix}.$$

7. If A is any 2×2 matrix show that

   (a) $A = a\begin{pmatrix} 1 & 0 \\ 0 & 0 \end{pmatrix} + b\begin{pmatrix} 0 & 0 \\ 1 & 0 \end{pmatrix} + c\begin{pmatrix} 0 & 1 \\ 0 & 0 \end{pmatrix} + d\begin{pmatrix} 0 & 0 \\ 0 & 1 \end{pmatrix}$
   for some real numbers a, b, c, and d;

   (b) $A = e\begin{pmatrix} 1 & 0 \\ 0 & 1 \end{pmatrix} + f\begin{pmatrix} 1 & 1 \\ 0 & 0 \end{pmatrix} + g\begin{pmatrix} 1 & 0 \\ 1 & 0 \end{pmatrix} + h\begin{pmatrix} 0 & 1 \\ 1 & 0 \end{pmatrix}$
   for some real numbers e, f, g, and h.

6.3 Product

The addition and the scalar multiplication of matrices are both quite simple and straightforward; they are defined by elementwise addition and elementwise multiplication respectively. The multiplication of matrices is somehow more complicated and some preparation is necessary.

Let $\mathbf{R}^p$ be the p-dimensional vector space. For any two vectors $\mathbf{a} = [a_1, a_2, \cdots, a_p]$ and $\mathbf{b} =]b_1, b_2, \cdots, b_p]$ of $\mathbf{R}^p$, their *dot product* is defined as the real number

$$\mathbf{a} \cdot \mathbf{b} = a_1 b_1 + a_2 b_2 + \cdots + a_p b_p \ .$$

Obviously the dot product of vectors of $\mathbf{R}^p$ and the dot product of vectors of $\mathbf{R}^2$ or $\mathbf{R}^3$ have the same formal properties. For example

(1) $\mathbf{a} \cdot \mathbf{b} = \mathbf{b} \cdot \mathbf{a}$;
(2) $(r\mathbf{a}) \cdot \mathbf{b} = r(\mathbf{a} \cdot \mathbf{b})$;
(3) $(\mathbf{a} + \mathbf{b}) \cdot \mathbf{c} = \mathbf{a} \cdot \mathbf{c} + \mathbf{b} \cdot \mathbf{c}$;
(4) $\mathbf{a} \cdot \mathbf{a} \geq 0$ and $\mathbf{a} \cdot \mathbf{a} = 0$ if and only if $\mathbf{a} = \mathbf{0}$.

Similarly if $\mathbf{a} \cdot \mathbf{a} = 1$, we say that $\mathbf{a}$ is a *unit vector* and if $\mathbf{a} \cdot \mathbf{b} = 0$ then we say that $\mathbf{a}$ and $\mathbf{b}$ are *orthogonal*.

For the sum $\mathbf{a} + \mathbf{b}$ and the dot product $\mathbf{a} \cdot \mathbf{b}$ of two vectors $\mathbf{a}$ and $\mathbf{b}$ to be defined, the number of components of $\mathbf{a}$ has to be the same as the number of components of $\mathbf{b}$. As the product AB of two matrices A and B is to be defined in terms of dot products of row vectors of A and column vectors of B, we must require the number of columns of A to be the same as the number of rows of B. In other words the order of A has to be (m, p) and the order of B has to be (p, n).

6.3.1 DEFINITION *Let A be an $m \times p$-matrix and B a $p \times n$-matrix. The product AB is the $m \times n$-matrix C with elements*

$$c_{ij} = \mathbf{r}_i(A) \cdot \mathbf{c}_j(B) = a_{i1} b_{1j} + a_{i2} b_{2j} + \cdots + a_{ip} b_{pj} = \sum_{k=1}^{p} a_{ik} b_{kj}$$

for all $i = 1, 2, \cdots, m$; $j = 1, 2, \cdots, n$.

For example if A is a 4×2-matrix and B is a 2×3-matrix then the product AB is the 4×3-matrix

$$\begin{pmatrix} a_{11}b_{11} + a_{12}b_{21} & a_{11}b_{12} + a_{12}b_{22} & a_{11}b_{13} + a_{12}b_{23} \\ a_{21}b_{11} + a_{22}b_{21} & a_{21}b_{12} + a_{22}b_{22} & a_{21}b_{13} + a_{22}b_{23} \\ a_{31}b_{11} + a_{32}b_{21} & a_{31}b_{12} + a_{32}b_{22} & a_{31}b_{13} + a_{32}b_{23} \\ a_{41}b_{11} + a_{42}b_{21} & a_{41}b_{12} + a_{42}b_{22} & a_{41}b_{13} + a_{42}b_{23} \end{pmatrix} \ .$$

6.3.2 EXAMPLE   Readers who encounter the multiplication of matrices for the first time may find the above definition somewhat unusual or even unnatural. Perhaps this example would provide a relevant explanation. Let

$\mathbf{x}_1, \mathbf{x}_2, \cdots, \mathbf{x}_m;\ \mathbf{y}_1, \mathbf{y}_2, \cdots, \mathbf{y}_p;\ \mathbf{z}_1, \mathbf{z}_2, \cdots, \mathbf{z}_n$ be vectors of the vector space $\mathbf{R}^q$. We know that if each

$$\mathbf{x}_i = a_{i1}\mathbf{y}_1 + a_{i2}\mathbf{y}_2 + \cdots + a_{ip}\mathbf{y}_p \quad i = 1, 2, \cdots, m$$

is a linear combination of the $\mathbf{y}'s$ and each

$$\mathbf{y}_k = b_{k1}\mathbf{z}_1 + b_{k2}\mathbf{z}_2 + \cdots + b_{kn}\mathbf{z}_n \quad k = 1, 2, \cdots, p$$

is a linear combination of the $\mathbf{z}'s$, then each $\mathbf{x}_i$ is also a linear combination of the $\mathbf{z}'s$. Indeed it follows from

$$\mathbf{x}_i = \sum_{k=1}^{p} a_{ik}\mathbf{y}_k = \sum_{k=1}^{p} a_{ik}\left(\sum_{j=1}^{n} b_{kj}\mathbf{z}_j\right) = \sum_{j=1}^{n}\left(\sum_{k=1}^{p} a_{ik}b_{kj}\right)\mathbf{z}_j$$

that if

$$c_{ij} = a_{i1}b_{1j} + a_{i2}b_{2j} + \cdots + a_{ip}b_{pj}$$

then

$$\mathbf{x}_i = c_{i1}\mathbf{z}_1 + c_{i2}\mathbf{z}_2 + \cdots + c_{in}\mathbf{z}_n \ .$$

On the other hand, writing the coefficients of the linear combinations in matrix form, we obtain

$$A = \begin{pmatrix} a_{11} & \cdots & a_{1p} \\ \cdots\cdots\cdots\cdots\cdots \\ a_{m1} & \cdots & a_{mp} \end{pmatrix}, \ B = \begin{pmatrix} b_{11} & \cdots & b_{1n} \\ \cdots\cdots\cdots\cdots \\ b_{p1} & \cdots & b_{pn} \end{pmatrix},$$

$$C = \begin{pmatrix} c_{11} & \cdots & c_{1n} \\ \cdots\cdots\cdots\cdots \\ c_{m1} & \cdots & c_{mn} \end{pmatrix}.$$

Using Definition 6.3.1, we would have

$$AB = C$$

expressing the relationship between the coefficients.

By definition, the element of the product AB on the i-th row and the j-th columns is the dot product of the i-th row vector of A and the j-th column vector of B. Between the numbers of rows and columns of the product AB and its factors A and B we have

number of rows of AB = number of rows of A ,

number of columns of A = number of rows of B ,

number of columns of B = number of columns of AB .

It follows that for both products AB and BA to be defined, A has to be of order (m, p) and B of order (p, m). In this case AB is an $m \times m$-matrix and BA a $p \times p$-matrix. Therefore if $m \neq p$, then the two products AB and BA are never equal. Even for the case $m = p$, where all four matrices A, B, AB and BA would be of order (m, m), the two products need not be equal. Take for example

$$A = \begin{pmatrix} 0 & 1 \\ 0 & 0 \end{pmatrix} \quad \text{and} \quad B = \begin{pmatrix} 1 & 0 \\ 2 & 1 \end{pmatrix},$$

then
$$AB = \begin{pmatrix} 2 & 1 \\ 0 & 0 \end{pmatrix} \quad \text{and} \quad BA = \begin{pmatrix} 0 & 1 \\ 0 & 2 \end{pmatrix}.$$

Therefore like the formation of the cross product of two vectors of $\mathbf{R}^3$, the *multiplication of matrices is not commutative*. Nevertheless *the multiplication of matrices is associative*, i.e. $A(BC) = (AB)C$. To prove this, let A be of order (m, p), B of order (p, q) and C of order (q, n). If $BC = D$, $A(BC) = AD = F$ and $AB = G$, $(AB)C = GC = H$. Then

$$D \text{ is a } p \times n\text{-matrix},$$
$$F \text{ an } m \times n\text{-matrix},$$
$$G \text{ an } m \times q\text{-matrix},$$
$$H \text{ an } m \times n\text{-matrix} \quad \text{and}$$

$$d_{rj} = \sum_{s=1}^{q} b_{rs} c_{sj} \qquad r = 1, \cdots, p; \; j = 1, \cdots, n \;;$$

$$f_{ij} = \sum_{r=1}^{p} a_{ir} d_{rj} \qquad i = 1, \cdots, m; \; j = 1, \cdots, n \;;$$

$$g_{is} = \sum_{r=1}^{p} a_{ir} b_{rs} \qquad i = 1, \cdots, m; \; s = 1, \cdots, q \;;$$

$$h_{ij} = \sum_{s=1}^{q} g_{is} c_{sj} \qquad i = 1, \cdots, m; \; j = 1, \cdots, n \;.$$

Therefore

$$f_{ij} = \sum_r a_{ir} \left(\sum_s b_{rs} c_{sj} \right) = \sum_{r,s} a_{ir} b_{rs} c_{sj} = \sum_s \left(\sum_r a_{ir} b_{rs} \right) c_{sj} = h_{ij} \;.$$

Hence $A(BC) = (AB)C$.

6.3.3 THEOREM *For matrices A, B, C of appropriate orders, the following statements hold:*

(1) $A(BC) = (AB)C$;

(2) $(rA)B = A(rB) = r(AB)$;

(3) $A(B + C) = AB + AC$;

(4) $(A + B)C = AC + BC$;

(5) $(AB)^t = B^t A^t$.

We take note that in (5) above the order in which A and B appear in the right hand side is the reverse of that on the left hand-side.

EXERCISES

1. Compute the following products:

(a) $(1\ 3) \begin{pmatrix} 2 \\ 7 \end{pmatrix}$; (b) $\begin{pmatrix} 1 & 2 \\ 0 & 1 \end{pmatrix} \begin{pmatrix} 2 & 4 \\ 8 & -1 \end{pmatrix}$; (c) $\begin{pmatrix} 2 & 4 \\ 8 & -1 \end{pmatrix} \begin{pmatrix} 1 & 2 \\ 0 & 1 \end{pmatrix}$.

2. For each of the following matrices, compute A^2 and A^3.

(a) $\begin{pmatrix} 2 & 0 \\ 0 & 1 \end{pmatrix}$; (b) $\begin{pmatrix} 1 & 1 \\ 1 & 1 \end{pmatrix}$; (c) $\begin{pmatrix} a & c \\ b & d \end{pmatrix}$.

3. Compute the following products:

(a) $\begin{pmatrix} 2 & 7 & -4 \\ 1 & 1 & 1 \end{pmatrix} \begin{pmatrix} 1 & 0 \\ 2 & 1 \\ 3 & 0 \end{pmatrix}$; (b) $\begin{pmatrix} 4 & 7 \\ 5 & 8 \\ 6 & 9 \end{pmatrix} \begin{pmatrix} 1 & 0 \\ 2 & 3 \end{pmatrix}$;

(c) $\begin{pmatrix} 1 & 1 & 0 \\ 0 & 1 & 1 \\ 1 & 0 & 1 \end{pmatrix} \begin{pmatrix} 1 & 0 & 0 \\ 2 & 1 & 0 \\ 3 & 4 & 1 \end{pmatrix}$.

4. Compute A^2 and A^3 for real numbers a, b, and c:

(a) $A = \begin{pmatrix} 0 & a & b \\ 0 & 0 & c \\ 0 & 0 & 0 \end{pmatrix}$; (b) $A = \begin{pmatrix} 1 & 0 & a \\ 0 & 1 & 0 \\ 0 & 0 & 1 \end{pmatrix}$.

5. Show that $\begin{pmatrix} a & 1+a \\ 1-a & -a \end{pmatrix}^2 = \begin{pmatrix} 1 & 0 \\ 0 & 1 \end{pmatrix}$ for all real numbers a.

6. Let $A = \begin{pmatrix} 3 & 4 \\ 1 & 2 \end{pmatrix}$, $B = \begin{pmatrix} 2 & 3 \\ 5 & 6 \end{pmatrix}$ and $C = \begin{pmatrix} 4 & 1 \\ 2 & 2 \end{pmatrix}$. Compute AC,

BC, and AB and verify that
(a) $(A + B)C = AC + BC$;
(b) $(AB)^t = B^t A^t$.

7. Find 2×2-matrices A, B and C such that $AB = AC$ but $B \neq C$.

8. Find 2×2-matrices A, B such that $AB = 0$ but $A \neq 0$, $B \neq 0$.

9. Find 2×2-matrices A, B such that $(AB)^2 \neq A^2 B^2$.

10. (a) Find 2×2-matrices A and B such that

$$(A + B)^2 \neq A^2 + 2AB + B^2 .$$

(b) Show that for any 2×2-matrices A and B such that $AB = BA$, then

$$(A + B)^2 = A^2 + 2AB + B^2 .$$

11. Let $M = \begin{pmatrix} \cos\theta & \sin\theta \\ -\sin\theta & \cos\theta \end{pmatrix}$. Show that $M^2 = \begin{pmatrix} \cos 2\theta & \sin 2\theta \\ -\sin 2\theta & \cos 2\theta \end{pmatrix}$;

hence by mathematical induction that $M^n = \begin{pmatrix} \cos n\theta & \sin n\theta \\ -\sin n\theta & \cos n\theta \end{pmatrix}$ for all positive integers n.

12. If $A = \begin{pmatrix} a & c \\ b & d \end{pmatrix}$, show that $A^2 = \begin{pmatrix} 0 & 1 \\ 0 & 0 \end{pmatrix}$ is not possible for any real numbers a, b, c and d.

13. Let $E = \begin{pmatrix} e & 0 \\ 0 & e \end{pmatrix}$, where e is an arbitrary real number. Show that $EA = AE$ for any 2×2-matrix A. Conversely show that if E is a matrix for which $AE = EA$ for any 2×2-matrix A, then E has the above form for some real number e.

14. (a) If $A = \begin{pmatrix} x & 0 \\ y & z \end{pmatrix}$ for some real numbers x, y and z, find A^4.

(b) Suppose $B = \begin{pmatrix} c - 2a & 0 \\ 2b - c & a - 2b \end{pmatrix}$. Find integers a, b and c such

that $B^4 = \begin{pmatrix} 16 & 0 \\ -20 & 1 \end{pmatrix}$.

(Hint: Put $x = c - 2a$, $y = 2b - c$, $z = a - 2b$, and use (a).)

15. Let $A = \begin{pmatrix} a & c \\ b & d \end{pmatrix}$ and $I = \begin{pmatrix} 1 & 0 \\ 0 & 1 \end{pmatrix}$, with $a + d \neq 0$.

(a) Prove that

$$A = \frac{1}{a+d}[A^2 + (ad - bc)I].$$

(b) If $A^2 = \begin{pmatrix} p & r \\ q & s \end{pmatrix}$, show by (a) that $p + s = (a + d)^2 - 2(ad - bc)$
and $ps - qr = (ad - bc)^2$.

(c) Hence or otherwise, find all possible matrices A such that $A^2 = \begin{pmatrix} 1 & 0 \\ 0 & 1 \end{pmatrix}$ and $a + d \neq 0$.

6.4 Square matrix

Any matrix of order (n, n) is called a *square matrix of order n* or an *n-square* matrix. Coordinate transformations of $\mathbf{R}^2$ and $\mathbf{R}^3$ can be expressed in terms of square matrices. For example the equations

$$x = x' \cos \theta - y' \sin \theta$$
$$y = x' \sin \theta + y' \cos \theta$$

of the coordinate transformation by a rotation of coordinate axes give rise to a square matrix

$$T = \begin{pmatrix} \cos \theta & -\sin \theta \\ \sin \theta & \cos \theta \end{pmatrix}$$

of order 2. Using the square matrix T and the column matrix $X = \begin{pmatrix} x \\ y \end{pmatrix}$ of coordinates of a point $P = (x, y)$ in the xy-plane and the column matrix $X' = \begin{pmatrix} x' \\ y' \end{pmatrix}$ of coordinates of the same point $P = ((x', y'))$ in the $x'y'$-plane , we can write the equations of transformation in matrix form as follows:

$$\begin{pmatrix} x \\ y \end{pmatrix} = \begin{pmatrix} \cos\theta & -\sin\theta \\ \sin\theta & \cos\theta \end{pmatrix} \begin{pmatrix} x' \\ y' \end{pmatrix} \quad \text{or} \quad X = TX' \ .$$

Similarly from the equations of transformation in reverse

$$x' = x\cos\theta + y\sin\theta$$
$$y' = -x\sin\theta + y\cos\theta \ ,$$

we have a 2-square matrix

$$S = \begin{pmatrix} \cos\theta & \sin\theta \\ -\sin\theta & \cos\theta \end{pmatrix}$$

such that

$$\begin{pmatrix} x' \\ y' \end{pmatrix} = \begin{pmatrix} \cos\theta & \sin\theta \\ -\sin\theta & \cos\theta \end{pmatrix} \begin{pmatrix} x \\ y \end{pmatrix} \quad \text{or} \quad X' = SX \ .$$

Therefore it follows that

$$X = TX' = TSX \quad \text{and} \quad X' = SX = STX' \ .$$

On the other hand

$$TS = \begin{pmatrix} \cos\theta & -\sin\theta \\ \sin\theta & \cos\theta \end{pmatrix} \begin{pmatrix} \cos\theta & \sin\theta \\ -\sin\theta & \cos\theta \end{pmatrix} = \begin{pmatrix} 1 & 0 \\ 0 & 1 \end{pmatrix} \ ,$$

$$ST = \begin{pmatrix} \cos\theta & \sin\theta \\ -\sin\theta & \cos\theta \end{pmatrix} \begin{pmatrix} \cos\theta & -\sin\theta \\ \sin\theta & \cos\theta \end{pmatrix} = \begin{pmatrix} 1 & 0 \\ 0 & 1 \end{pmatrix} \ .$$

Denoting this identical product of matrices by I, we have

$$X = IX \quad \text{and} \quad X' = IX'.$$

More generally if A is any $m \times 2$-matrix and B any $2 \times n$ matrix, then

$$AI = \begin{pmatrix} a_{11} & a_{12} \\ a_{21} & a_{22} \\ \vdots & \vdots \\ a_{m1} & a_{m2} \end{pmatrix} \begin{pmatrix} 1 & 0 \\ 0 & 1 \end{pmatrix} = \begin{pmatrix} a_{11} & a_{12} \\ a_{21} & a_{22} \\ \vdots & \vdots \\ a_{m1} & a_{m2} \end{pmatrix} = A \ ,$$

$$IB = \begin{pmatrix} 1 & 0 \\ 0 & 1 \end{pmatrix} \begin{pmatrix} b_{11} & b_{12} & \cdots & b_{1n} \\ b_{21} & b_{22} & \cdots & b_{2n} \end{pmatrix} = \begin{pmatrix} b_{11} & b_{12} & \cdots & b_{1n} \\ b_{21} & b_{22} & \cdots & b_{2n} \end{pmatrix} = B.$$

In other words the square matrix I of order 2 acts as the multiplicative identity in the formation of products.

In general the square matrix of order n

$$I = \begin{pmatrix} 1 & 0 & \cdots & \cdots & 0 \\ 0 & 1 & \cdots & \cdots & 0 \\ & & \ddots & & \\ & & & \ddots & \\ 0 & \cdots & \cdots & 0 & 1 \end{pmatrix} = \begin{pmatrix} \delta_{11} & \delta_{12} & \cdots & \delta_{1n} \\ \delta_{21} & \delta_{22} & \cdots & \delta_{2n} \\ \cdots\cdots\cdots\cdots\cdots \\ \cdots\cdots\cdots\cdots\cdots \\ \cdots\cdots\cdots\cdots\cdots \\ \delta_{n1} & \delta_{n2} & \cdots & \delta_{nn} \end{pmatrix}$$

is called the *identity matrix of order* n. Here the Kronecker symbol δ_{ij} has the value 1 if $i = j$ and 0 if $i \neq j$. The i-th row vector and the i-th column vector of I are both the i-th unit coordination vector e_i of $\mathbf{R}^n$. This matrix has the characteristic property that

$$AI = A \quad \text{and} \quad IB = B$$

for any $m \times n$-matrix A and any $n \times m$-matrix B. The identity matrix is usually written as

$$\begin{pmatrix} 1 & & & 0 \\ & \ddots & & \\ & & \ddots & \\ 0 & & & 1 \end{pmatrix}$$

where all the unspecified elements on the diagonal are presumed to be 1 and all those off the diagonal to be zero.

Let us consider some other special square matrices. Recall that in Section 5.4, we have used elementary transformations on a matrix to find its rank. Applying these transformations on the identity matrix I, we obtain some rather special square matrices. For example, using an elementary row transformation of the first type $T : \mathbf{r}_i(I) \to s\mathbf{r}_i(I)$ on I with some $s \neq 0$, we get

$$E = \begin{pmatrix} 1 & & & & & & 0 \\ & \ddots & & & & & \\ & & s & & & & \\ & & & \ddots & & & \\ & & & & \ddots & & \\ & & & & & \ddots & \\ 0 & & & & & & 1 \end{pmatrix} \quad i\text{-th row}$$

i.e. $T(I) = E$. Similarly from an elementary row transformation of the second type $T' : \mathbf{r}_i(I) \to \mathbf{r}_i(I) + s\mathbf{r}_j(I)$, we get

$$E' = \begin{pmatrix} 1 & & & & & & 0 \\ & \ddots & & & & & \\ & & 1 & & s & & \\ & & & \ddots & & & \\ & & & & 1 & & \\ & & & & & \ddots & \\ 0 & & & & & & 1 \end{pmatrix} \quad \begin{matrix} i\text{-th row} \\ \\ j\text{-th row} \end{matrix}$$

i.e. $T'(I) = E'$. Finally with an elementary row transformation of the third type $T'' : \mathbf{r}_i(I) \leftrightarrow \mathbf{r}_j(I)$, we obtain

$$E'' = \begin{pmatrix} 1 & & & & & & 0 \\ & \ddots & & & & & \\ & & 0 & & 1 & & \\ & & & \ddots & & & \\ & & 1 & & 0 & & \\ & & & & & \ddots & \\ 0 & & & & & & 1 \end{pmatrix} \quad \begin{matrix} i\text{-th row} \\ \\ j\text{-th row} \end{matrix}$$

i.e. $T''(I) = E''$.

Let us examine these special square matrices. For any $n \times m$-matrix B, the products EB, $E'B$, $E''B$ which are obtained by multiplication of the above matrices E, E', E'' from the left have the following row vectors.

$$\mathbf{r}_i(EB) = s\mathbf{r}_i(B); \quad \mathbf{r}_k(EB) = \mathbf{r}_k(B), \quad k \neq i \ ;$$

$$\mathbf{r}_i(E'B) = \mathbf{r}_i(B) + s\mathbf{r}_j(B); \quad \mathbf{r}_k(E'B) = \mathbf{r}_k(B), \quad k \neq i ;$$
$$\mathbf{r}_i(E''B) = \mathbf{r}_j(B), \mathbf{r}_j(E''B) = \mathbf{r}_i(B); \quad \mathbf{r}_k(E''B) = \mathbf{r}_k(B), \quad k \neq i, j .$$

But these are precisely the row vectors of the matrices $T(B)$, $T'(B)$, $T''(B)$ respectively. Therefore

$$EB = T(B), \quad E'B = T'(B), \quad E''B = T''(B)$$

In other words there is a one-to-one correspondence between the elementary row transformations and the left multiplications by the above special square matrices. This suggests that these matrices be called *elementary matrices of order n*.

Similarly applying to the identity matrix I the elementary column transformations $S : \mathbf{c}_i(I) \rightarrow s\mathbf{c}_i(I)$ with $s \neq 0$, $S' : \mathbf{c}_j(I) \rightarrow \mathbf{c}_j(I) + s\mathbf{c}_i(I)$, and $S'' : \mathbf{c}_i(I) \leftrightarrow \mathbf{c}_j(I)$ we obtain the same elementary matrices E, E' and E''. We notice that the roles of i and j in S' are reverses of those in T'. Moreover for any $m \times n$-matrix A, we have

$$AE = S(A), \quad AE' = S'(A), \quad AE'' = S''(A)$$

which establishes a one-to-one correspondence between the elementary column transformations and the right multiplications by the elementary matrices E, E', E''. We observe that as a consequence of the change of roles of i and j, the left multiplication by E' changes the i-th row while the right multiplication by the same matrix E' changes the j-th column. We shall have many applications of this correspondence between elementary transformations and multiplications by elementary matrices.

EXERCISES

1. In this section, we have shown that if A and B are $n \times n$ matrices and A is obtained from B by means of an elementary row operation, then $A = EB$, where E is the elementary matrix obtained from I by the same elementary row operation. In each of the following pairs of matrices A and B, find an elementary matrix E such that $A = EB$, and verify your result.

(a) $A = \begin{pmatrix} 5 & 1 \\ 4 & 8 \end{pmatrix}$, $\qquad B = \begin{pmatrix} 4 & 8 \\ 5 & 1 \end{pmatrix}$.

(b) $A = \begin{pmatrix} 1 & 0 \\ 2 & 3 \end{pmatrix}$, $\qquad B = \begin{pmatrix} 7 & 9 \\ 2 & 3 \end{pmatrix}$.

(c) $A = \begin{pmatrix} 1 & -2 & 1 \\ 3 & 0 & 2 \\ 2 & 4 & 6 \end{pmatrix}$, $\qquad B = \begin{pmatrix} 1 & -2 & 1 \\ 3 & 0 & 2 \\ 1 & 2 & 3 \end{pmatrix}$.

2. Again if A and B are $n \times n$ matrices and B is obtained from A by means of an elementary column operation, then $B = AE$, where E is the elementary matrix obtained from I by the same elementary column operation. In each of the following pairs of matrices A and B, find an elementary matrix E such that $B = AE$, and verify your result.

(a) $A = \begin{pmatrix} 1 & 3 \\ 2 & 4 \end{pmatrix}$, $\qquad B = \begin{pmatrix} 3 & 1 \\ 4 & 2 \end{pmatrix}$.

(b) $A = \begin{pmatrix} 1 & 4 \\ -1 & 5 \end{pmatrix}$, $\qquad B = \begin{pmatrix} 2 & 4 \\ -2 & 5 \end{pmatrix}$.

(c) $A = \begin{pmatrix} 1 & -2 & 1 \\ 3 & 0 & 2 \\ 2 & 4 & 6 \end{pmatrix}$, $\qquad B = \begin{pmatrix} 1 & -1 & 1 \\ 3 & 2 & 2 \\ 2 & 10 & 6 \end{pmatrix}$.

3. Convert each of the following sequences of row operations on a 2×2 matrix B into a product of elementary matrices and B.
   (a) $r_1 \leftrightarrow r_2$, $r_1 \rightarrow r_1 + 2r_2$.
   (b) $r_2 \rightarrow r_2 - r_1$, $r_3 \rightarrow r_3 + 2r_1$, $r_1 \rightarrow r_1 - 3r_2$.

4. For any square matrix A of order n, A is called *symmetric* if $A^t = A$, and A is called *anti-symmetric* if $A^t = -A$.
   (a) Show that $A + A^t$ is symmetric for any square matrix A of order n.
   (b) Show that $A - A^t$ is anti-symmetric for any square matrix A of order n.
   (c) Hence deduce that every square matrix is the sum of a symmetric and an anti-symmetric matrix.

5. Let A be a square matrix of order n. The *trace* of A, denoted $\mathrm{tr}(A)$, is the sum of the diagonal elements, that is, $a_{11} + a_{22} + \cdots + a_{nn}$.
   (a) If A and B are square matrices of order n, show that
   (i) $\mathrm{tr}(A + B) = \mathrm{tr}(A) + \mathrm{tr}(B)$;

(ii) $\text{tr}(kA) = k\,\text{tr}(A)$ for any real number k;

(iii) $\text{tr}(A^t) = \text{tr}(A)$;

(iv) $\text{tr}(AB) = \text{tr}(BA)$;

(v) $\text{tr}(AA^t) =$ sum of the squares of all entries of A.

(b) Find 2×2 matrices A and B such that $\text{tr}(AB) \neq \text{tr}(A) \cdot \text{tr}(B)$.

6. A square matrix A is called an *idempotent* if $A^2 = A$. Prove the following statements.

(a) $\begin{pmatrix} 1 & 1 \\ 0 & 0 \end{pmatrix}$, $\begin{pmatrix} 1 & 0 \\ 1 & 0 \end{pmatrix}$ and $\begin{pmatrix} 0 & 0 \\ -1 & 1 \end{pmatrix}$ are idempotents.

(b) If A is an idempotent, then $I - A$ is an idempotent.

(c) If A is an idempotent, so is A^t.

(d) If A is an idempotent, then $B = A + PA - APA$ is also an idempotent for any square matrix P of the same order.

(e) If A is of order $n \times m$ and B is $m \times n$, and if $AB = I_n$, identity of order n, then BA is an idempotent.

6.5 Invertible matrix

A square matrix A of order n is called an *invertible matrix* or a *non-singular matrix* if there exists a square matrix A' of the same order such that

$$A'A = AA' = I \, .$$

In this case the matrix A' is uniquely determined by the above condition. Indeed if A' and A'' are such that

$$A'A = AA' = I \quad \text{and} \quad A''A = AA'' = I$$

then $A' = A'I = A'(AA'') = (A'A)A'' = IA'' = A''$. Therefore we may call A' *the inverse* of A and denote it by A^{-1}. Because of $AA^{-1} = A^{-1}A = I$, A^{-1} is also invertible and $(A^{-1})^{-1} = A$.

Obviously the identity matrix I is invertible with $I^{-1} = I$ and so are all elementary matrices. The matrices T and S of coordinate transformations of the last section are also invertible and inverses of each other. We take note that an invertible matrix is a special kind of square matrix; therefore invertibility is not a property of non-square matrices.

If A and B are invertible matrices of the same order, then the product AB and the transposes A^t and B^t are also invertible. Indeed

$$(B^{-1}A^{-1})(AB) = B^{-1}(A^{-1}A)B = I, \ (AB)(B^{-1}A^{-1}) = A(BB^{-1})A^{-1} = I$$
$$(A^{-1})^t A^t = (AA^{-1})^t = I^t = I, \ A^t(A^{-1})^t = (A^{-1}A)^t = I .$$

6.5.1 THEOREM *Let A and B be invertible matrices, and r a non-zero real number. Then the following statements hold.*

    (1) $(A^{-1})^{-1} = A$.
    (2) $(A^t)^{-1} = (A^{-1})^t$.
    (3) $(AB)^{-1} = B^{-1}A^{-1}$.
    (4) $(rA)^{-1} = \frac{1}{r}A^{-1}$.

In general, given a square matrix A, is there an effective method to find out if A is invertible? Furthermore if A is found to be invertible, is there an easy way of evaluating its inverse? The following theorem which gives a necessary and sufficient condition for invertibility in terms of rank is a preliminary step toward a satisfactory answer to the first question.

6.5.2 THEOREM *A square matrix A of order n is invertible if and only if the rank of A is n.*

PROOF Suppose that A is invertible. Let $B = A^{-1}$. Then it follows from $I = BA$ that

$$\delta_{ij} = \sum_{k=1}^{n} b_{ik} a_{kj} \quad \text{for} \quad i, j = 1, 2, \cdots, n .$$

Therefore for all $i = 1, 2, \cdots, n$, the unit coordinate vector

$$\begin{aligned}
\mathbf{e}_i &= [\delta_{i1}, \delta_{i2}, \cdots, \delta_{in}] \\
&= [\sum_k b_{ik} a_{k1}, \sum_k b_{ik} a_{k2}, \cdots, \sum_k b_{ik} a_{kn}] \\
&= \sum_k [b_{ik} a_{k1}, b_{ik} a_{k2}, \cdots, b_{ik} a_{kn}] \\
&= \sum_k b_{ik} [a_{k1}, a_{k2}, \cdots, a_{kn}] \\
&= b_{i1} \mathbf{r}_1(A) + b_{i2} \mathbf{r}_2(A) + \cdots + b_{in} \mathbf{r}_n(A)
\end{aligned}$$

of $\mathbf{R}^n$ is a linear combination of the row vectors $\mathbf{r}_1(A), \mathbf{r}_2(A), \cdots, \mathbf{r}_n(A)$ of A and hence a vector of the subspace $\langle \mathbf{r}_1(A), \mathbf{r}_2(A), \cdots, \mathbf{r}_n(A) \rangle$ generated by the row vectors. From this it follows that rank $A = n$ because $n \geq$ rank $A = \dim \langle \mathbf{r}_1(A), \mathbf{r}_2(A), \cdots, \mathbf{r}_n(A) \rangle \geq \dim \langle \mathbf{e}_1, \mathbf{e}_2, \cdots, \mathbf{e}_n \rangle = n$.

Conversely suppose that the rank of A is n. Then the subspace $\langle \mathbf{r}_1(A), \mathbf{r}_2(A), \cdots, \mathbf{r}_n(A) \rangle$ of $\mathbf{R}^n$ generated by the row vectors of A has dimension n and contains all unit coordinate vectors of $\mathbf{R}^n$. Therefore for $i = 1, 2, \cdots, n$,

$$\mathbf{e}_i = b_{i1} \mathbf{r}_1(A) + b_{i2} \mathbf{r}_2(A) + \cdots + b_{in} \mathbf{r}_n(A)$$

for certain n^2 scalars b_{ij} $i, j = 1, 2, \cdots, n$. Retracing the steps of the previous calculation backward, we conclude $BA = I$. Similarly working with the column vectors $\mathbf{c}_j(A)$ of A or with the row vectors $\mathbf{r}_i(A^t)$ of A^t, we find a square matrix C of order n such that $AC = I$. Finally $C = IC = (BA)C = B(AC) = BI = B$, proving that B is the inverse of A. Now the proof of the theorem is complete.

We are now in a good position to answer affirmatively the first question. Clearly we shall make use of elementary row transformations to evaluate the rank of the given square A of order n. Using column transformations would, of course, lead to the same result. According to the procedure of Section 5.3 and 5.4, the matrix is invertible if and only if by a series of elementary row transformations of the first and the second types A can be transformed into an $n \times n$-matrix H whose n row vectors are linearly independent and whose n column vectors are exactly the n distinct unit coordinate vectors $\mathbf{e}_1, \mathbf{e}_2, \cdots, \mathbf{e}_n$ of $\mathbf{R}^n$. By then the vectors $\mathbf{e}_1, \mathbf{e}_2, \cdots, \mathbf{e}_n$ also appear in some order as the n row vectors of H. For example

$$\begin{pmatrix} 1 & 0 & 1 \\ -1 & 3 & -11 \\ 2 & 1 & -1 \end{pmatrix} \rightarrow \begin{pmatrix} 1 & 0 & 1 \\ 0 & 3 & -10 \\ 0 & 1 & -3 \end{pmatrix} \rightarrow \begin{pmatrix} 1 & 0 & 1 \\ 0 & 0 & -1 \\ 0 & 1 & -3 \end{pmatrix}$$

$$\rightarrow \begin{pmatrix} 1 & 0 & 1 \\ 0 & 0 & 1 \\ 0 & 1 & -3 \end{pmatrix} \rightarrow \begin{pmatrix} 1 & 0 & 0 \\ 0 & 0 & 1 \\ 0 & 1 & 0 \end{pmatrix}.$$

We can now rearrange the rows of H by a series of row transformations of the third type and transform H into the identity matrix I of order n. We have therefore prove the following Corollary.

6.5.3 COROLLARY *A square matrix A of order n is invertible if and only if it can be transformed into the identity matrix I of order n by a series of elementary row (column) transformations.*

Let us now consider the second question of finding the inverse A^{-1} of a non-singular matrix A. It turns out that to obtain A^{-1} we may actually apply to I the same elementary row transformations that bring A into I. Suppose the square matrix A has rank n. Then by 6.5.3, we can actually find p elementary row transformations $T_1, T_2, \cdots, T_p$ that transform A into I:

$$T_p(T_{p-1}(\cdots T_1(A)\cdots)) = I .$$

But for each row transformation T_i there is a unique elementary matrix $T_i(I) = E_i$ so that $E_i B = T_i(B)$ for any square matrix B of order n. Therefore it follows from $T_p(T_{p-1}(\cdots T_1(A)\cdots)) = I$ that

$$I = E_p E_{p-1} \cdots E_1 A .$$

Multiplying both sides of the above equation by the inverse A^{-1} from the right, we obtain

$$A^{-1} = E_p E_{p-1} \cdots E_1 I .$$

Converting left multiplications by elementary matrix back as elementary row transformations, we obtain

$$A^{-1} = T_p(T_{p-1}(\cdots T_1(I)\cdots)) .$$

6.5.4 THEOREM *Elementary row (column) transformations $T_1, T_2, \cdots, T_p$ that transform an invertible matrix A into I, will take I to A^{-1}.*

In practice we can actually carry out the two procedures, one on A and the other on I, at the same time.

6.5.5 EXAMPLE

$$A = \begin{pmatrix} 1 & 0 & 1 \\ -1 & 3 & -11 \\ 2 & 1 & -1 \end{pmatrix} \rightarrow \begin{pmatrix} 1 & 0 & 1 \\ 0 & 3 & -10 \\ 0 & 1 & -3 \end{pmatrix} \rightarrow \begin{pmatrix} 1 & 0 & 1 \\ 0 & 0 & -1 \\ 0 & 1 & -3 \end{pmatrix} \rightarrow$$

$$\begin{pmatrix} 1 & 0 & 1 \\ 0 & 0 & 1 \\ 0 & 1 & -3 \end{pmatrix} \rightarrow \begin{pmatrix} 1 & 0 & 0 \\ 0 & 0 & 1 \\ 0 & 1 & 0 \end{pmatrix} \rightarrow \begin{pmatrix} 1 & 0 & 0 \\ 0 & 1 & 0 \\ 0 & 0 & 1 \end{pmatrix} = I$$

$$I = \begin{pmatrix} 1 & 0 & 0 \\ 0 & 1 & 0 \\ 0 & 0 & 1 \end{pmatrix} \rightarrow \begin{pmatrix} 1 & 0 & 0 \\ 1 & 1 & 0 \\ -2 & 0 & 1 \end{pmatrix} \rightarrow \begin{pmatrix} 1 & 0 & 0 \\ 7 & 1 & -3 \\ -2 & 0 & 1 \end{pmatrix} \rightarrow$$

$$\begin{pmatrix} 1 & 0 & 0 \\ -7 & -1 & 3 \\ -2 & 0 & 1 \end{pmatrix} \rightarrow \begin{pmatrix} 8 & 1 & -3 \\ -7 & -1 & 3 \\ -23 & -3 & 10 \end{pmatrix} \rightarrow \begin{pmatrix} 8 & 1 & -3 \\ -23 & -3 & 10 \\ -7 & -1 & 3 \end{pmatrix} = A^{-1}$$

The elementary row transformations on both A and I are as follows:

$$(r_2 \rightarrow r_2 + r_1) \; ; \; (r_3 \rightarrow r_3 - 2r_1) \; ; \; (r_2 \rightarrow r_2 - 3r_3) \; ; \; (r_2 \rightarrow -r_2) \; ;$$
$$(r_1 \rightarrow r_1 - r_2) \; ; \; (r_3 \rightarrow r_3 + 3r_2) \; ; \; (r_2 \leftrightarrow r_3) \; .$$

Indeed we have

$$AA^{-1} = \begin{pmatrix} 1 & 0 & 1 \\ -1 & 3 & -11 \\ 2 & 1 & -1 \end{pmatrix} \begin{pmatrix} 8 & 1 & -3 \\ -23 & -3 & 10 \\ -7 & -1 & 3 \end{pmatrix} = \begin{pmatrix} 1 & 0 & 0 \\ 0 & 1 & 0 \\ 0 & 0 & 1 \end{pmatrix} = I$$

$$A^{-1}A = \begin{pmatrix} 8 & 1 & -3 \\ -23 & -3 & 10 \\ -7 & -1 & 3 \end{pmatrix} \begin{pmatrix} 1 & 0 & 1 \\ -1 & 3 & -11 \\ 2 & 1 & -1 \end{pmatrix} = \begin{pmatrix} 1 & 0 & 0 \\ 0 & 1 & 0 \\ 0 & 0 & 1 \end{pmatrix} = I$$

Readers may find it convenient to put the two matrices A and I side by side and carry out the row transformations on the resulting $n \times 2n$-matrix:

$$\left(\begin{array}{ccc|ccc} 1 & 0 & 1 & 1 & 0 & 1 \\ -1 & 3 & 11 & 0 & 1 & 0 \\ 2 & 1 & -1 & 0 & 0 & 1 \end{array} \right) \rightarrow \left(\begin{array}{ccc|ccc} 1 & 0 & 1 & 1 & 0 & 1 \\ 0 & 3 & -10 & 1 & 1 & 0 \\ 0 & 1 & -3 & -2 & 0 & 1 \end{array} \right)$$

$$\rightarrow \left(\begin{array}{ccc|ccc} 1 & 0 & 1 & 1 & 0 & 1 \\ 0 & 0 & -1 & 7 & 1 & -3 \\ 0 & 1 & -3 & -2 & 0 & 1 \end{array} \right) \rightarrow \left(\begin{array}{ccc|ccc} 1 & 0 & 1 & 1 & 0 & 1 \\ 0 & 0 & 1 & -7 & -1 & 3 \\ 0 & 1 & -3 & -2 & 0 & 1 \end{array} \right)$$

$$\rightarrow \left(\begin{array}{ccc|ccc} 1 & 0 & 0 & 8 & 1 & -3 \\ 0 & 0 & 1 & -7 & -1 & 3 \\ 0 & 1 & 0 & -23 & -3 & 10 \end{array} \right) \rightarrow \left(\begin{array}{ccc|ccc} 1 & 0 & 0 & 8 & 1 & -3 \\ 0 & 1 & 0 & -23 & -3 & 10 \\ 0 & 0 & 1 & -7 & -1 & 3 \end{array} \right)$$

EXERCISES

1. Find the inverses of the following matrices.

(a) $\begin{pmatrix} 1 & 2 \\ 1 & 3 \end{pmatrix}$;   (b) $\begin{pmatrix} -2 & 3 \\ 3 & -5 \end{pmatrix}$;   (c) $\begin{pmatrix} 2 & 3 \\ 5 & 4 \end{pmatrix}$.

2. Find the inverses of the following matrices if the matrix is invertible.

(a) $\begin{pmatrix} 1 & 2 & 3 \\ 2 & 0 & 1 \\ -1 & 1 & 0 \end{pmatrix}$;   (b) $\begin{pmatrix} 3 & 1 & 5 \\ 2 & 4 & 1 \\ -4 & 2 & -9 \end{pmatrix}$;   (c) $\begin{pmatrix} 2 & 2 & 3 \\ 1 & -1 & 0 \\ -1 & 2 & 1 \end{pmatrix}$.

3. Consider the matrix $A = \begin{pmatrix} 2 & 5 \\ 0 & 1 \end{pmatrix}$.

  (a) Find elementary matrices E_1 and E_2 such that $E_2 E_1 A = I$.

  (b) Hence express A^{-1} as a product of two elementary matrices.

4. Suppose $A = \begin{pmatrix} a & c \\ b & d \end{pmatrix}$ is invertible with $A^{-1} = \begin{pmatrix} e & g \\ f & h \end{pmatrix}$. Let $B = \begin{pmatrix} b & d \\ a & c \end{pmatrix}$, that is, B is obtained from A by a row interchange. Show that $B^{-1} = \begin{pmatrix} g & e \\ h & f \end{pmatrix}$, that is, B^{-1} is obtained by a column interchange from A^{-1}.

5. Show that the matrix

$$A = \begin{pmatrix} 1 & 0 & 0 \\ 0 & \cos\theta & \sin\theta \\ 0 & -\sin\theta & \cos\theta \end{pmatrix}$$

  is invertible for all values of θ and find A^{-1}.

6. Let A be an invertible matrix and B a square matrix of the same order. Show that for all positive integers n,

$$(ABA^{-1})^n = AB^n A^{-1} .$$

7. Let $A_1, A_2, \cdots, A_n$ be any n invertible matrices of the same order. Show that

$$(A_1 \cdots A_n)^{-1} = A_n^{-1} \cdots A_1^{-1} .$$

8. Find the inverses of the following matrices, where a, b, c, d and e are all non-zero real numbers.

(a) $\begin{pmatrix} a & 0 & 0 & 0 \\ 0 & b & 0 & 0 \\ 0 & 0 & c & 0 \\ 0 & 0 & 0 & d \end{pmatrix}$ (b) $\begin{pmatrix} 0 & 0 & 0 & a \\ 0 & 0 & b & 0 \\ 0 & c & 0 & 0 \\ d & 0 & 0 & 0 \end{pmatrix}$ (c) $\begin{pmatrix} e & 0 & 0 & 0 \\ 1 & e & 0 & 0 \\ 0 & 1 & e & 0 \\ 0 & 0 & 1 & e \end{pmatrix}$.

9. A square matrix A satisfies

$$A^2 - 2A + I = 0 .$$

Show that A is invertible and find A^{-1}. (Hint: Consider the definition of invertible matrix.)

10. By a similar argument employed in Exercise 9, show that if square matrix A satisfying $A^3 + 2A^2 - 3A + 4I = 0$, then A is invertible and find A^{-1}. Can it be extended to a general result?

11. Let A be a square matrix for which there is exactly one matrix B such that $AB = I$. By considering $BA + B - I$, show that A is invertible.

12. Let A and B be invertible matrices of order n.
    (a) Find examples to show that $A + B$ may not be invertible.
    (b) If $A + B$ is invertible, show that

$$[A(A + B)^{-1}B]^{-1} = A^{-1} + B^{-1} .$$

13. Given two square matrices A and B such that $AB = 0$. Prove the following statements.
    (a) If one of A and B is invertible, the other is zero.
    (b) It is impossible for both A and B to have inverses.
    (c) $(BA)^2 = 0$.

14. Let A, B and C be square matrices of order n. Prove the following statements.
    (a) If A and AB are both invertible, then B is invertible.
    (b) If AB and BA are both invertible, then A and B are both invertible.
    (c) If A, C and ABC are all invertible, then B is invertible.

15. Let A and B be square matrices of order n.
    (a) If $AB = BA$, show that $(AB)^2 = A^2 B^2$.
    (b) If A and B are invertible and $(AB)^2 = A^2 B^2$, show that $AB = BA$.
    (c) If $A = \begin{pmatrix} 0 & 0 \\ 0 & 1 \end{pmatrix}$ and $B = \begin{pmatrix} 1 & 1 \\ 0 & 0 \end{pmatrix}$, show that $(AB)^2 = A^2 B^2$ but $AB \neq BA$.

16. Prove that every non-singular square matrix is a product of elementary matrices.

6.6 Determinant of order 2

We recall that in Theorem 1.3.2 we have found $a_1 b_2 - a_2 b_1 = 0$ to be a necessary and sufficient condition for two given vectors $\mathbf{a} = [a_1, a_2]$, $\mathbf{b} = [b_1, b_2]$ of $\mathbf{R}^2$ to be linearly dependent. The expression $a_1 b_2 - a_2 b_1$ in the components of $\mathbf{a}$, $\mathbf{b}$ is called the *determinant*

$$\begin{vmatrix} a_1 & b_1 \\ a_2 & b_2 \end{vmatrix} \quad \text{of the matrix} \quad A = \begin{pmatrix} a_1 & b_1 \\ a_2 & b_2 \end{pmatrix}$$

whose column vectors are $\mathbf{a}$ and $\mathbf{b}$. This determinant of the second order is also denoted by $\det A$, $|A|$ or $|\mathbf{a}\ \mathbf{b}|$, indicating that the enclosed vectors are column vectors. The following properties of the determinant are easily verified.

6.6.1 THEOREM *Let A be a square matrix of order 2 whose column vectors are $\mathbf{a}$ and $\mathbf{b}$. Then the determinant $\det A = |\mathbf{a}\ \mathbf{b}|$ of the second order has the following properties:*

(1) $\det A^t = \det A$;

(2) $\det A$ *is linear in each column (row) of A;*

(3) $\det A$ *changes signs if the columns (rows) of A interchange positions;*

(4) $|\det A|$ *is the area of the parallelogram of which the column (row) vectors are coterminous edges;*

(5) $|\mathbf{e}_1\ \mathbf{e}_2| = \det I = 1$, *if $\mathbf{e}_1$ and $\mathbf{e}_2$ are the unit coordinate vectors of $\mathbf{R}^2$;*

(6) $|\mathbf{a}\ \mathbf{b}| = 0$ *if and only if $\mathbf{a}$ and $\mathbf{b}$ are linearly dependent.*

By direct computation we can also verify that the product of the determinants of two matrices equals the determinant of the product of the two matrices.

6.6.2 THEOREM *Let A and B be two square matrices of order 2. Then $\det AB = (\det A)(\det B)$.*

EXERCISES

1. Evaluate the following determinant.

(a) $\begin{vmatrix} 1 & 0 \\ 0 & 1 \end{vmatrix}$; 　　　(b) $\begin{vmatrix} 1 & 2 \\ -1 & 4 \end{vmatrix}$; 　　　(c) $\begin{vmatrix} 6 & 3 \\ 4 & 2 \end{vmatrix}$.

2. By putting $A = \begin{pmatrix} a_1 & a_3 \\ a_2 & a_4 \end{pmatrix}$ and $B = \begin{pmatrix} b_1 & b_3 \\ b_2 & b_4 \end{pmatrix}$, prove Theorem 6.6.2.

3. For each of the following pairs of vectors, write down a determinant which represents the area of the parallelogram of which the vectors are coterminous edges. Hence find the area.

   (a) $[1, 6]$, $[-2, 3]$;

   (b) $[1, -15]$, $[4, 11]$.

4. Let $A = \begin{pmatrix} a_1 & b_1 \\ a_2 & b_2 \end{pmatrix}$.

   (a) Write down A^t and show that $\det A^t = \det A$.

   (b) Prove that $\det A$ changes signs if the columns (or rows) of A are interchanged.

   (c) Show that

   $$|r\mathbf{a} + s\mathbf{c} \ \mathbf{b}| = r|\mathbf{a} \ \mathbf{b}| + s|\mathbf{c} \ \mathbf{b}|$$

   and hence show that $\det A$ is linear in each column (row).

5. Suppose that $\begin{vmatrix} a & b \\ c & d \end{vmatrix} = 7$. By using Exercise 4, evaluate the following determinants by inspection.

   (a) $\begin{vmatrix} 3a & 3b \\ c & d \end{vmatrix}$;　　　(b) $\begin{vmatrix} c & d \\ a & b \end{vmatrix}$;　　　(c) $\begin{vmatrix} a + 2c & b + 2d \\ c & d \end{vmatrix}$.

6. Consider examples of 2×2 matrices. Show that the following relations may not be true.

   (a) $\det A = 0$ if and only if $A = 0$.

   (b) $\det(A + B) \leq \det A + \det B$.

6.7 Determinant of order 3

The *determinant* of a square matrix A of order 3 whose column vectors are $\mathbf{a}$, $\mathbf{b}$, $\mathbf{c}$ is defined as the real number

$$\det A = |A| = |\mathbf{a} \ \mathbf{b} \ \mathbf{c}| = \begin{vmatrix} a_1 & b_1 & c_1 \\ a_2 & b_2 & c_2 \\ a_3 & b_3 & c_3 \end{vmatrix}$$

$$= a_1 b_2 c_3 + a_2 b_3 c_1 + a_3 b_1 c_2 - a_1 b_3 c_2 - a_2 b_1 c_3 - a_3 b_2 c_1 .$$

Clearly by definition we have $|\mathbf{e}_1 \ \mathbf{e}_2 \ \mathbf{e}_3| = \det I = 1$ for the unit coordinate vectors $\mathbf{e}_1, \mathbf{e}_2, \mathbf{e}_3$ of $\mathbf{R}^3$.

The determinant $|\mathbf{a}\ \mathbf{b}|$ of second order can be easily remembered by means of the diagram

$$
\begin{array}{cc}
a_1 & b_1 \\
a_2 & b_2
\end{array}
$$

in which the broken lines and the signs show how the terms of the determinant are obtained. The diagrams

$$
\begin{array}{ccc}
a_1 & b_1 & c_1 \\
a_2 & b_2 & c_2 \\
a_3 & b_3 & c_3
\end{array}
\qquad\qquad
\begin{array}{ccc}
a_1 & b_1 & c_1 \\
a_2 & b_2 & c_2 \\
a_3 & b_3 & c_3
\end{array}
$$

fulfill the same purpose for the determinant $|\mathbf{a}\ \mathbf{b}\ \mathbf{c}|$ of the third order.

Using this scheme of products and signs on the transpose A^t of A we find that $\det A^t = \det A$.

Arranging the terms of the determinant into

$$
|\mathbf{a}\ \mathbf{b}\ \mathbf{c}| = a_1(b_2c_3 - b_3c_2) + a_2(b_3c_1 - b_1c_3) + a_3(b_1c_2 - b_2c_1)
$$

we recognize that

$$
|\mathbf{a}\ \mathbf{b}\ \mathbf{c}| = \mathbf{a} \cdot (\mathbf{b} \times \mathbf{c})
$$

expressing the determinant as a "mix-product" of the column vectors. Therefore it follows from Theorem 2.7.7 that $|\det A|$ is indeed the volume of the parallelepiped with $\mathbf{a}$, $\mathbf{b}$, $\mathbf{c}$ as its coterminous edges.

Because both the dot product and the cross product are linear in each factor, it follows that the determinant is linear in each column vector. For example

$$
|\mathbf{a}\ s\mathbf{b} + s'\mathbf{b}'\ \mathbf{c}| = s|\mathbf{a}\ \mathbf{b}\ \mathbf{c}| + s'|\mathbf{a}\ \mathbf{b}'\ \mathbf{c}| \ .
$$

Again by Theorem 2.7.7 we get:

$$
\mathbf{a} \cdot (\mathbf{b} \times \mathbf{c}) = \mathbf{b} \cdot (\mathbf{c} \times \mathbf{a}) = \mathbf{c} \cdot (\mathbf{a} \times \mathbf{b})
$$
$$
= -\mathbf{a} \cdot (\mathbf{c} \times \mathbf{b}) = -\mathbf{b} \cdot (\mathbf{a} \times \mathbf{c}) = -\mathbf{c} \cdot (\mathbf{b} \times \mathbf{a})
$$

from which we see the effect of a permutation on the column vectors of $|\mathbf{a}\ \mathbf{b}\ \mathbf{c}|$ as follows.

$$|\mathbf{a} \ \mathbf{b} \ \mathbf{c}| = |\mathbf{b} \ \mathbf{c} \ \mathbf{a}| = |\mathbf{c} \ \mathbf{a} \ \mathbf{b}|$$
$$= -|\mathbf{a} \ \mathbf{c} \ \mathbf{b}| = -|\mathbf{c} \ \mathbf{b} \ \mathbf{a}| = -|\mathbf{b} \ \mathbf{a} \ \mathbf{c}| \ .$$

In other words the determinant changes signs if any two columns interchange positions.

We have now successfully prove the first five statements of the following theorem which is the order 3 counterpart of Theorem 6.6.1.

6.7.1 THEOREM *Let A be a square matrix of order 3 whose column vectors are* **a**, **b**, **c**. *Then the determinant* $\det A = |\mathbf{a} \ \mathbf{b} \ \mathbf{c}|$ *of the third order has the following properties.*

(1) $\det A^t = \det A$;

(2) $\det A$ *is linear in each column (row) vector of A;*

(3) $\det A$ *changes signs if any two column (row) vectors interchange positions;*

(4) $|\det A|$ *is the volume of the parallelepiped of which the column (row) vectors of A are coterminous edges;*

(5) $|\mathbf{e}_1 \ \mathbf{e}_2 \ \mathbf{e}_3| = 1$ *if* $\mathbf{e}_1$, $\mathbf{e}_2$, $\mathbf{e}_3$ *are the unit coordinate vectors of* $\mathbf{R}^3$;

(6) $|\mathbf{a} \, \mathbf{b} \, \mathbf{c}| = 0$ *if and only if* **a**, **b**, **c** *are linearly dependent.*

PROOF OF (6) Suppose that the vectors are linearly dependent. Then one of them is a linear combination of the others, say $\mathbf{c} = r\mathbf{a} + s\mathbf{b}$. Then

$$|\mathbf{a} \ \mathbf{b} \ \mathbf{c}| = (\mathbf{a} \times \mathbf{b}) \cdot \mathbf{c} = (\mathbf{a} \times \mathbf{b}) \cdot (r\mathbf{a} + s\mathbf{b})$$
$$= r(\mathbf{a} \times \mathbf{b}) \cdot \mathbf{a} + s(\mathbf{a} \times \mathbf{b}) \cdot \mathbf{b} = 0$$

because $\mathbf{a} \times \mathbf{b}$ is orthogonal to both **a** and **b**.

Conversely suppose that $|\mathbf{a} \ \mathbf{b} \ \mathbf{c}| = 0$. If any two among **a**, **b**, **c** are linearly dependent, then all three vectors are linearly dependent. Assume that **a** and **b** are linearly independent. Then it follows from $|\mathbf{a} \ \mathbf{b} \ \mathbf{c}| = (\mathbf{a} \times \mathbf{b}) \cdot \mathbf{c} = 0$ that **c** is orthogonal to $\mathbf{a} \times \mathbf{b}$. But this also means that **c** is on the plane of the vectors **a** and **b**. Therefore **a**, **b**, **c** are linearly dependent.

Before we study the determinant of a product of two matrices, we write down some easy consequences of 6.7.1(6).

6.7.2 COROLLARY *Let A be a square matrix of order 3. Then the following statements hold.*

(a) $\det A \neq 0$ if and only if A has rank 3.

(b) $\det A \neq 0$ if and only if A is invertible.

(c) $\det A = 0$ if A has two identical rows or two identical column.

Let us now turn to the product formula $\det AB = \det A \det B$ of determinants. For two square matrices

$$A = \begin{pmatrix} a_{11} & a_{12} & a_{13} \\ a_{21} & a_{22} & a_{23} \\ a_{31} & a_{32} & a_{33} \end{pmatrix} \quad \text{and} \quad B = \begin{pmatrix} b_{11} & b_{12} & b_{13} \\ b_{21} & b_{22} & b_{23} \\ b_{31} & b_{32} & b_{33} \end{pmatrix}$$

both of order $n = 3$, each element of the product $AB = C$ is a sum of $n = 3$ terms, e.g. $c_{12} = a_{11}b_{12} + a_{12}b_{22} + a_{13}b_{32}$. Now $\det AB$ has $n! = 6$ summands, each being a product of $n = 3$ elements of AB; hence it is a sum of $(n!)(n^n) = 162$ terms. On the other side of the product formula, we have $\det A \det B$ which is the product of two sums of $n! = 6$ terms each; we have a sum of $(n!)^2 = 36$ terms. If we were to prove $\det AB = \det A \det B$ by brute force, we would have to examine $(n!)^2 + (n!)(n^n) = 198$ terms each of which is a product of $2n = 6$ elements among the a_{ij} and b_{ij}. This is certainly something that we ought to avoid.

As a viable atternative, we consider special cases of the product formula where one of the factors is a special matrix and try to reduce the general case to the special cases by elementary transformations. We have already seen in 6.7.1(3) that an elementary row (column) transformation of the third type only changes the sign of the determinant. In general, we have the following easy theorem.

6.7.3 THEOREM *Let A be a square matrix of order 3 and T an elementary column (row) transformation. Then*

(i) $\det T(A) = s \det A$ *if T is of the first type that replaces one column (row) vector of A by its scalar multiple with scalar $s \neq 0$;*

(ii) $\det T(A) = \det A$ *if T is of the second type;*

(iii) $\det T(A) = -\det A$ *if T is of the third type.*

PROOF Let $\det A = |\mathbf{a\,b\,c}|$. For statement (i), suppose that T replaces $\mathbf{a}$ by $s\mathbf{a}$. Then $\det T(A) = |s\mathbf{a\,b\,c}| = s|\mathbf{a\,b\,c}| = s \det A$ by 6.7.1(2). Similarly (i) holds for all column transformations of the first type. For statement (ii), suppose T replaces $\mathbf{b}$ by $\mathbf{b} + s\mathbf{c}$. Then $\det T(A) = |\mathbf{a\,b} + s\mathbf{c\,c}| =$

$|\mathbf{a}\,\mathbf{b}\,\mathbf{c}| + s|\mathbf{a}\,\mathbf{c}\,\mathbf{c}| = |\mathbf{a}\,\mathbf{b}\,\mathbf{c}| = \det A$ by 6.7.1 and 6.7.2. Similarly (ii) holds for all column transformations of the second type. As (iii) has been proved, the theorem holds for all column transformations T. But clearly it also holds for all row transformations since $\det A = \det A^t$. The proof is now complete.

Applying elementary transformations to the identity I which has $\det I = 1$ we obtain the following corollary.

6.7.4 COROLLARY *Let E be an elementary matrix of order 3. Then*
  (i) $\det E = s$ *if E is of the first type with one element on the diagonal equal to $s \neq 0$;*
  (ii) $\det E = 1$ *if E is of the second type;*
  (iii) $\det E = -1$ *if E is of the third type.*

Combining 6.7.3 and 6.7.4, and by the correspondence between elementary transformations and multiplications by elementary matrix: $T(A) = EA$ for row transformations T or $T(A) = AE$ for column transformations T, we see that for any square matrix A and any elementary matrix E, the product formula holds:

$$\det EA = \det E \det A = \det AE \ .$$

But the product formula also holds for other special matrices. A matrix D is called a *diagonal matrix* if all its elements off the diagonal are zero: $d_{ij} = 0$ for all $i \neq j$. Clearly the product formula holds if one of the factors is a diagonal matrix. We shall now make use of the following special case 6.7.5 of the product formula to prove the general case in 6.7.6.

6.7.5 THEOREM *If X is an elementary matrix or a diagonal matrix, then*

$$\det XA = \det X \det A = \det AX$$

for all square matrices A of order 3.

6.7.6 THEOREM *Let A and B be two square matrices of order 3. Then* $\det AB = \det A \det B$.

PROOF We have seen in Sections 5.4 and 5.5 that we can transform the matrix A by elementary row and column transformations of the first and the second types into a matrix H in which all non-zero rows are distinct unit coordinate vectors. Now we extend the procedure by elementary transformations of the third type to transform H into a diagonal matrix J. Converting each row transformation into a left multiplication and each column transformation into a right multiplication by elementary matrix we obtain elementary matrices $E_1, E_2, \cdots, E_p$ and $F_1, F_2, \cdots, F_q$ and a diagonal matrix J so that

$$E_p \cdots E_2 E_1 A F_1 F_2 \cdots F_q = J \ .$$

Multiplying the inverses of the elementary matrices one after the other to both sides of the equations, we get

$$A = E_1^{-1} E_2^{-1} \cdots E_p^{-1} J F_q^{-1} \cdots F_2^{-1} F_1^{-1} \ .$$

But the inverse of an elementary matrix is itself an elementary; therefore A and similarly B are both products of a diagonal matrix and elementary matrices. But then AB as a product of two such products is also a product of diagonal matrices and elementary matrices. Now the theorem follows from repeated applications of the special product formula 6.7.5. The proof is now complete.

The method of elementary transformation not only saves us from pages of tiresome computation of the 198 terms for the present case of order 3 determinants, it is also a method that can be used for determinants of higher order. Moreover it also affords us a convenient way of evaluating determinants.

6.7.7 EXAMPLE

$$\begin{vmatrix} -1 & 2 & -2 \\ 8 & -9 & 16 \\ 4 & 5 & 8 \end{vmatrix} = \begin{vmatrix} -1 & 2 & -2 \\ 0 & 7 & 0 \\ 0 & 13 & 0 \end{vmatrix} = 0 \ .$$

6.7.8 EXAMPLE

$$\begin{vmatrix} 3 & 0 & 2 \\ -1 & 3 & -11 \\ 2 & 1 & -1 \end{vmatrix} = \begin{vmatrix} 0 & 9 & -31 \\ -1 & 3 & -11 \\ 0 & 7 & -23 \end{vmatrix} = \begin{vmatrix} 0 & 2 & -8 \\ -1 & 3 & -11 \\ 0 & 7 & -23 \end{vmatrix}$$

$$= 2 \begin{vmatrix} 0 & 1 & -4 \\ -1 & 3 & -11 \\ 0 & 7 & -23 \end{vmatrix} = 2 \begin{vmatrix} 0 & 1 & -4 \\ -1 & 0 & 1 \\ 0 & 0 & 5 \end{vmatrix} = -2 \begin{vmatrix} -1 & 0 & 1 \\ 0 & 1 & -4 \\ 0 & 0 & 5 \end{vmatrix}$$

$$= (-2)(-1)(1)(5) = 10 \ .$$

Here we observe that if a matrix is in *triangular form* [in which all elements below (or all elements above) the diagonal are zero], then its determinant is simply the product of its diagonal elements.

Before we conclude our discussion on determinants of order 3 and move on to determinants of higher order in the next section, we outline two new approaches to our subject. Both approaches will lead to the same numerical value of the 3×3-determinant $\det A$ and they are susceptible to straightforward generalization to a higher order.

Consider first the following rearrangement of the six terms of the determinant:

$$\det A = \begin{vmatrix} a_{11} & a_{12} & a_{13} \\ a_{21} & a_{22} & a_{23} \\ a_{31} & a_{32} & a_{33} \end{vmatrix} = \mathbf{c}_1(A) \cdot (\mathbf{c}_2(A) \times \mathbf{c}_3(A))$$

$$= a_{11} \begin{vmatrix} a_{22} & a_{23} \\ a_{32} & a_{33} \end{vmatrix} - a_{21} \begin{vmatrix} a_{12} & a_{13} \\ a_{32} & a_{33} \end{vmatrix} + a_{31} \begin{vmatrix} a_{12} & a_{13} \\ a_{22} & a_{23} \end{vmatrix}.$$

This expresses the determinant $\det A$ of order 3 in terms of determinants of order 2. We observe that the 2×2-determinant of the first term is obtained by deleting from $\det A$ the first row and the first column on which the element a_{11} is situated. In general the 2×2-determinant obtained by deleting from the 3×3-determinant the i-th row and the j-th column is called the *minor of the element* a_{ij}. We multiply the minor of a_{ij} by $(-1)^{i+j}$ to obtain the *cofactor of the element* a_{ij} and denote it by A_{ij}:

$$A_{11} = \begin{vmatrix} a_{22} & a_{23} \\ a_{32} & a_{33} \end{vmatrix} \qquad A_{12} = -\begin{vmatrix} a_{21} & a_{23} \\ a_{31} & a_{33} \end{vmatrix} \qquad A_{13} = \begin{vmatrix} a_{21} & a_{22} \\ a_{31} & a_{32} \end{vmatrix}$$

$$A_{21} = -\begin{vmatrix} a_{12} & a_{13} \\ a_{32} & a_{33} \end{vmatrix} \qquad A_{22} = \begin{vmatrix} a_{11} & a_{13} \\ a_{31} & a_{33} \end{vmatrix} \qquad A_{23} = -\begin{vmatrix} a_{11} & a_{12} \\ a_{31} & a_{32} \end{vmatrix}$$

$$A_{31} = \begin{vmatrix} a_{12} & a_{13} \\ a_{22} & a_{23} \end{vmatrix} \qquad A_{32} = -\begin{vmatrix} a_{11} & a_{13} \\ a_{21} & a_{23} \end{vmatrix} \qquad A_{33} = \begin{vmatrix} a_{11} & a_{12} \\ a_{21} & a_{22} \end{vmatrix}$$

We therefore see that we may define $\det A$ by

$$\det A = a_{11}A_{11} + a_{21}A_{21} + a_{31}A_{31}$$

which is called the *expansion of* $\det A$ *by the first column*. It is not difficult to prove the following expansion formulas:

$$\det A = a_{1j}A_{1j} + a_{2j}A_{2j} + a_{3j}A_{3j} \,, \qquad j = 1, 2, 3$$
$$\det A = a_{i1}A_{i1} + a_{i2}A_{i2} + a_{i3}A_{i3} \,, \qquad i = 1, 2, 3 \,,$$

i.e. det A *can be obtained by expansion by any column and any row.*

Since each element a_{ij} of the 3×3-matrix A has a cofactor A_{ij}, we use the cofactors A_{ij} to form a new matrix

$$\begin{pmatrix} A_{11} & A_{12} & A_{13} \\ A_{21} & A_{22} & A_{23} \\ A_{31} & A_{32} & A_{33} \end{pmatrix}$$

with A_{ij} in the same position as a_{ij}. To make good use of the above expansion formulas and to rewrite them in terms of matrix multiplication, we call the transpose of this matrix the *adjoint matrix* of A and denote it by $\mathrm{ad}\, A$:

$$\mathrm{ad}\, A = \begin{pmatrix} A_{11} & A_{21} & A_{31} \\ A_{12} & A_{22} & A_{32} \\ A_{13} & A_{23} & A_{33} \end{pmatrix} \,.$$

The expansion formulas now become

$$\det A = \mathbf{r}_1(\mathrm{ad}\, A) \cdot \mathbf{c}_1(A) = \mathbf{r}_2(\mathrm{ad}\, A) \cdot \mathbf{c}_2(A) = \mathbf{r}_3(\mathrm{ad}\, A) \cdot \mathbf{c}_3(A) \,;$$
$$\det A = \mathbf{r}_1(A) \cdot \mathbf{c}_1(\mathrm{ad}\, A) = \mathbf{r}_2(A) \cdot \mathbf{c}_2(\mathrm{ad}\, A) = \mathbf{r}_3(A) \cdot \mathbf{c}_3(\mathrm{ad}\, A) \,.$$

In other words all diagonal elements of the products $\mathrm{ad}\, A \cdot A$ and $A \cdot \mathrm{ad}\, A$ equal det A. What about the off-diagonal elements of these two products? Take for example the element of $\mathrm{ad}\, A \cdot A$ on the first row and the second column:

$$\mathbf{r}_1(\mathrm{ad}\, A) \cdot \mathbf{c}_2(A) = A_{11}a_{12} + A_{21}a_{22} + A_{31}a_{32} = a_{12}A_{11} + a_{22}A_{21} + a_{32}A_{31} \,.$$

Applying the expansion by the second column on the matrix obtained by replacing the first column of A by its second column, we get

$$0 = \begin{vmatrix} a_{12} & a_{12} & a_{13} \\ a_{22} & a_{22} & a_{23} \\ a_{32} & a_{32} & a_{33} \end{vmatrix} = a_{12}\begin{vmatrix} a_{22} & a_{23} \\ a_{32} & a_{33} \end{vmatrix} - a_{22}\begin{vmatrix} a_{12} & a_{13} \\ a_{32} & a_{33} \end{vmatrix} + a_{23}\begin{vmatrix} a_{12} & a_{13} \\ a_{22} & a_{23} \end{vmatrix}$$
$$= a_{12}A_{11} + a_{22}A_{21} + a_{32}A_{31} = \mathbf{r}_1(\mathrm{ad}\, A) \cdot \mathbf{c}_2(A) \,.$$

Similarly we get

$$0 = \begin{vmatrix} a_{11} & a_{12} & a_{13} \\ a_{11} & a_{12} & a_{13} \\ a_{21} & a_{22} & a_{23} \end{vmatrix} = a_{11} \begin{vmatrix} a_{12} & a_{13} \\ a_{22} & a_{23} \end{vmatrix} - a_{12} \begin{vmatrix} a_{11} & a_{13} \\ a_{21} & a_{23} \end{vmatrix} + a_{13} \begin{vmatrix} a_{11} & a_{12} \\ a_{21} & a_{22} \end{vmatrix}$$

$$= a_{11} A_{31} + a_{12} A_{32} + a_{13} A_{33} = \mathbf{r}_1(A) \cdot \mathbf{c}_3(\text{ad}\, A),$$

and similarly all elements off the diagonal of the products $\text{ad}\, A \cdot A$ and $A \cdot \text{ad}\, A$ equal zero. We have therefore proved the following theorem from which Cramer's rule follows.

6.7.9 THEOREM *Let A be a square matrix of order 3 and $\text{ad}\, A$ its adjoint matrix. Then*

$$\text{ad}\, A \cdot A = A \cdot \text{ad}\, A = \det A \cdot I.$$

6.7.10 CRAMER'S RULE *If A is an invertible matrix of order 3, then*

$$A^{-1} = \frac{1}{\det A} \text{ad}\, A = \frac{1}{|A|} \begin{pmatrix} A_{11} & A_{21} & A_{31} \\ A_{12} & A_{22} & A_{32} \\ A_{13} & A_{23} & A_{33} \end{pmatrix}.$$

While both the original definition 3×3-determinant and the alternative definition in terms of 2×2-determinants are explicit and computational, we shall now derive a second alternative definition which is implicit and conceptual. Let A be a square matrix of order 3 with arbitrary column vectors $\mathbf{a}$, $\mathbf{b}$, $\mathbf{c}$. We now regard the 3×3-determinant $\det A = |\mathbf{a}\,\mathbf{b}\,\mathbf{c}|$ as a real-valued function in the three vector variables $\mathbf{a}, \mathbf{b}, \mathbf{c}$. If we denote this function by D, then the domain of D is the triple cartesian product $\mathbf{R}^3 \times \mathbf{R}^3 \times \mathbf{R}^3$ and the range of D is simply the system $\mathbf{R}$ of all real numbers, i.e. $D : \mathbf{R}^3 \times \mathbf{R}^3 \times \mathbf{R}^3 \to \mathbf{R}$. It follows from the first part of the present section that D has in particular properties similar to (2), (5) of 6.7.1 and (c) of 6.7.2. Now it turns out that these three fundamental properties actually characterizes the 3×3-determinant. More precisely we put down this characterization of the 3×3-determinant as a theroem.

6.7.11 THEOREM *Let $D : \mathbf{R}^3 \times \mathbf{R}^3 \times \mathbf{R}^3 \to \mathbf{R}$ be a real-valued function in three vector variables. Then $D(\mathbf{a}, \mathbf{b}, \mathbf{c}) = |\mathbf{a}\,\mathbf{b}\,\mathbf{c}|$ for any three vectors $\mathbf{a}$, $\mathbf{b}$, $\mathbf{c}$ of $\mathbf{R}^3$ if the following conditions are satisfied.*

    (i) *D is linear in each variable, i.e. for any six vectors $\mathbf{a}$, $\mathbf{a}'$, $\mathbf{b}$, $\mathbf{b}'$, $\mathbf{c}$, $\mathbf{c}'$ of $\mathbf{R}^3$ and any two real numbers r and s, the following identities hold:*

$$D(r\mathbf{a} + s\mathbf{a'}, \mathbf{b}, \mathbf{c}) = rD(\mathbf{a}, \mathbf{b}, \mathbf{c}) + sD(\mathbf{a'}, \mathbf{b}, \mathbf{c}) \ ;$$
$$D(\mathbf{a}, r\mathbf{b} + s\mathbf{b'}, \mathbf{c}) = rD(\mathbf{a}, \mathbf{b}, \mathbf{c}) + sD(\mathbf{a}, \mathbf{b'}, \mathbf{c}) \ ;$$
$$D(\mathbf{a}, \mathbf{b}, r\mathbf{c} + s\mathbf{c'}) = rD(\mathbf{a}, \mathbf{b}, \mathbf{c}) + sD(\mathbf{a}, \mathbf{b}, \mathbf{c'}) \ .$$

(ii) $D(\mathbf{a}, \mathbf{b}, \mathbf{c}) = 0$ if any two among the vectors $\mathbf{a}$, $\mathbf{b}$, $\mathbf{c}$ are equal.

(iii) $D(\mathbf{e}_1, \mathbf{e}_2, \mathbf{e}_3) = 1$ if $\mathbf{e}_1$, $\mathbf{e}_2$, $\mathbf{e}_3$ are the unit coordinate vectors of $\mathbf{R}^3$.

PROOF Applying conditions (i) and (ii) to $D(\mathbf{a} + \mathbf{b}, \mathbf{a} + \mathbf{b}, \mathbf{c})$, we get

$$\begin{aligned}
0 &= D(\mathbf{a} + \mathbf{b}, \mathbf{a} + \mathbf{b}, \mathbf{c}) \\
&= D(\mathbf{a}, \mathbf{a}, \mathbf{c}) + D(\mathbf{a}, \mathbf{b}, \mathbf{c}) + D(\mathbf{b}, \mathbf{a}, \mathbf{c}) + D(\mathbf{b}, \mathbf{b}, \mathbf{c}) \\
&= D(\mathbf{a}, \mathbf{b}, \mathbf{c}) + D(\mathbf{b}, \mathbf{a}, \mathbf{c}) \ .
\end{aligned}$$

Therefore $D(\mathbf{a}, \mathbf{b}, \mathbf{c}) = -D(\mathbf{b}, \mathbf{a}, \mathbf{c})$. Similarly we get $D(\mathbf{a}, \mathbf{b}, \mathbf{c}) = -D(\mathbf{a}, \mathbf{c}, \mathbf{b})$ $= -D(\mathbf{c}, \mathbf{b}, \mathbf{a})$, i.e. D changes signs if two of its arguments interchange positions. Thus D also has property (3) of 6.7.1. Now for $\mathbf{a} = [a_1, a_2, a_3]$, $\mathbf{b} = [b_1, b_2, b_3]$, $\mathbf{c} = [c_1, c_2, c_3]$ we apply condition (i) to

$$D(\mathbf{a}, \mathbf{b}, \mathbf{c}) = D(a_1\mathbf{e}_1 + a_2\mathbf{e}_2 + a_3\mathbf{e}_3, b_1\mathbf{e}_1 + b_2\mathbf{e}_2 + b_3\mathbf{e}_3, c_1\mathbf{e}_1 + c_2\mathbf{e}_2 + c_3\mathbf{e}_c)$$

to get a sum of 27 terms of the form $a_i b_j c_k D(\mathbf{e}_i, \mathbf{e}_j, \mathbf{e}_k)$. But by condition (ii) only 6 of these 27 terms can be non-zero, namely those with 3 distinct indices i, j, k. Therefore

$$\begin{aligned}
D(\mathbf{a}, \mathbf{b}, \mathbf{c}) = \ &a_1 b_2 c_3 D(\mathbf{e}_1, \mathbf{e}_2, \mathbf{e}_3) + a_2 b_3 c_1 D(\mathbf{e}_2, \mathbf{e}_3, \mathbf{e}_1) + a_3 b_1 c_2 D(\mathbf{e}_3, \mathbf{e}_1, \mathbf{e}_2) \\
&+ a_2 b_1 c_3 D(\mathbf{e}_2, \mathbf{e}_1, \mathbf{e}_3) + a_1 b_3 c_2 D(\mathbf{e}_1, \mathbf{e}_3, \mathbf{e}_2) + a_3 b_2 c_1 D(\mathbf{e}_3, \mathbf{e}_2, \mathbf{e}_1) \ .
\end{aligned}$$

Using condition (iii) and the above rule of sign change:

$$D(\mathbf{e}_1, \mathbf{e}_2, \mathbf{e}_3) = D(\mathbf{e}_2, \mathbf{e}_3, \mathbf{e}_1) = D(\mathbf{e}_3, \mathbf{e}_1, \mathbf{e}_2) = 1 \ ,$$
$$D(\mathbf{e}_2, \mathbf{e}_1, \mathbf{e}_3) = D(\mathbf{e}_1, \mathbf{e}_3, \mathbf{e}_2) = D(\mathbf{e}_3, \mathbf{e}_2, \mathbf{e}_1) = -1 \ ,$$

we finally obtained

$$\begin{aligned}
D(\mathbf{a}, \mathbf{b}, \mathbf{c}) &= a_1 b_2 a_3 + a_2 b_3 c_1 + a_3 b_1 c_2 - a_2 b_1 c_3 - a_1 b_3 c_2 - a_3 b_2 c_1 \\
&= |\mathbf{a}\,\mathbf{b}\,\mathbf{c}| \ .
\end{aligned}$$

The proof is now complete.

In general a function $D : \mathbf{R}^3 \times \mathbf{R}^3 \times \mathbf{R}^3 \to \mathbf{R}$ is said to be *trilinear* if it satisfies condition (i) of 6.7.11, *alternating* if it satisfies (ii) and *normalized* if it satisfies (iii). Therefore we have proved in 6.7.11 that the determinant is the unique normalized alternating trilinear function of $\mathbf{R}^3$, providing a second alternative definition of 3×3-determinant.

EXERCISES

1. Evaluate the following determinant.

(a) $\begin{vmatrix} 2 & 3 & -2 \\ 3 & 5 & 2 \\ 4 & 1 & 0 \end{vmatrix}$;   (b) $\begin{vmatrix} 7 & 4 & 5 \\ 3 & 0 & 1 \\ 2 & 0 & 6 \end{vmatrix}$;   (c) $\begin{vmatrix} 3 & 7 & -2 \\ 2 & 1 & -3 \\ 4 & 0 & 6 \end{vmatrix}$.

2. Find all values of a for which $\det A = 0$.

(a) $A = \begin{pmatrix} a-1 & -2 \\ 1 & a-4 \end{pmatrix}$;   (b) $A = \begin{pmatrix} a-6 & 0 & 0 \\ 0 & a & -1 \\ 0 & 4 & a-4 \end{pmatrix}$.

3. The results in this question is useful in evaluating determinants. They are another version of Theorem 6.7.1(2). For any real number k, prove that

(a) $\begin{vmatrix} ka_1 & kb_1 & kc_1 \\ a_2 & b_2 & c_2 \\ a_3 & b_3 & c_3 \end{vmatrix} = k \begin{vmatrix} a_1 & b_1 & c_1 \\ a_2 & b_2 & c_2 \\ a_3 & b_3 & c_3 \end{vmatrix}$;

(b) $\begin{vmatrix} a_1 + ka_2 & b_1 + kb_2 & c_1 + kc_2 \\ a_2 & b_2 & c_2 \\ a_3 & b_3 & c_3 \end{vmatrix} = \begin{vmatrix} a_1 & b_1 & c_1 \\ a_2 & b_2 & c_2 \\ a_3 & b_3 & c_3 \end{vmatrix}$.

In fact (a) and (b) are true for any row or column.

4. By Exercise 3 and Theorem 6.7.1(3), evaluate

(a) $\begin{vmatrix} 2 & 3 & 7 \\ 0 & 0 & -3 \\ 1 & -2 & 7 \end{vmatrix}$;   (b) $\begin{vmatrix} 2 & -4 & 8 \\ -2 & 7 & -2 \\ 0 & 1 & 5 \end{vmatrix}$.

5. Suppose that $\begin{vmatrix} a & b & c \\ d & e & f \\ g & h & i \end{vmatrix} = 4.$ Find

(a) $\begin{vmatrix} d & e & f \\ g & h & i \\ a & b & c \end{vmatrix}$;   (b) $\begin{vmatrix} -a & -b & -c \\ 5d & 5e & 5f \\ 3g & 3h & 3i \end{vmatrix}$;   (c) $\begin{vmatrix} a+g & b+h & c+i \\ a-d & b-e & c-f \\ g & h & i \end{vmatrix}$.

6. (a) Express $\begin{vmatrix} a_1 + b_1 & c_1 + d_1 \\ a_2 + b_2 & c_2 + d_2 \end{vmatrix}$

as a sum of four determinants whose entries contain no sums.

(b) Express $\begin{vmatrix} a_1 + b_1 & c_1 + d_1 & e_1 + f_1 \\ a_2 + b_2 & c_2 + d_2 & e_2 + f_2 \\ a_3 + b_3 & c_3 + d_3 & e_3 + f_3 \end{vmatrix}$

as a sum of eight determinants whose entries contain no sums.

7. Use Corollary 6.7.2 to determine which of the following matrices are invertible.

(a) $\begin{pmatrix} 2 & 1 & 0 \\ 1 & 0 & 3 \\ 0 & 3 & 2 \end{pmatrix}$;   (b) $\begin{pmatrix} 1 & 2 & 4 \\ 2 & 3 & 1 \\ 3 & 4 & 2 \end{pmatrix}$;   (c) $\begin{pmatrix} -2 & 1 & -4 \\ 1 & 1 & 2 \\ 3 & 1 & 6 \end{pmatrix}$.

8. By using Cramer's rule, find the inverse of each invertible matrix in Exercise 7.

9. Show that $\begin{pmatrix} \sin^2 a & \sin^2 b & \sin^2 c \\ \cos^2 a & \cos^2 b & \cos^2 c \\ 1 & 1 & 1 \end{pmatrix}$
is not invertible for any values of a, b and c.

10. Find the values of x such that $\begin{vmatrix} 1 & x & x \\ x & 1 & x \\ x & x & 1 \end{vmatrix} = 0.$

11. Given that the roots of the equation $x^3 + qx + r = 0$ are a, b and c,
express the value of $\begin{vmatrix} 1+a & 1 & 1 \\ 1 & 1+b & 1 \\ 1 & 1 & 1+c \end{vmatrix}$ in terms of q and r.

12. Show that $\begin{vmatrix} (b+c)^2 & a^2 & a^2 \\ b^2 & (c+a)^2 & b^2 \\ c^2 & c^2 & (a+b)^2 \end{vmatrix} = 2abc(a+b+c)^3.$

13. (a) Evaluate $\begin{vmatrix} 0 & c & b \\ c & 0 & a \\ b & a & 0 \end{vmatrix}^2$.

    (b) Hence show that $\begin{vmatrix} b^2+c^2 & ab & ca \\ ab & c^2+a^2 & bc \\ ca & bc & a^2+b^2 \end{vmatrix} = 4a^2b^2c^2.$

14. If A is invertible, show that $\det(A^{-1}) = (\det A)^{-1}$.

15. Prove that for any real number k, $\det(kA) = k^3 \det A$.

16. Prove that if AB is invertible, then both A and B are invertible.

17. Prove that if B is invertible, then

$$\det(I - B^{-1}AB) = \det(I - A) \; ,$$

    where I is the identity matrix of order 3.

18. What can be said about $\det A$ and $\det B$ if:
    (a) $A^2 = A$,
    (b) $A^t = -A$,
    (c) $A^2 = kA$ for some real number k,
    (d) $A^2 + I = 0$,
    (e) $AB + BA = 0$?

6.8 Determinant of higher order

Let A be square matrix of order 4. Then the 4×4-determinant $\det(A) = |A|$ is a sum of $4! = 24$ terms, each being a product of the form $\pm a_{1i}a_{2j}a_{3k}a_{4\ell}$. However, to write down this sum precisely, we shall need to specify that the term $a_{1i}a_{2j}a_{3k}a_{4\ell}$ is given a positive sign if i, j, k, ℓ is an even permutation of the digits 1, 2, 3, 4 and a negative sign if it is an odd permutation. With such a complicated definition, it would be very difficult for us to study the properties of det A and prove the counterparts of 6.7.1. etc.

Instead we may use the two alternative definitions of 3×3-determinants as models to define det A. For example we can define det A by an analogous expansion formula:

$$|A| = a_{11}A_{11} + a_{12}A_{12} + a_{13}A_{13} + a_{14}A_{14}$$

$$= a_{11}\begin{vmatrix} a_{22} & a_{23} & a_{24} \\ a_{32} & a_{33} & a_{34} \\ a_{42} & a_{43} & a_{44} \end{vmatrix} - a_{12}\begin{vmatrix} a_{21} & a_{23} & a_{24} \\ a_{31} & a_{33} & a_{34} \\ a_{41} & a_{43} & a_{44} \end{vmatrix} + a_{13}\begin{vmatrix} a_{21} & a_{22} & a_{24} \\ a_{31} & a_{32} & a_{34} \\ a_{41} & a_{42} & a_{44} \end{vmatrix}$$

$$- a_{14}\begin{vmatrix} a_{21} & a_{22} & a_{23} \\ a_{31} & a_{32} & a_{33} \\ a_{41} & a_{42} & a_{43} \end{vmatrix}.$$

Using this definition, we shall have no difficulty in proving det $A = $ det A^t. With some patience and care the other properties of det A can be proved to be valid as well.

A second alternative would be to define 4×4-determinant as a function $D : \mathbf{R}^4 \times \mathbf{R}^4 \times \mathbf{R}^4 \times \mathbf{R}^4 \to \mathbf{R}$ which satisfies

  (i) D is linear in each vector argument, i.e. it is a *multilinear* function;

  (ii) D is alternating, i.e. it is zero if two vector arguments are equal;

  (iii) D is normalized, i.e. $D(\mathbf{e}_1, \mathbf{e}_2, \mathbf{e}_3, \mathbf{e}_4) = 1$.

Under this definition, it would be much easier to prove the majority of the properties. However some effort is necessary to show that one such unique normalized alternating multilinear function does exist.

It is now clear that determinants of still higher order can be defined analogously and that they have properties similar to those of 3×3-determinant.

CHAPTER SEVEN

LINEAR EQUATIONS

In this final chapter, we shall apply results of the last two chapters to investigate systems of linear equations in several unknowns. A necessary and sufficient condition will be given in terms of the ranks of certain matrices and a general method of solution is described. Readers will find that this method is essentially the classical successive eliminations of unknowns but given in terms of elementary row transformations.

7.1 Terminology

A *linear equation* in n unknowns $x_1, x_2, \cdots, x_n$ is an expression of the form

$$a_1 x_1 + a_2 x_2 + \cdots + a_n x_n = b \tag{1}$$

where the *coefficients* $a_1, a_2, \cdots, a_n$ and the *constant term* b are all fixed real numbers such that at least one coefficient should be non-zero. An ordered n-tuple $(t_1, t_2, \cdots, t_n)$ of real numbers is called a *solution* of the equation (1) if after substituting each t_i for x_i, we get

$$a_1 t_1 + a_2 t_2 + \cdots + a_n t_n = b \ .$$

In this case the n-tuple $(t_1, t_2, \cdots, t_n)$ or the vector $\mathbf{t}$ is said to *satisfy* the equation (1) and we also say that $x_1 = t_1$, $x_2 = t_2, \cdots, x_n = t_n$ constitute a *solution* to (1). Alternatively if we denote by $\mathbf{t} \in \mathbf{R}^n$ the vector $[t_1, t_2, \cdots, t_n]$ and by $\mathbf{a} \in \mathbf{R}^n$ the vector $[a_1, a_2, \cdots, a_n]$, then $\mathbf{t}$ is a solution of the equation (1) if and only if the dot product

$$\mathbf{a} \cdot \mathbf{t} = a_1 t_1 + a_2 t_2 + \cdots + a_n t_n = b \ .$$

For example

$$2x_1 + 3x_2 + 0x_3 - 2x_4 = 7 \quad \text{or} \quad 2x_1 + 3x_2 - 2x_4 = 7$$

is a linear equation in the 4 unknowns x_1, x_2, x_3, x_4. The 4-tuples $(1, 1, -1, -1)$ and $(0, 1, 0, -2)$ are both solutions of the equations while $(0, 0, 2, 0)$, $(1, 2, -4, 7)$ are not. In particular we observe that the equation (1) always has solutions. For example if $a_1 \neq 0$, then $x_1 = b/a_1$, $x_2 = x_3 = \cdots = x_n = 0$ constitute a solution to (1).

Besides considering one linear equation at a time, we also study simultaneously several linear equations

$$a_{i1} x_1 + a_{i2} x_2 + \cdots a_{in} x_n = b_i, \quad i = 1, 2, \cdots, m \tag{2}$$

and their common solutions. These m linear equations constitute *a system of m linear equations in n unknowns* $x_1, x_2, \cdots, x_n$:

$$\begin{aligned} a_{11} x_1 + a_{12} x_2 + \cdots + a_{1n} x_n &= b_1 \\ a_{21} x_1 + a_{22} x_2 + \cdots + a_{2n} x_n &= b_2 \\ &\cdots\cdots\cdots\cdots\cdots \\ a_{m1} x_1 + a_{m2} x_2 + \cdots + a_{mn} x_n &= b_m \ . \end{aligned} \tag{3}$$

An ordered n-tuple $(t_1, t_2, \cdots, t_n)$ is called a *solution* of the system (3) if after substituting each t_j for x_j, we get

$$\begin{aligned} a_{11} t_1 + a_{12} t_2 + \cdots + a_{1n} t_n &= b_1 \\ a_{21} t_1 + a_{22} t_2 + \cdots + a_{2n} t_n &= b_2 \\ &\cdots\cdots\cdots\cdots\cdots \\ a_{m1} t_1 + a_{m2} t_2 + \cdots + a_{mn} t_n &= b_m \ . \end{aligned}$$

Alternatively if $\mathbf{a}_i = [a_{i1}, a_{i2}, \cdots, a_{in}]$ for $i = 1, 2, \cdots, m$, then $\mathbf{t} = [t_1, t_2, \cdots, t_n]$ is a solution of the system (2) if and only if $\mathbf{a}_i \cdot \mathbf{t} = b_i$ for $i = 1, 2, \cdots, m$.

Therefore a solution to the system (3) is a common solution to all equations of the system and vice versa. Our chief concern with a given system of linear equations is clearly that we would like to know if solutions exist, and if they do exist then we would like to have an effective method of evaluating them and a clear and precise way to present them.

282

EXERCISES

1. Show that $(1, 0, -1, 1)$, $(-1, 1, -1, 1)$, and $(-9, 5, -1, 1)$ are all solutions of the system of linear equations

$$
\begin{aligned}
3x_1 + 6x_2 - x_3 + x_4 &= 5 \\
-2x_1 - 4x_2 + x_3 &= -3 \\
x_3 + x_4 &= 0.
\end{aligned}
$$

2. Verify each of the following:
   (a) $\left(-\frac{1}{2}t - 4, \frac{1}{2}t - 6, t\right)$ is a solution of the system of linear equations

$$
\begin{aligned}
x_1 - x_2 + x_3 &= 2 \\
3x_1 - x_2 + 2x_3 &= -6 \\
3x_1 + x_2 + x_3 &= -18
\end{aligned}
$$

   for any real number t.
   (b) $(-s - t, s, -t, 0, t)$ is a solution of the system of linear equations

$$
\begin{aligned}
2x_1 + 2x_2 - x_3 + x_5 &= 0 \\
-x_1 - x_2 + 2x_3 - 3x_4 + x_5 &= 0 \\
x_1 + x_2 - 2x_3 - x_5 &= 0
\end{aligned}
$$

   for any real numbers s and t.

3. Given that (s_1, t_1, u_1) and (s_2, t_2, u_2) are two solutions of the system of linear equations

$$
\begin{aligned}
a_1x_1 + a_2x_2 + a_3x_3 &= 0 \\
b_1x_1 + b_2x_2 + b_3x_3 &= 0 \\
c_1x_1 + c_2x_2 + c_3x_3 &= 0,
\end{aligned}
$$

   show that for any real numbers m and n, $(ms_1 + ns_2, mt_1 + nt_2, mu_1 + nu_2)$ is a solution of the system.

7.2 Condition for consistency

While every single linear equation always admits solutions, some systems of linear equatoins do not. For example the system

$$
\begin{aligned}
x_1 + x_2 &= 1 \\
x_1 + x_2 &= 2
\end{aligned}
$$

certainly has no solution. In general we say that a system of linear equations is *consistent* or *solvable* if it admits a solution. This leads us to the first important problem of this chapter. That is to find a necessary and sufficient condition for a given system of linear equations to be consistent. We shall find that matrices would provide a most suitable language to express such a condition.

Let S:

$$a_{11}x_1 + a_{12}x_2 + \cdots + a_{1n}x_n = b_1$$
$$a_{21}x_1 + a_{22}x_2 + \cdots + a_{2n}x_n = b_2$$
$$\cdots\cdots\cdots\cdots\cdots\cdots\cdots\cdots\cdots\cdots$$
$$a_{m1}x_1 + a_{m2}x_2 + \cdots + a_{mn}x_n = b_m$$

be a system of linear equations in the unknowns $x_1, x_2, \cdots, x_n$. After detaching the coefficients from the unknowns, we obtain an $m \times n$-matrix

$$A_0 = \begin{pmatrix} a_{11} & a_{12} & \cdots & a_{1n} \\ a_{22} & a_{22} & \cdots & a_{2n} \\ \cdots\cdots\cdots\cdots\cdots\cdots \\ a_{m1} & a_{m2} & \cdots & a_{mn} \end{pmatrix}$$

which is called the *coefficient matrix* of the system S. Adjoining to A_0 an additional column that consists of the constant terms of the system, we further obtain an $m \times (n+1)$-matrix

$$A = \begin{pmatrix} a_{11} & a_{12} & \cdots & a_{1n} & b_1 \\ a_{21} & a_{21} & \cdots & a_{2n} & b_2 \\ \cdots\cdots\cdots\cdots\cdots\cdots\cdots \\ a_{m1} & a_{m2} & \cdots & a_{mn} & b_m \end{pmatrix}$$

which is called the *augmented matrix* of the system S.

Thus every system of linear equations has a unique augmented matrix and hence a unique coefficient matrix; conversely every $m \times (n+1)$-matrix determines a unique system of m linear equations in n unknowns. In fact the system S of linear equations can be written as a single matrix equation

$$\begin{pmatrix} a_{11} & a_{12} & \cdots & a_{1n} \\ a_{21} & a_{22} & \cdots & a_{2n} \\ \cdots\cdots\cdots\cdots\cdots\cdots \\ a_{m1} & a_{m2} & \cdots & a_{mn} \end{pmatrix} \begin{pmatrix} x_1 \\ x_2 \\ \vdots \\ x_n \end{pmatrix} = \begin{pmatrix} b_1 \\ b_2 \\ \vdots \\ b_m \end{pmatrix}.$$

If the two column matrices are denoted by X and B respectively, then the equation becomes

$$A_0 X = B .$$

Therefore we can make use of the results of the last two chapters to find a condition for consistency in terms of vectors and matrices.

Suppose that the system S is consistent and let $(t_1, t_2, \cdots, t_n)$ be a solution. Then

$$\begin{aligned}
a_{11}t_1 + a_{12}t_2 + \cdots + a_{1n}t_n &= b_1 \\
a_{21}t_1 + a_{22}t_2 + \cdots + a_{2n}t_n &= b_2 \\
\cdots\cdots\cdots\cdots\cdots\cdots\cdots\cdots\cdots \\
a_{m1}t_1 + a_{m2}t_2 + \cdots + a_{mn}t_n &= b_m .
\end{aligned}$$

In terms of vectors of the vector space $\mathbf{R}^m$, this can be written into

$$t_1[a_{11}, a_{21}, \cdots, a_{m1}] + t_2[a_{12}, a_{22}, \cdots, a_{m2}]$$
$$+ \cdots + t_n[a_{1n}, a_{2n}, \cdots, a_{mn}]$$
$$= [b_1, b_2, \cdots, b_m]$$

or
$$t_1 c_1(A) + t_2 c_2(A) + \cdots + t_n c_n(A) = c_{n+1}(A).$$

Therefore the last column vector $c_{n+1}(A)$ of the augmented matrix A is a linear combination of the first n column vectors $c_1(A), c_2(A), \cdots, c_n(A)$ of A. The latter n vectors being the column vectors of the coefficient matrix A_0, it follows that rank A_0 = rank A. Therefore a necessary condition for a system S of linear equations to be consistent is that *the rank of the coefficient matrix of S equals the rank of the augmented matrix of S.*

This condition turns out to be also sufficient. Let r = rank A_0 = rank A. Then there are r linear independent vectors among the column vectors $c_1(A_0) = c_1(A), c_2(A_0) = c_2(A), \cdots, c_n(A_0) = c_n(A)$ of A_0 such that all column vectors of A_0 are linear combinations of them. Since the augmented A also has rank r by assumption, the said r linearly independent column vectors of A_0 also form a maximum set of r linearly independent column vectors of A. Hence all column vectors of A are also linear combinations of these same r linear independent column vectors of A_0 and hence of all n column vectors of A_0. In particular, for the last column vector $c_{n+1}(A) = \mathbf{b}$ of A, we get

$$[b_1, b_2, \cdots, b_n] = t_1[a_{11}, a_{21}, \cdots, a_{m1}] + t_2[a_{12}, a_{22}, \cdots, a_{m2}]$$
$$+ \cdots + t_n[a_{1n}, a_{2n}, \cdots, a_{mn}]$$

for some scalars $t_1, t_2, \cdots, t_n$. But this also means that $(t_1, t_2, \cdots, t_n)$ is a solution of the system S. Hence S is consistent, proving that the condition is sufficient. We now put this very important result in a theorem.

7.2.1 THEOREM *A system S of linear equations is consistent if and only if the coefficient matrix A_0 of S has the same rank as the augmented matrix A of S, i.e. rank A_0 = rank A.*

7.2.2 EXAMPLE The system

$$x_1 + x_2 = 1$$
$$x_1 + x_2 = 2$$

is inconsistent because

$$\text{rank} \begin{pmatrix} 1 & 1 \\ 1 & 1 \end{pmatrix} = 1 \quad \text{and rank} \begin{pmatrix} 1 & 1 & 1 \\ 1 & 1 & 2 \end{pmatrix} = 2 \ .$$

7.2.3 EXAMPLE Determine whether the following system of linear equations is solvable.

$$x_1 + x_2 + x_3 = 4$$
$$x_1 + 2x_2 - x_3 = 5$$
$$x_1 - x_2 + 5x_3 = 2 \ .$$

SOLUTION Apply appropriate elementary *row* transformations to the augmented matrix A of the system to get

$$\begin{pmatrix} 1 & 1 & 1 & 4 \\ 1 & 2 & -1 & 5 \\ 1 & -1 & 5 & 2 \end{pmatrix} \longrightarrow \begin{pmatrix} 1 & 1 & 1 & 4 \\ 0 & 1 & -2 & 1 \\ 0 & -2 & 4 & -2 \end{pmatrix} \longrightarrow \begin{pmatrix} 1 & 0 & 3 & 3 \\ 0 & 1 & -2 & 1 \\ 0 & 0 & 0 & 0 \end{pmatrix} .$$

Thus rank $A = 2$. But the same row transformations on the coefficient matrix A_0 would bring it to

$$\begin{pmatrix} 1 & 0 & 3 \\ 0 & 1 & -2 \\ 0 & 0 & 0 \end{pmatrix} .$$

Therefore rank $A_0 = 2$. Hence the system is solvable. In fact for every real number t, the triple $(3 - 3t, 1 + 2t, t)$ is a solution.

In the above method only elementary row transformations are used. Every row transformation on A has a corresponding row transformation on A_0; therefore the ranks of A and A_0 can be obtained simultaneously. Because of the absence of a fourth column in A_0 of the system in 7.2.3, this is no more true of the elementary column transformations on A. For example the column transformation $c_1(A) \rightarrow c_1(A) + c_4(A)$ on A has no corresponding column transformation on A_0. Therefore the use of column transformations should be avoided in the simultaneous evaluation of the ranks of the coefficient matrix and the augmented matrix of a system of equations. Later on we shall see yet another compelling reason for disallowing the use of column transformations when working with linear equations.

7.2.4 EXAMPLE For any values of b_1, b_2, b_3, the following system of linear equations is consistent.

$$
\begin{aligned}
2x_1 + 4x_2 + 6x_3 &= b_1 \\
-3x_1 + x_2 + 2x_3 &= b_2 \\
5x_1 + 3x_2 + x_3 &= b_3 \,.
\end{aligned}
$$

PROOF The determinant of the coefficient matrix is

$$
\begin{vmatrix}
2 & 4 & 6 \\
-3 & 1 & 2 \\
5 & 3 & 1
\end{vmatrix} = -42 \neq 0 \,.
$$

Therefore rank $A_0 = 3$. Since rank $A_0 \le$ rank $A \le 3$, rank $A_0 =$ rank A. The system is therefore consistent for all b_1, b_2, b_3.

We now have a necessary and sufficient conditon (7.2.1) for consistency. Moreover, given any system S of linear equations, we may use our trusted warhorse of elementary row transformations on the augmented matrix A to test if S is consistent. Obviously following the test, we would like to obtain all solutions of the system S if it turns out to be consistent. We shall however delay this consideration for the moment and study the special case of linear equations with vanishing constant terms because the results that we shall obtain for the special case will be very valuable for the general case.

EXERCISES

1. Show that every single linear equation must have solutions.
2. Write down the augmented matrix for each of the following systems of linear equations.

   (a) $\quad 5x_1 - 2x_2 = 1$
   $\qquad x_1 + 4x_2 = 6$.

   (b) $\quad x_1 - 3x_2 = 4$
   $\qquad\qquad x_2 = 1$.

   (c) $\quad x_1 - x_2 + x_3 = 4$
   $\qquad x_1 \qquad - x_3 = -1$
   $\qquad\quad x_2 + 2x_3 = 1$.

3. For each of the following augmented matrices, write down a corresponding system of linear equations.

   (a) $\begin{pmatrix} 1 & 2 & -4 & 2 \\ 0 & 1 & -2 & -1 \\ 3 & 1 & 0 & 7 \end{pmatrix}$.

   (b) $\begin{pmatrix} 1 & 0 & 4 & 7 & 10 \\ 0 & 1 & -3 & -4 & -2 \\ 0 & 0 & 1 & 1 & 2 \end{pmatrix}$.

4. Determine whether the following systems of linear equations are solvable.

   (a) $\quad x_1 - 2x_2 = 1$
   $\qquad -x_1 + 4x_2 = -2$.

   (b) $\quad x_1 - x_2 + x_3 = 1$
   $\qquad x_1 + 2x_2 - x_3 = 7$
   $\qquad -x_1 + 4x_2 - 3x_3 = 4$.

   (c) $\quad x_1 + x_2 + 2x_3 = 8$
   $\qquad x_1 + 2x_2 - 3x_3 = -1$
   $\qquad 3x_1 - 7x_2 + 4x_3 = 10$.

5. Find the value of k such that the following system of equations is solvable.

$$2x_1 + x_2 - x_3 = 1$$
$$x_1 + x_2 + x_3 = 3$$
$$3x_1 + 2x_2 \qquad = k .$$

6. Find conditions on the numbers a, b, c and d such that the following system of linear equations is consistent.

$$ax_1 + cx_2 = 1$$
$$bx_1 + dx_2 = 0 .$$

7. Show that for any real numbers b_1, b_2 and b_3, the following systems of linear equations are consistent.

(a) $\quad 3x_1 + 5x_2 = b_1$
$\quad\quad 2x_1 - \ x_2 = b_2$.

(b) $\quad x_1 + \ x_2 + \ x_3 = b_1$
$\quad\quad 2x_1 \quad\quad\ + 2x_3 = b_2$
$\quad\quad\quad\ 3x_2 + 4x_3 = b_3$.

7.3 Homogeneous linear equations

A linear equation is said to be *homogeneous* if its constant term is zero. A system of linear equations whose constant terms are all zero is similarly said to be *homogeneous*. Such a system S_0

$$a_{11}x_1 + \ a_{12}x_2 + \cdots + \ a_{1n}x_n = 0$$
$$a_{21}x_1 + \ a_{22}x_2 + \cdots + \ a_{2n}x_n = 0$$
$$\cdots\cdots\cdots\cdots\cdots\cdots\cdots\cdots\cdots\cdots\cdots\cdots$$
$$a_{m1}x_1 + a_{m2}x_2 + \cdots + a_{mn}x_n = 0$$

is always consistent, because $(0, 0, \cdots, 0)$ is trivially a solution of the system. Regarding every solution $(t_1, t_2, \cdots, t_n)$ of the system as a vector $\mathbf{t} = [t_1, t_2, \cdots, t_n]$ of $\mathbf{R}^n$, we shall consider the set X_0 of all solutions of the system S_0 as a subset of the vector space $\mathbf{R}^n$. By Example 5.2.7, X_0 is a subspace of $\mathbf{R}^n$. In view of its importance in the theory of linear equations, we shall formulate this result as a theorem.

7.3.1 THEOREM *Let* S_0

$$a_{11}x_1 + \ a_{12}x_2 + \cdots + \ a_{1n}x_n = 0$$
$$a_{21}x_1 + \ a_{22}x_2 + \cdots + \ a_{2n}x_n = 0$$
$$\cdots\cdots\cdots\cdots\cdots\cdots\cdots\cdots\cdots\cdots\cdots\cdots$$
$$a_{m1}x_1 + a_{m2}x_2 + \cdots + a_{mn}x_n = 0$$

be a homogeneous system and X_0 *the set of solutions of* S_0. *Then* X_0 *is a subspace of the vector space* $\mathbf{R}^n$.

PROOF Let $\mathbf{u}$ and $\mathbf{v}$ be solutions of S_0. Converting them into column matrices U and V respectively. Then

$$A_0 U = 0 \quad \text{and} \quad A_0 V = 0 .$$

Therefore

$$A_0(U + V) = A_0 U + A_0 V = 0 .$$

Similarly for any scalar r

$$A_0(rU) = r(A_0 U) = 0 .$$

Therefore both $\mathbf{u} + \mathbf{v}$ and $r\mathbf{u}$ are solutions of S_0. Hence the set X_0 which also contains the trivial solution 0 is a subspace of $\mathbf{R}^n$.

By the above theorem it is now legitimate to call X_0 the *solution space* of the system S_0. Clearly we would want to find out the dimension of the solution space X_0 and develop an effective algorithm for finding a base of X_0 from which all solutions of S_0 would be generated.

Meanwhile we consider a simple example that would illustrate the main idea of the subsequent discussion.

7.3.2 EXAMPLE   The usual way of solving the following system

$$\begin{aligned}
x_1 + 2x_2 - 3x_3 &= 0 \\
2x_1 + 5x_2 - 2x_2 &= 0
\end{aligned} \tag{1}$$

is by the method of successive eliminations of unknowns. For example to eliminate the unknown x_1 from the second equation, we subtract twice the first equation from the second to get

$$x_2 + 4x_3 = 0 .$$

Then replace the second equation of (1) by the new equation to obtain a new system

$$\begin{aligned}
x_1 + 2x_2 - 3x_3 &= 0 \\
x_2 + 4x_3 &= 0
\end{aligned} \tag{2}$$

which has the same solutions as (1). Then we eliminate the unknown x_2 from the first equation by subtracting twice the second equation of (2) from the first equation to get

$$x_1 - 11x_3 = 0 .$$

Replace now the first equation of (2) by this new equation to obtain another system

$$\begin{aligned}
x_1 \quad - 11x_3 &= 0 \\
x_2 + 4x_3 &= 0
\end{aligned} \tag{3}$$

which has the same solutions as (2). But all solutions of (3) and hence all solutions of (1) are of the form $(11t, -4t, t)$ for arbitrary values of t. This means that the vector $[11, -4, 1]$ by itself is a base of the solution space of (1). Hence the solution space has dimension 1. On the other hand the rank of the coefficient matrix of (1) is 2. Therefore in this particular case the dimension of the solution space is equal to the number of unknowns minus the rank of the coefficient matrix.

In general two systems of linear equations are said to be *equivalent* if they have the same solutions. The systems (1), (2) and (3) in the above example are therefore all equivalent. We also observe that the transformation of the given system (1) into the equivalent system (2) corresponds to an elementary row transformation

$$\begin{pmatrix} 1 & 2 & -3 \\ 2 & 5 & -2 \end{pmatrix} \xrightarrow{\ r_2 \to r_2 - 2r_1\ } \begin{pmatrix} 1 & 2 & -3 \\ 0 & 1 & 4 \end{pmatrix}$$

of the coefficient matrix of (1) into that of (2). Similarly the transformation of (2) into (3) corresponds to

$$\begin{pmatrix} 1 & 2 & -3 \\ 0 & 1 & 4 \end{pmatrix} \xrightarrow{\ r_1 \to r_1 - 2r_2\ } \begin{pmatrix} 1 & 0 & -11 \\ 0 & 1 & 4 \end{pmatrix}.$$

In the subsequent discussion we shall see that the method by elementary row transformations used in the example can be applied to any system of homogeneous equations.

Let S_0:

$$\begin{aligned}
a_{11}x_1 + a_{12}x_2 + \cdots + a_{1n}x_n &= 0 \\
a_{21}x_1 + a_{22}x_2 + \cdots + a_{2n}x_n &= 0 \\
&\cdots\cdots\cdots \\
a_{m1}x_1 + a_{m2}x_2 + \cdots + a_{mn}x_n &= 0
\end{aligned}$$

be a homogeneous system with coefficient matrix A_0. Then it is quite easy to see that if T is any elementary row transformation, then the homogeneous system S_0' with coefficient matrix $A_0' = T(A_0)$ is equivalent to the given system S_0. For example if T is $r_1(A_0) \to r_1(A_0) + sr_2(A_0) = r_1(A_0')$, then S_0' is the homogeneous system

$$\begin{aligned}
(a_{11} + sa_{21})x_1 + (a_{12} + sa_{22})x_2 + \cdots + (a_{1n} + sa_{2n})x_n &= 0 \\
a_{21}x_1 + a_{22}x_2 + \cdots + a_{2n}x_n &= 0 \\
&\cdots\cdots\cdots \\
a_{m1}x_1 + a_{m2}x_2 + \cdots + a_{mn}x_n &= 0.
\end{aligned}$$

If $\mathbf{u}$ is a vector of the solution space X_0 of S_0, then $\mathbf{a}_i \cdot \mathbf{u} = 0$ for all row vectors $\mathbf{a}_i$ $(i = 1, 2, \cdots, m)$ of A_0. Since $(\mathbf{a}_1 + s\,\mathbf{a}_2) \cdot \mathbf{u} = \mathbf{a}_1 \cdot \mathbf{u} + s\mathbf{a}_2 \cdot \mathbf{u} = 0$, the vector $\mathbf{u}$ also belongs to the solution space X_0' of S_0'. Conversely if $\mathbf{v}$ is a vector of X_0' then $(\mathbf{a}_1 + s\,\mathbf{a}_2) \cdot \mathbf{v} = 0$ and $\mathbf{a}_i \cdot \mathbf{v} = 0$ for $i = 2, 3, \cdots, m$. But $\mathbf{a}_1 \cdot \mathbf{v} = (\mathbf{a}_1 + s\,\mathbf{a}_2) \cdot \mathbf{v} - s\,\mathbf{a}_2 \cdot \mathbf{v} = 0$; therefore $\mathbf{v}$ belongs to X_0. Hence $X_0 = X_0'$ and therefore S_0 and S_0' are equivalent. Similar argument applies to other elementary row transformations T on the coefficient matrix. We have therefore proved the following important theorem.

7.3.3 THEOREM *Let S_0 be a system of homogeneous linear equations with coefficient matrix A_0. If T is an elementary row transformation, then the homogeneous system with coefficient matrix $T(A_0)$ is equivalent to the system S_0.*

By the above theorem we can use elementary row transformations to bring a given system of linear equations into an equivalent but much simplified system. For this purpose we shall adopt some slight modifications to the procedure of Section 5.4 by which we evaluate the row rank of a matrix. Recall that the procedure consists of a number of stages at each of which a certain non-zero element on some column is made into 1 and chosen as the pivot to eliminate other elements on the same column, so that the said column vector is changed into a unit coordinate vector. No preference is given to any particular column on which the pivot is to be selected. At present each column index j of the coefficient matrix is tied to the index of one definite unknown x_j. It is therefore very much in our interest to devise a systematic selection of the column indices $j_1, j_2, \cdots, j_r$ from whose corresponding columns the pivots are chosen. Here is a description of the modified procedure of r stages where $r = \text{rank } A_0$.

STAGE 1. Let j_1 be the least column index so that the j_1-th column $\mathbf{c}_{j_1}(A_0)$ is non-zero. In other words, $a_{ij} = 0$ for all $i = 1, 2, \cdots, m$ and $j = 1, 2, \cdots, j_1 - 1$, but $a_{ij_1} \neq 0$ for at least one row index i. Pick a non-zero component, say $a_{i_1 j_1}$, of this column vector, change it to 1 by a row transformation of the first type and eliminate all other components of this column vector by row transformations of the second type. Consequently this column vector is changed into the unit coordinate vector e_{i_1} of $\mathbf{R}^m$ and the rows of A_0 are changed accordingly while the first $j_1 - 1$ zero columns remain unchanged. Now use an elementary row transformation of the third type to interchange the positions of the first row and the i_1-th row. Now the j_1-th column vector of the new matrix B_0 is the first unit coordinate vector e_1 of $\mathbf{R}^m$ and the system of linear equations with coefficient matrix B_0 is equivalent to original one with A_0. For the rows of the $m \times n$-matrix B_0, we have

(a) $1 \le j_1 \le n$;

(b_1) $b_{1j} = 0$ for all $j < j_1$; $b_{1j_1} = 1$;

(c) $b_{ij} = 0$ for all $i = 2, 3, \cdots, m$ and $j \le j_1$.

STAGE 2. Now examine the column vectors of the matrix B_0. Let j_2 be the least column index so that the j_2-th column vector $c_{j_2}(B_0)$ has, disregarding its first component, a non-zero component. In other words, $b_{ij} = 0$ for all $i = 2, 3, \cdots, m$ and $j = 1, 2, \cdots, j_2 - 1$, but $a_{ij_2} \neq 0$ for at least one row index $i \ge 2$. Then $j_1 < j_2$. Pick any one such non-zero component $b_{i_2 j_2}$ ($i_2 \ge 2$), change it to 1 and use it as the pivot to change the j_2-th column vector into a unit coordinate vector e_{i_2} of $\mathbf{R}^m$. Rows of B_0 are changed accordingly while the first $j_2 - 1$ columns of B_0 remain unchanged. Interchange the second row and the i_2-th row of B_0 to make the j_2-th column vector into e_2. Now the new matrix C_0 corresponds to a system equivalent to the original S_0 and has rows of the following special form.

(a) $1 \le j_1 < j_2 \le n$;

(b_1) $c_{1j} = 0$ for all $j < j_1$; $c_{1j_1} = 1$; $c_{1j_2} = 0$;

(b_2) $c_{2j} = 0$ for all $j < j_2$ and $c_{2j_2} = 1$;

(c) $c_{ij} = 0$ for $i = 3, 4, \cdots, m$ and $j \le j_2$.

Clearly we can carry on in the same manner for a total of r stages where $r = \text{rank } A_0 = \text{rank } B_0 = \text{rank } C_0 = \cdots$. At the conclusion of the procedure, we would then have transformed A_0 into a final matrix D_0 which corresponds to a system equivalent to the given S_0 and has rows of the following very special form.

(a) $1 \le j_1 < j_2 < \cdots < j_r \le n$;

(b_1) $d_{1j} = 0$ for all $j < j_1$; $d_{1j_1} = 1$; $d_{1j_2} = d_{1j_3} = \cdots = d_{1j_r} = 0$;

(b_2) $d_{2j} = 0$ for all $j < j_2$; $d_{2j_2} = 1$; $d_{2j_3} = d_{2j_4} = \cdots = d_{2j_r} = 0$;

. . . .

(b$_r$) $d_{rj} = 0$ for all $j < j_r$; $d_{rj_r} = 1$;

(c) $d_{ij} = 0$ for all $i > r$, i.e. $\mathbf{r}_{r+1}(D_0) = \mathbf{r}_{r+2}(D_0) = \cdots = \mathbf{r}_m(D_0) = 0$.

In general any $m \times n$-matrix whose elements satisfy the above specifications (a), (b$_1$), (b$_2$), $\cdots$, (b$_r$), (c) is called an *echelon matrix*. The r specified column indices $j_1, j_2, \cdots, j_r$ of (a) are called the *constrained column indices* of D_0; for these r constrained column indices, we have $\mathbf{c}_{j_1}(D_0) = \mathbf{e}_1, \mathbf{c}_{j_2}(D_0) = \mathbf{e}_2, \cdots, \mathbf{c}_{j_r}(D_0) = \mathbf{e}_r$. All other $n - r$ column indices of D_0 are called *unconstrained column indices*. The specified 1's and 0's form a block that resembles a flight of steps:

$$D_0 = \begin{pmatrix} 0 \cdots 0\ 1\ d_{1j_1+1} \cdots 0 \cdots\cdots 0 \cdots\cdots 0 \cdots\cdots d_{1n} \\ 0 \cdots\cdots\cdots 0\ \ 1\ d_{2j_2+1} \cdots 0 \cdots\cdots 0 \cdots\cdots d_{2n} \\ 0 \cdots\cdots\cdots\cdots\cdots 0\ \ 1\ d_{3j_3+1} \cdots 0 \cdots\cdots d_{3n} \\ \cdots\cdots\cdots\cdots\cdots\cdots\cdots\cdots\cdots\cdots\cdots \\ 0 \cdots\cdots\cdots\cdots\cdots\cdots\cdots\cdots 0\ \ 1\ d_{rj_r+1} \cdots d_{rn} \\ 0 \cdots\cdots\cdots\cdots\cdots\cdots\cdots\cdots\cdots\cdots 0 \\ \cdots\cdots\cdots\cdots\cdots\cdots\cdots\cdots\cdots\cdots \\ 0 \cdots\cdots\cdots\cdots\cdots\cdots\cdots\cdots\cdots\cdots 0 \end{pmatrix}$$

We put the result of the above discussion into the following lemma.

7.3.4 LEMMA *Any $m \times n$-matrix can be brought into echelon form by elementary row transformations.*

7.3.5 EXAMPLE Find all solutions to the following system S_0 of homogeneous linear equations.

$$\begin{aligned} -x_2 - 4x_3 + 5x_4 + 42x_5 + x_6 - 5x_7 &= 0 \\ x_2 + 4x_3 + x_4 + 6x_5 + 2x_6 - 10x_7 &= 0 \\ 2x_2 + 8x_3 + x_4 + 4x_5 + 3x_6 - 18x_7 &= 0 \\ -3x_2 - 12x_3 + 4x_4 + 38x_5 - 2x_6 + 13x_7 &= 0 \ . \end{aligned}$$

SOLUTION We use row transformations to bring the coefficient matrix into echelon form:

$$\begin{pmatrix} 0 & -1 & -4 & 5 & 42 & 1 & -5 \\ 0 & 1 & 4 & 1 & 6 & 2 & -10 \\ 0 & 2 & 8 & 1 & 4 & 3 & -18 \\ 0 & -3 & -12 & 4 & 38 & -2 & 13 \end{pmatrix} \longrightarrow \begin{pmatrix} 0 & 0 & 0 & 6 & 48 & 3 & -15 \\ 0 & 1 & 4 & 1 & 6 & 2 & -10 \\ 0 & 0 & 0 & -1 & -8 & -1 & 2 \\ 0 & 0 & 0 & 7 & 56 & 4 & -17 \end{pmatrix}$$

$$\begin{pmatrix} 0 & 1 & 4 & 1 & 6 & 2 & -10 \\ 0 & 0 & 0 & 6 & 48 & 3 & -15 \\ 0 & 0 & 0 & -1 & -8 & -1 & 2 \\ 0 & 0 & 0 & 7 & 56 & 4 & -17 \end{pmatrix} \longrightarrow \begin{pmatrix} 0 & 1 & 4 & 0 & -2 & 1 & -8 \\ 0 & 0 & 0 & 0 & 0 & -3 & -3 \\ 0 & 0 & 0 & 1 & 8 & 1 & -2 \\ 0 & 0 & 0 & 0 & 0 & -3 & -3 \end{pmatrix} \longrightarrow$$

$$\begin{pmatrix} 0 & 1 & 4 & 0 & -2 & 0 & -9 \\ 0 & 0 & 0 & 1 & 8 & 1 & -2 \\ 0 & 0 & 0 & 0 & 0 & -3 & -3 \\ 0 & 0 & 0 & 0 & 0 & -3 & -3 \end{pmatrix} \longrightarrow \begin{pmatrix} 0 & 1 & 4 & 0 & -2 & 0 & -9 \\ 0 & 0 & 0 & 1 & 8 & 0 & -3 \\ 0 & 0 & 0 & 0 & 0 & 1 & 1 \\ 0 & 0 & 0 & 0 & 0 & 0 & 0 \end{pmatrix} .$$

Here the constrained column indices are $2, 4, 6$ and the unconstrained column indices are $1, 3, 5, 7$. Therefore rank A_o = number of constrained column indices = 3. By Theorem 7.3.3 the given system is equivalent the following new one:

$$\begin{aligned} x_2 + 4x_3 \quad - 2x_5 \quad - 9x_7 &= 0 \\ x_4 + 8x_5 \quad - 3x_7 &= 0 \\ x_6 + \quad x_7 &= 0 . \end{aligned}$$

Each equation begins with an unknown with a different contrained index. Call the unknowns x_2, x_4, x_6 *constrained unknowns*, the other unknowns x_1, x_3, x_5, x_7 *unconstrained*. Then each of the three equations above expresses one constrained unknown in terms of the uncontrained unknowns. Assign to each uncontrained unknown x_k a free parameter λ_k, and express each constrained unknown in terms of the parameters by one equation to get:

$$\begin{aligned} x_2 &= -4\lambda_3 + 2\lambda_5 + 9\lambda_7 ; \\ x_4 &= \qquad - 8\lambda_5 + 3\lambda_7 ; \\ x_6 &= \qquad\qquad - \lambda_7 . \end{aligned}$$

Therefore for arbitrary $\lambda_1, \lambda_3, \lambda_5, \lambda_7$

$$\mathbf{t} = [\lambda_1, -4\lambda_3 + 2\lambda_5 + 9\lambda_7, \lambda_3, -8\lambda_5 + 3\lambda_7, \lambda_5, -\lambda_7, \lambda_7]$$

is a vector of the solution space X_0 of S_0. Every time we put one of the parameters equal to 1 and the others zero, we obtain a vector of the solution space X_0:

$$\begin{aligned} \mathbf{u}_1 &= [1, \quad 0, 0, \quad 0, 0, \quad 0, 0] \ ; \\ \mathbf{u}_3 &= [0, \ -4, 1, \quad 0, 0, \quad 0, 0] \ ; \\ \mathbf{u}_5 &= [0, \quad 2, 0, -8, 1, \quad 0, 0] \ ; \\ \mathbf{u}_7 &= [0, \quad 0, 0, \quad 3, 0, -1, 1] \ . \end{aligned}$$

Here the components of the vector $\mathbf{u}_k$ are simply the coefficients of the parameter λ_k in the components of the general solution $\mathbf{t}$. It follows from

$\mathbf{t} = \lambda_1\mathbf{u}_1 + \lambda_3\mathbf{u}_3 + \lambda_5\mathbf{u}_5 + \lambda_7\mathbf{u}_7$ that the solution space X_0 of S_0 is generated by the four vectors $\mathbf{u}_k$. On the other hand the 4×7-matrix

$$U = \begin{pmatrix} 1 & 0 & 0 & 0 & 0 & 0 & 0 \\ 0 & -4 & 1 & 0 & 0 & 0 & 0 \\ 0 & 2 & 0 & -8 & 1 & 0 & 0 \\ 0 & 0 & 0 & 3 & 0 & -1 & 1 \end{pmatrix}$$

with row vectors $\mathbf{u}_1$, $\mathbf{u}_3$, $\mathbf{u}_5$, $\mathbf{u}_7$ has maximum rank 4 because among its column vectors we find $\mathbf{c}_1(u) = \mathbf{e}_1$, $\mathbf{c}_3(u) = \mathbf{e}_2$, $\mathbf{c}_5(u) = \mathbf{e}_3$, $\mathbf{c}_7(u) = \mathbf{e}_4$ of $\mathbf{R}^4$. Therefore $\dim X_0 =$ number of unconstrained unknowns $=$ number of unknowns minus rank A_o.

After working through the above example in detail, we use it as a model to prove the last important theorem on systems of homogeneous linear equations.

7.3.6 THEOREM *Let S_0 be a system of homogeneous linear equations in n unknowns, A_0 its coefficient matrix and X_0 its solution space. Then $\dim X_0 = n - \text{rank } A_0$.*

PROOF By 7.3.3 and 7.3.4, S_0 is equivalent to a system $\overline{S_0}$ whose coefficient matrix D_0 is in echelon form. Let $1 \leq j_1 < j_2 < \cdots < j_r \leq n$ be the constrained column indices. Then rank $A_0 = $ rank $D_0 = r$. On the first row of D_0 we find $d_{1j} = 0$ for all $j < j_1$, $d_{1j_1} = 1$ and $d_{1j_2} = \cdots = d_{1j_r} = 0$. Therefore the first constrained unknown x_{j_1} can be expressed in terms of the unconstrained unknowns. Similarly we can express all r constrained unknowns $x_{j_1}, x_{j_2}, \cdots, x_{j_r}$ in terms of the $n - r$ unconstrained unknowns x_k where $k \neq j_1, j_2, \cdots, j_r$:

$$x_{j_1} = -\sum_k d_{1k} x_k$$

$$x_{j_2} = -\sum_k d_{2k} x_k$$

$$\cdots\cdots\cdots\cdots$$

$$x_{j_r} = -\sum_k d_{rk} x_k$$

where the coefficients d_{ik} are elements of the echelon matrix D_0 and the summations are taken over all the unconstrained column indices $k \neq j_1$, $j_2, \cdots, j_r$. For each unconstrained unknown x_k, let λ_k be a parameter. Putting

$$t_k = \lambda_k \quad \text{for} \quad k \neq j_1, j_2, \cdots, j_r, \text{ and}$$

$$t_{j_1} = -\sum_k d_{1k}\lambda_k$$

$$\cdots \cdots \cdots \cdots \cdots$$

$$t_{j_2} = -\sum_k d_{rk}\lambda_k$$

we obtain a general solution

$$\mathbf{t} = [t_1, t_2, \cdots, t_n]$$

of the system $\overline{S_0}$ and hence of S_0. Then the $n - r$ vectors $\mathbf{u}_k$ ($k \neq j_1$, $j_2, \cdots, j_r$) whose components are the coefficients of λ_k in $\mathbf{t} = [t_1, t_2, \cdots, t_n]$ form a base of the solution space X_0 of S_0. Therefore $\dim X_0 = n - r$.

7.3.7 EXAMPLE Find a parametric representation for the solutions of the homogeneous system

$$
\begin{aligned}
x_2 + x_3 + 5x_4 + 3x_5 &= 0 \\
x_1 - 2x_2 - x_3 - 4x_4 \phantom{{}+ 3x_5} &= 0 \\
-x_1 + 3x_2 + 2x_3 + x_4 - 5x_5 &= 0 \\
3x_1 - x_2 + 2x_3 + 6x_4 + 8x_5 &= 0 .
\end{aligned}
$$

SOLUTION Apply appropriate elementary row transformations to the coefficient matrix

$$
\begin{pmatrix}
0 & 1 & 1 & 5 & 3 \\
1 & -2 & -1 & -4 & 0 \\
-1 & 3 & 2 & 1 & -5 \\
3 & -1 & 2 & 6 & 8
\end{pmatrix}
\longrightarrow
\begin{pmatrix}
1 & -2 & -1 & -4 & 0 \\
0 & 1 & 1 & 5 & 3 \\
-1 & 3 & 2 & 1 & -5 \\
3 & -1 & 2 & 6 & 8
\end{pmatrix}
\longrightarrow
$$

$$
\begin{pmatrix}
1 & -2 & -1 & -4 & 0 \\
0 & 1 & 1 & 5 & 3 \\
0 & 1 & 1 & -3 & -5 \\
0 & 5 & 5 & 18 & 8
\end{pmatrix}
\longrightarrow
\begin{pmatrix}
1 & 0 & 1 & 6 & 6 \\
0 & 1 & 1 & 5 & 3 \\
0 & 0 & 0 & -8 & -8 \\
0 & 0 & 0 & -7 & -7
\end{pmatrix}
\longrightarrow
$$

$$
\begin{pmatrix}
1 & 0 & 1 & 6 & 6 \\
0 & 1 & 1 & 5 & 3 \\
0 & 0 & 0 & 1 & 1 \\
0 & 0 & 0 & 1 & 1
\end{pmatrix}
\longrightarrow
\begin{pmatrix}
1 & 0 & 1 & 0 & 0 \\
0 & 1 & 1 & 0 & -2 \\
0 & 0 & 0 & 1 & 1 \\
0 & 0 & 0 & 0 & 0
\end{pmatrix}
$$

to obtain an equivalent system

$$
\begin{aligned}
x_1 \phantom{{}+ x_2} + x_3 \phantom{{}- 2x_5} &= 0 \\
x_2 + x_3 \phantom{{}+} - 2x_5 &= 0 \\
x_4 + x_5 &= 0 .
\end{aligned}
$$

The constrained unknowns are x_1, x_2, x_4 and the unconstrained ones are x_3, x_5. Take $x_3 = \lambda$ and $x_5 = \mu$ as free independent parameters to get solutions

$$(-\lambda, -\lambda + 2\mu, \lambda, -\mu, \mu)$$

of the given system in parametric form.

REMARK 7.3.8   At the end of Section 7.2, we found that elementary column transformations should not be used simultaneously in evaluating the ranks of the coefficient matrix and the augmented matrix of a system of linear equations. Here again in this section we have also refrained from using column transformations. The reason for this is that while row transformation would transform a system into an equivalent system, column transformations would not do so. For example the last system

$$\begin{aligned}
x_1 \quad + x_3 \qquad\qquad &= 0 \\
x_2 + x_3 \qquad - 2x_5 &= 0 \\
x_4 + \quad x_5 &= 0
\end{aligned}$$

of 7.3.7 with general solution $(-\lambda, -\lambda + 2\mu, \lambda, -\mu, \mu)$ can be readily transformed by column transformations into

$$\begin{aligned}
x_1 \qquad\qquad &= 0 \\
x_2 \quad &= 0 \\
x_4 &= 0
\end{aligned}$$

with general solution $(0, 0, \lambda, 0, \mu)$. The two systems are clearly not equivalent.

EXERCISES

1. Which of the following matrices are echelon ones?

(a) $\begin{pmatrix} 1 & 0 & 0 \\ 0 & 1 & 0 \\ 0 & 0 & 0 \end{pmatrix}$ (b) $\begin{pmatrix} 0 & 0 & 0 \\ 1 & 0 & 1 \\ 0 & 1 & 0 \end{pmatrix}$ (c) $\begin{pmatrix} 1 & 0 & 1 & 0 \\ 0 & 1 & 1 & 2 \\ 0 & 1 & 1 & 0 \end{pmatrix}$

(d) $\begin{pmatrix} 1 & 3 & 0 & -2 & 4 \\ 0 & 0 & 1 & 3 & 7 \\ 0 & 0 & 0 & 0 & 0 \end{pmatrix}$.

For those non-echelon ones, can we change them to echelon ones?

2. By using elementary row transformations, reduce the following matrices to echelon ones.

(a) $\begin{pmatrix} 3 & 6 \\ -1 & 2 \end{pmatrix}$ (b) $\begin{pmatrix} 3 & 1 & 1 & 1 \\ 5 & -1 & 1 & -1 \end{pmatrix}$ (c) $\begin{pmatrix} 2 & -1 & 0 & 1 \\ -1 & 2 & -3 & 1 \\ 1 & -2 & 0 & -1 \\ 0 & 1 & 1 & 1 \end{pmatrix}$.

3. Find all solutions of the following systems of linear equations.

(a) $2x_1 + 5x_2 + 6x_3 = 0$
    $x_1 - 2x_2 + x_3 = 0$.

(b) $3x_1 + x_2 + x_3 + x_4 = 0$
    $5x_1 - x_2 + x_3 - x_4 = 0$.

4. Solve

$$2x_1 + 3x_2 + x_3 = 0$$
$$4x_1 - x_2 + 5x_3 = 0$$
$$3x_1 + 6x_2 - 7x_3 = 0.$$

5. Solve

$$x_1 + x_2 + x_3 = 0$$
$$2x_1 + 3x_2 + 2x_3 = 0$$
$$4x_1 + 5x_2 + 4x_3 = 0.$$

6. Solve

$$2x_1 - x_2 \qquad + x_4 = 0$$
$$-x_1 + 2x_2 - 3x_3 + x_4 = 0$$
$$x_1 - 2x_2 \qquad - x_4 = 0$$
$$x_2 + x_3 + x_4 = 0.$$

7. Use Theorem 7.3.6 to prove that a system of homogeneous linear equations with coefficient matrix A_0 of order $n \times n$ has non-trivial solution if and only if $\det A_0 = 0$.

8. By evaluating suitable determinants, show that the only solution of

$$x_1 - 2x_2 = ax_1$$
$$x_1 - x_2 = ax_2$$

is $x_1 = 0$, $x_2 = 0$ for all real numbers a.

9. Find all real values of a for which the following system of linear equations has a non-trivial solution

$$x_1 - 2x_2 = ax_1$$
$$-3x_1 + 2x_2 = ax_2.$$

Hence find the solutions.

299

10. For what value of k will the system

$$x + y + kz = 0$$
$$x + y - z = 0$$
$$kx + y + z = 0$$

    have non-trivial solutions? Hence find the solutions.

In Exercises 11-13, by considering suitable system of linear equations, determine whether the given vectors are linearly independent or not.

11. $[1, 0, 2]$, $[1, -1, 1]$, $[-2, 1, 1]$.

12. $[1, 2, 3, 4]$, $[2, 1, 0, -1]$, $[1, -1, -2, 1]$.

13. $[3, 1, -1, 4]$, $[2, 0, 0, 3]$, $[4, 7, -2, 2]$, $[1, -4, -1, 4]$.

14. The notion of orthogonal vectors in $\mathbf{R}^2$ and $\mathbf{R}^3$ can be extended to $\mathbf{R}^n$ for any n, by $[a_1, a_2, \cdots, a_n] \cdot [b_1, b_2, \cdots, b_n] = a_1 b_1 + \cdots + a_n b_n = 0$. Find all the vectors in $\mathbf{R}^4$ which are orthogonal to both $[1, -1, 7, -1]$ and $[2, 3, -8, 1]$.

15. For each of the following systems of homogeneous linear equations, find a base for the solution space.

    (a) $x_1 + 2x_2 + 3x_3 = 0$ (b) $2x_1 \quad\quad -3x_3 + x_4 = 0$
            $x_2 + 3x_3 = 0$. $x_2 \quad\quad + x_4 = 0$.

16. Extend the bases in the above question to a base in $\mathbf{R}^3$ and a base in $\mathbf{R}^4$ respectively. (Hint: You may consider the unit coordinate vectors.)

7.4 Inhomogeneous system

We shall see here that the effective method of solving homogeneous linear equations is also applicable to inhomogeneous equations with some slight modification.

Let S:

$$a_{11}x_1 + a_{12}x_2 + \cdots + a_{1n}x_n = b_1$$
$$a_{21}x_1 + a_{22}x_2 + \cdots + a_{2n}x_n = b_2$$
$$\cdots\cdots\cdots\cdots\cdots\cdots\cdots\cdots\cdots\cdots\cdots\cdots$$
$$a_{m1}x_1 + a_{m2}x_2 + \cdots + a_{mn}x_n = b_m$$

be a system of linear equations in n unknowns where the constant terms $b_1, b_2, \cdots, b_m$ need not be all equal to zero. As usual we denote by A_0 the coefficient matrix and by A the augmented matrix of the system S. Putting all constant terms of S equal to zero, we get a homogeneous system S_0:

$$a_{11}x_1 + a_{12}x_2 + \cdots + a_{1n}x_n = 0$$
$$a_{21}x_1 + a_{22}x_2 + \cdots + a_{2n}x_n = 0$$
$$\cdots\cdots\cdots\cdots\cdots\cdots\cdots\cdots\cdots\cdots\cdots\cdots$$
$$a_{m1}x_1 + a_{m2}x_2 + \cdots + a_{mn}x_n = 0 \, .$$

We shall call S_0 the *associated homogeneous system* of the system S. Then both S and S_0 have the same coefficient matrix A_0.

Suppose that rank A = rank A_0. Then by 7.2.1 the system S is consistent. Under this assumption we shall now examine the relationship between the set X of all solutions of S and the solution space X_0 of S_0. Let $\mathbf{u} \in X$ be a fixed solution of S. Then $\mathbf{a}_i \cdot \mathbf{u} = b_i$ for all $i = 1, 2, \cdots, m$. If $\mathbf{t} \in X_0$ is any solution of the associated system S_0, then $\mathbf{a}_i \cdot \mathbf{t} = 0$. It follows from $\mathbf{a}_i \cdot (\mathbf{u} + \mathbf{t}) = \mathbf{a}_i \cdot \mathbf{u} + \mathbf{a}_i \cdot \mathbf{t} = b_i + 0 = b_i$ that the sum $\mathbf{u} + \mathbf{t}$ is also a solution of S. Conversely if $\mathbf{v} \in X$ is any other solution of S, then $\mathbf{a}_i \cdot (\mathbf{v} - \mathbf{u}) = \mathbf{a}_i \cdot \mathbf{v} - \mathbf{a}_i \cdot \mathbf{u} = b_i - b_i = 0$. Therefore $\mathbf{t} = \mathbf{v} - \mathbf{u} \in X_0$ and $\mathbf{v} = \mathbf{u} + \mathbf{t}$. Hence we have proved the following relation between X and X_0.

7.4.1 THEOREM *Let S be a consistent system of linear equations and S_0 its associated homogeneous system. If $\mathbf{u}$ is any fixed solution of S, then $\mathbf{u} + \mathbf{t}$ is a solution of S for every solution $\mathbf{t}$ of S_0. Conversely every solution $\mathbf{v}$ of S can be written into $\mathbf{v} = \mathbf{u} + \mathbf{t}$ for some solution $\mathbf{t}$ of S_0.*

In other words, if S is a consistent system with associated homogeneous system S_0, then the set X of solutions of S has the form $X = \mathbf{u} + X_0 = \{\mathbf{u} + \mathbf{t} | \mathbf{t} \in X_0\}$ where $\mathbf{u}$ is any solution of S and X_0 the solution space of S_0. In general X is not a subspace but just a subset of $\mathbf{R}^n$; therefore it is referred to as the *solution set* but not the solution space of S. The following corollaries are direct consequences of the above theorem and Theorem 7.3.6.

7.4.2. COROLLARY *A system S of linear equations in n unknowns has one and only one solution if and only if both the coefficient matrix and the augmented matrix of S have the same rank equal to n.*

7.4.3 COROLLARY *A system S of n linear equations in n unknowns has one and only one solution if and only if the coefficient matrix of S has a rank equal to n.*

In the last section we have seen that elementary row transformations on the coefficient matrix can be used effectively for solving a homogeneous system. This method is also effective on inhomogeneous systems. It is easy to see that if T is any elementary row transformation and $A' = T(A)$ is the transform of the augmented matrix A of the system S under T, then the system S'

$$
\begin{aligned}
a'_{11}x_1 + a'_{12}x_2 + \cdots + a'_{1n}x_n &= b'_1 \\
a'_{21}x_1 + a'_{22}x_2 + \cdots + a'_{2n}x_n &= b'_2 \\
&\cdots\cdots\cdots\cdots \\
a'_{m1}x_1 + a'_{m2}x_2 + \cdots + a'_{mn}x_n &= b'_m
\end{aligned}
$$

will have the same solutions as S. Therefore S and S' are equivalent systems and $X = X'$ for their solution sets.

Consequently as the first step towards a complete solution of system S, we change the augmented matrix A into an echelon matrix

$$
D = \begin{pmatrix}
0 \cdots 0\ 1\ d_{1j_1+1} \cdots 0 \cdots\cdots\cdots\cdots\cdots\cdots d_{1n}\ f_1 \\
0 \cdots\cdots\cdots\cdots 0\ 1\ d_{2j_2+1} \cdots\cdots\cdots d_{2n}\ f_2 \\
\cdots\cdots\cdots\cdots\cdots\cdots\cdots\cdots\cdots\cdots\cdots\cdots \\
0 \cdots\cdots\cdots\cdots\cdots 0\ 1\ d_{rj_r+1} \cdots d_{rn}\ f_r \\
0 \cdots\cdots\cdots\cdots\cdots\cdots\cdots\cdots\cdots\cdots\cdots 0 \\
\cdots\cdots\cdots\cdots\cdots\cdots\cdots\cdots\cdots\cdots\cdots \\
0 \cdots\cdots\cdots\cdots\cdots\cdots\cdots\cdots\cdots\cdots\cdots 0
\end{pmatrix}
$$

by a sequence of elementary row transformations. At the same time, the coefficient matrix A_0 consisting of the first n columns of the augmented matrix A is transformed into an echelon matrix D_0 consisting of the first n columns of D by the same sequence of elementary row transformations. Let us examine the constrained column indices of $D : 1 \le j_1 < j_2 < \cdots < j_r \le n+1$. Depending on the value of the last constrained column index there are two cases: $j_r = n+1$ and $j_r \le n$.

Case 1. $j_r = n + 1$.

The echelon matrix D has r constrained column indices and the echelon matrix D_0 has just $r - 1$ constrained column indices. This means that D has r linearly independent row vectors and $m - r$ zero row vectors while D_0 has $r - 1$ linearly independent row vectors and $m - r + 1$ zero row vectors. Therefore $r = \operatorname{rank} D = \operatorname{rank} A$ and $r - 1 = \operatorname{rank} D_0 = \operatorname{rank} A_0$. Hence the given system is inconsistent and has no solutions.

Case 2. $j_r \leq n$.

Both echelon matrices D and D_0 have the same number of constrained column indices and hence the same rank. As rank $A =$ rank $D =$ rank $D_0 =$ rank A_0, the given system is consistent. The given system S is now equivalent to the new system $\overline{S}$:

$$x_{j_1} + d_{1j_1+1}x_{j_1+1} + \cdots\cdots\cdots + d_{1n}x_n = f_1$$
$$x_{j_2} + d_{2j_2+1}x_{j_2+1} + \cdots\cdots + d_{2n}x_n = f_2$$
$$\cdots\cdots\cdots\cdots\cdots\cdots\cdots\cdots$$
$$x_{j_r} + d_{rj_r+1}x_{j_r+1} + \cdots + d_{rn}x_n = f_r$$

with augmented matrix D and coefficient matrix D_0. Furthermore their associated homogeneous systems S_0 and $\overline{S}_0$ are also equivalent and have identical solution space $X_0 = \overline{X}_0$ of dimension $n - r$. Now by Theorem 7.4.1 it is sufficient to find a particular solution $\mathbf{u}$ of $\overline{S}$ and a general solution $\mathbf{t}$ of $\overline{S}_0$ to express all solutions of $\overline{S}$ and hence of S in the form $\mathbf{u} + \mathbf{t}$.

For the components of the particular solution

$$\mathbf{u} = [u_1, u_2, \cdots, u_n]$$

of $\overline{S}$ and hence of S, we may put $u_k = 0$ if k is not one of the $n - r$ unconstrained column indices of D and $u_{j_1} = f_1, u_{j_2} = f_2, \cdots, u_{j_r} = f_r$ for the r constrained column indices. The general solution

$$\mathbf{t} = [t_1, t_2, \cdots, t_n]$$

of $\overline{S}_0$ and hence of S_0 is the same as that of Theorem 7.3.6. Therefore among the components of a general solution

$$\mathbf{v} = \mathbf{u} + \mathbf{t} = [v_1, v_2, \cdots, v_n]$$

of S we find the $n - r$ unconstrained components to be

$$v_k = \lambda_k \qquad k \neq j_1, j_2, \cdots, j_r$$

with independent parameters λ_k, and the r constrained components of the form

$$v_{j_i} = f_i - \sum_k d_{ik}\lambda_k \qquad i = 1, 2, \cdots, r$$

where d_{ik} are elements of D_0 and the summation is taken over all unconstrained indices.

7.4.4 EXAMPLE Solve the system of inhomogeneous linear equations

$$\begin{array}{rcl} x_2 + x_3 + 5x_4 + 3x_5 &=& 2 \\ x_1 - 2x_2 - x_3 - 4x_4 &=& 2 \\ -x_1 + 3x_2 + 2x_3 + x_4 - 5x_5 &=& -8 \\ 3x_1 - x_2 + 2x_3 + 6x_4 + 8x_5 &=& 9 \ . \end{array}$$

SOLUTION The coefficient matrix is the same as the coefficient matrix of the homogeneous system of Example 7.3.7. Using the same sequence of elementary row transformations we get

$$\begin{pmatrix} 0 & 1 & 1 & 5 & 3 & 2 \\ 1 & -2 & -1 & -4 & 0 & 2 \\ -1 & 3 & 2 & 1 & -5 & -8 \\ 3 & -1 & 2 & 6 & 8 & 9 \end{pmatrix} \longrightarrow \begin{pmatrix} 1 & 0 & 1 & 0 & 0 & 0 \\ 0 & 1 & 1 & 0 & -2 & -3 \\ 0 & 0 & 0 & 1 & 1 & 1 \\ 0 & 0 & 0 & 0 & 0 & 0 \end{pmatrix} .$$

Therefore

$$\begin{array}{rcl} x_1 + x_3 &=& 0 \\ x_2 + x_3 - 2x_5 &=& -3 \\ x_4 + x_5 &=& 1 \ . \end{array}$$

Hence $(0, -3, 0, 1, 0)$ is a particular solution of the inhomogeneous system while $(-\lambda, -\lambda+2\mu, \lambda, -\mu, \mu)$ is a general solution of the associated homogeneous system. Therefore $(-\lambda, -3 - \lambda + 2\mu, \lambda, 1 - \mu, \mu)$ is a general solution of the given system.

7.4.5 EXAMPLE Solve the system of inhomogeneous linear equations

$$\begin{array}{rcl} x_1 + 2x_2 - x_3 + 4x_5 &=& 2 \\ x_1 + 4x_2 - 5x_3 + x_4 + 3x_5 &=& 1 \\ 3x_1 + 2x_2 + 5x_3 + 2x_4 + 2x_5 &=& 12 \\ 2x_1 - 2x_2 + 10x_3 + x_4 - x_5 &=& 11 \ . \end{array}$$

SOLUTION Transformation of the augmented matrix:

$$\begin{pmatrix} 1 & 2 & -1 & 0 & 4 & 2 \\ 1 & 4 & -5 & 1 & 3 & 1 \\ 3 & 2 & 5 & 2 & 2 & 12 \\ 2 & -2 & 10 & 1 & -1 & 11 \end{pmatrix} \longrightarrow \begin{pmatrix} 1 & 2 & -1 & 0 & 4 & 2 \\ 0 & 2 & -4 & 1 & -1 & -1 \\ 0 & -4 & 8 & 2 & -10 & 6 \\ 0 & -6 & 12 & 1 & -9 & 7 \end{pmatrix} \longrightarrow$$

$$\begin{pmatrix} 1 & 0 & 3 & -1 & 5 & 3 \\ 0 & 2 & -4 & 1 & -1 & -1 \\ 0 & 0 & 0 & 4 & -12 & 4 \\ 0 & 0 & 0 & 4 & -12 & 4 \end{pmatrix} \longrightarrow \begin{pmatrix} 1 & 0 & 3 & -1 & 5 & 3 \\ 0 & 2 & -4 & 1 & -1 & -1 \\ 0 & 0 & 0 & 1 & -3 & 1 \\ 0 & 0 & 0 & 0 & 0 & 0 \end{pmatrix} \longrightarrow$$

$$\begin{pmatrix} 1 & 0 & 3 & 0 & 2 & 4 \\ 0 & 2 & -4 & 0 & 2 & -2 \\ 0 & 0 & 0 & 1 & -3 & 1 \\ 0 & 0 & 0 & 0 & 0 & 0 \end{pmatrix} \longrightarrow \begin{pmatrix} 1 & 0 & 3 & 0 & 2 & 4 \\ 0 & 1 & -2 & 0 & 1 & -1 \\ 0 & 0 & 0 & 1 & -3 & 1 \\ 0 & 0 & 0 & 0 & 0 & 0 \end{pmatrix} .$$

The desired equivalent system:

$$\begin{aligned} x_1 & & + 3x_3 & & + 2x_5 & = 4 \\ & x_2 & - 2x_3 & & + x_5 & = -1 \\ & & & x_4 & - 3x_5 & = 1 . \end{aligned}$$

The general solution:

$$(4 - 3\lambda - 2\mu, \ -1 + 2\lambda - \mu, \ \lambda, \ 1 + 3\mu, \ \mu) .$$

7.4.6 EXAMPLE Find a necessary and sufficient condition on the constants a and b such that the following system is consistent.

$$\begin{aligned} bx_1 + & \quad x_2 + & (ab + a + b)x_3 + & \quad (ab^2 + b^2)x_4 = a^2 \\ -ax_1 + & (a+1)x_2 + & (ab + a + b)x_3 + & (ab^2 - a^2b + b^2)x_4 = a^2 + ab^2 \\ & (b+1)x_2 + & (b^2 + ab + a + b)x_3 + & (b^3 + b^2)x_4 = b^3 + a^2 . \end{aligned}$$

SOLUTION Transformation of the augmented matrix:

$$\begin{pmatrix} b & 1 & ab+a+b & ab^2+b^2 & a^2 \\ -a & a+1 & ab+a+b & ab^2-a^2b+b^2 & a^2+ab^2 \\ 0 & b+1 & b^2+ab+a+b & b^3+b^2 & b^3+a^2 \end{pmatrix} \begin{matrix} \mathbf{r_2 \to r_2 - r_1} \\ \longrightarrow \\ \mathbf{r_3 \to r_3 - r_1} \end{matrix}$$

$$\begin{pmatrix} b & 1 & ab+a+b & ab^2+b^2 & a^2 \\ -a-b & a & 0 & -a^2b & ab^2 \\ -b & b & b^2 & -ab^2+b^3 & b^3 \end{pmatrix} \begin{matrix} \mathbf{r_2 \to r_2 - ar_1} \\ \longrightarrow \\ \mathbf{r_3 \to r_3 - br_1} \end{matrix}$$

$$\begin{pmatrix} b & 1 & (ab+a+b) & ab^2+b^2 & a^2 \\ -(ab+a+b) & 0 & -a(ab+a+b) & -ab(ab+a+b) & -a(a^2-b^2) \\ -b(b+1) & 0 & -ab(b+1) & -ab^2(b+1) & -b(a^2-b^2) \end{pmatrix}$$

Consider the coefficient matrix B_0 of the last augmented matrix B. The second row vector $\mathbf{r_2}(B_0) = -(ab + a + b)[1, 0, a, ab]$ and the third row vector $\mathbf{r_3}(B_0) = -b(b + 1)[1, 0, a, ab]$ are linearly dependent, being multiples of $[1, 0, a, ab]$. But the first row vector and the second row vector are linearly independent because their second components are 1 and 0 respectively.

Therefore rank $B_0 = 2$. It follows from $b(b+1)\mathbf{r}_2(B_0) = (ab + a + b)\mathbf{r}_3(B_0)$ that rank $B_0 = $ rank $B = 2$ if and only if

$$b(b+1)\mathbf{r}_2(B) = (ab + a + b)\mathbf{r}_3(B) .$$

A comparison of their last components yields

$$b(b+1)a(a^2 - b^2) = (ab + a + b)b(a^2 - b^2) .$$

Simplifying we get

$$b^2(a^2 - b^2) = 0 .$$

Therefore a necessary and sufficient condition for the given system to be consistent is that $b = 0$ or $b = \pm a$.

EXERCISES

In Exercises 1-4, find all solutions, if any, of the given system of linear equations.

1. $\begin{aligned} x_1 + 2x_2 + x_3 &= 6 \\ 2x_1 + 4x_2 + x_3 &= 7 \\ 3x_1 + 2x_2 + 9x_3 &= 14 . \end{aligned}$

2. $\begin{aligned} 2x_1 + x_2 - x_3 &= 1 \\ -x_1 + x_2 + x_3 &= 3 \\ 3x_1 - 2x_2 + 2x_3 &= 12 . \end{aligned}$

3. $\begin{aligned} x_1 - x_3 + x_4 &= 5 \\ 2x_1 + 6x_2 - 4x_3 + 2x_4 &= 4 \\ 3x_1 - 2x_2 + 2x_3 &= 2 . \end{aligned}$

4. $\begin{aligned} 2x_1 + 3x_2 - x_3 + x_4 &= 2 \\ x_1 + x_2 + x_3 - x_4 &= -3 \\ x_1 + 2x_3 + 3x_4 &= 10 \\ 2x_2 + x_3 - x_4 &= 1 . \end{aligned}$

5. When $x = -1, 1, 2$, the function $ax^2 + bx + c$ takes the values $9, 5, 12$. Find the value of the function when $x = \frac{1}{2}$.

6. Find all real numbers k such that the following system of linear equations is consistent. What then is the solution?

$$\begin{aligned} x_1 - x_2 + x_3 &= 7 \\ 4x_1 + 4x_2 + 2x_3 &= 9 \\ 5x_1 + 3x_2 + 3x_3 &= k \; . \end{aligned}$$

7. Show that the following system of linear equations

$$\begin{aligned} x_1 - x_2 &= a_1 \\ x_2 - x_3 &= a_2 \\ x_3 - x_4 &= a_3 \\ x_4 - x_5 &= a_4 \\ x_5 - x_1 &= a_5 \end{aligned}$$

is consistent if and only if $a_1 + a_2 + a_3 + a_4 + a_5 = 0$. Find the solution under such a condition.

8. The system of linear equations

$$\begin{aligned} 2x_1 + x_2 - x_3 &= 6 \\ x_1 - 2x_2 - 2x_3 &= 1 \\ x_1 - 12x_2 - 8x_3 &= -7 \end{aligned}$$

represents three planes in $\mathbf{R}^3$. Solve the system. What does this mean about the three planes?

9. Find a plane $a_1 x + a_2 y + a_3 z + a_4 = 0$ in $\mathbf{R}^3$ which passes through the points $(-1, -2, 2)$, $(4, 2, 1)$, $(0, 4, -5)$.

10. Consider the system of linear equations

$$\begin{aligned} 3x_1 + x_2 - 5x_3 &= a \\ 2x_1 - x_2 + 3x_3 &= b \\ -5x_1 - 5x_2 + 21x_3 &= c \; . \end{aligned}$$

(a) Show that the system is inconsistent if $c \neq 2b - 3a$.
(b) Solve the system for $a = 1$, $b = 2$, and $c = 1$.

11. Solve

$$\begin{aligned} kx_1 + x_2 + x_3 &= 1 \\ x_1 + kx_2 + x_3 &= k \\ x_1 + x_2 + kx_3 &= k^2 \; , \end{aligned}$$

when (i) $k \neq -2, 1$, (ii) $k = -2$, and (iii) $k = 1$.

12. Solve

$$x_1 + (k^2 + 1)x_2 + \qquad 2x_3 = k$$
$$kx_1 + \qquad kx_2 + (2k+1)x_3 = 0$$
$$x_1 + (2k+1)x_2 + \qquad 2x_3 = 2 \,,$$

when (i) $k \neq 0, 2$, (ii) $k = 0$, and (iii) $k = 2$.

13. Consider the system of linear equations

$$x_1 + 2x_2 - \quad x_3 = b$$
$$2x_1 + \quad x_2 + ax_3 = 0$$
$$ax_1 + \quad x_2 + 2x_3 = 0$$

for real numbers a and b. Show that the system has a unique solution if $a^2 \neq 4$. For what values of b does the system have a solution when $a^2 = 4$?

14. Given that

$$6x_1 - 4x_2 + 3x_3 = a$$
$$4x_1 + 2x_2 + 8x_3 = b$$
$$x_1 + 2x_2 + 3x_3 = c \,,$$

find x_1, x_2, x_3 in terms of a, b and c. Hence obtain the inverse of the matrix

$$\begin{pmatrix} 6 & -4 & 3 \\ 4 & 2 & 8 \\ 1 & 2 & 3 \end{pmatrix} .$$

15. How should the coefficients a, b and c be chosen so that the system of linear equations

$$ax_1 + \quad x_2 - \quad cx_3 = -5$$
$$cx_1 - bx_2 + 2x_3 = \quad 3$$
$$bx_1 + ax_2 + \quad x_3 = 12$$

has the solution $x_1 = -1$, $x_2 = 3$, $x_3 = 5$?

16. (a) Solve the system of linear equations

$$x_1 + \quad x_2 + 2x_3 - \quad x_4 = \quad 3$$
$$2x_1 + 6x_2 + 3x_3 - 4x_4 = -6$$
$$-3x_1 \qquad - \quad x_3 + 2x_4 = -7$$
$$5x_1 + \quad x_2 \qquad + \quad x_4 = \quad 1 \,.$$

(b) Hence express the vector $[3, -6, -7, 1]$ as a linear combination of the vectors $[1, 2, -3, 5]$, $[1, 6, 0, 1]$, $[2, 3, -1, 0]$ and $[-1, -4, 2, 1]$.

7.5 Cramer's rule

A linear equation

$$ax + by = k$$

in two unknowns x and y represents a straight line on the coordinate plane. Here the line consists of points (x, y) whose coordinates satisfy the equation. A system

$$a_1 x + b_1 y = k_1$$
$$a_2 x + b_2 y = k_2$$

therefore represents the intersection of the two lines of the individual equations. By 7.4.3 the two lines intersect at exactly one point if and only if the matrix

$$\begin{pmatrix} a_1 & b_1 \\ a_2 & b_2 \end{pmatrix}$$

has rank 2 or equivalently

$$|\mathbf{a}\ \mathbf{b}| = \begin{vmatrix} a_1 & b_1 \\ a_2 & b_2 \end{vmatrix} = a_1 b_2 - a_2 b_1 \neq 0 \ .$$

In this case, it is easily seen that the unique point of intersection (x, y) has coordinates

$$x = (k_1 b_2 - k_2 b_1)/(a_1 b_2 - a_2 b_1)$$
$$y = (a_1 k_2 - a_2 k_1)/(a_1 b_2 - a_2 b_1) \ .$$

Writing them in terms of determinants, we obtain

$$x = \frac{|\mathbf{k}\ \mathbf{b}|}{|\mathbf{a}\ \mathbf{b}|}, \quad y = \frac{|\mathbf{a}\ \mathbf{k}|}{|\mathbf{a}\ \mathbf{b}|} \ .$$

The common denominator is the determinant $|\mathbf{a}\ \mathbf{b}|$ of the coefficient matrix. In the numerator of x, we find that in the determinant $|\mathbf{a}\ \mathbf{b}|$, the column $\mathbf{a}$ is replaced by the column $\mathbf{k}$ of constant terms. Similarly in the numerator of y, $\mathbf{b}$ in $|\mathbf{a}\ \mathbf{b}|$ is replaced by $\mathbf{k}$. This expression of the solution was discovered by Gabriel Cramer in the eighteenth century.

Similarly a system

$$a_1 x + b_1 y + c_1 z = k_1$$
$$a_2 x + b_2 y + c_2 z = k_2$$
$$a_3 x + b_3 y + c_3 z = k_3$$

of linear equations in three unknowns x, y, z represents the intersection of the three planes in space defined by the individual equations of the system. Again by 7.4.3 the intersection is precisely a point (x, y, z) in space if and only if the coefficient matrix

$$M = \begin{bmatrix} a_1 & b_1 & c_1 \\ a_2 & b_2 & c_2 \\ a_3 & b_3 & c_3 \end{bmatrix}$$

has rank 3 or equivalently

$$\det M = |\mathbf{a}\ \mathbf{b}\ \mathbf{c}| = \begin{vmatrix} a_1 & b_1 & c_1 \\ a_2 & b_2 & c_2 \\ a_3 & b_3 & c_3 \end{vmatrix} \neq 0 \ .$$

To find Cramer's rule for the solution (x, y, z), we denote by A_1, A_2, A_3 the cofactors of a_1, a_2, a_3 in M respectively. Multiply the equations of the system by A_1, A_2, A_3 respectively and add the resulting equations to get

$$(a_1 A_1 + a_2 A_2 + a_3 A_3)x + (b_1 A_1 + b_2 A_2 + b_3 A_3)y$$
$$+ (c_1 A_1 + c_2 A_2 + c_3 A_3)z = k_1 A_1 + k_2 A_2 + k_3 A_3 \ .$$

For the coefficients and the constant term of this equation, we have

$$a_1 A_1 + a_2 A_2 + a_3 A_3 = |\mathbf{a}\ \mathbf{b}\ \mathbf{c}|$$
$$b_1 A_1 + b_2 A_2 + b_3 A_3 = |\mathbf{b}\ \mathbf{b}\ \mathbf{c}| = 0$$
$$c_1 A_1 + c_2 A_2 + c_3 A_3 = |\mathbf{c}\ \mathbf{b}\ \mathbf{c}| = 0$$
$$k_1 A_1 + k_2 A_2 + k_3 A_3 = |\mathbf{k}\ \mathbf{b}\ \mathbf{c}| \ .$$

On the assumption that $|\mathbf{a}\ \mathbf{b}\ \mathbf{c}| \neq 0$, we obtain

$$x = |\mathbf{k}\ \mathbf{b}\ \mathbf{c}| / |\mathbf{a}\ \mathbf{b}\ \mathbf{c}| \ .$$

Similarly we get

$$y = |\mathbf{a}\ \mathbf{k}\ \mathbf{c}| / |\mathbf{a}\ \mathbf{b}\ \mathbf{c}|$$
$$z = |\mathbf{a}\ \mathbf{b}\ \mathbf{k}| / |\mathbf{a}\ \mathbf{b}\ \mathbf{c}| \ .$$

Linear Equations

These expressions of the coordinates x, y, z of the unique point of intersection in terms of determinants are knowns as Cramer's rule.

7.5.1 CRAMER'S RULE A System

$$a_1 x + b_1 y = k_1$$
$$a_2 x + b_2 y = k_2$$

of two linear equations in two unknowns has exactly one solution if and only if

$$|M| = \begin{vmatrix} a_1 & b_1 \\ a_2 & b_2 \end{vmatrix} \neq 0 \; .$$

In this case the unique solution is given by

$$x = \frac{1}{|M|} \begin{vmatrix} k_1 & b_1 \\ k_2 & b_2 \end{vmatrix} , \qquad y = \frac{1}{|M|} \begin{vmatrix} a_1 & k_1 \\ a_2 & k_2 \end{vmatrix} .$$

7.5.2 CRAMER'S RULE A system

$$a_1 x + b_1 y + c_1 z = k_1$$
$$a_2 x + b_2 y + c_3 z = k_2$$
$$a_3 x + b_3 y + c_3 z = k_3$$

of three linear equations in three unknowns has exactly one solution if and only if

$$|M| = \begin{vmatrix} a_1 & b_1 & c_1 \\ a_2 & b_2 & c_2 \\ a_3 & b_3 & c_3 \end{vmatrix} \neq 0 \; .$$

In this case the unique solution is given by

$$x = \frac{1}{|M|} \begin{vmatrix} k_1 & b_1 & c_1 \\ k_2 & b_2 & c_2 \\ k_3 & b_3 & c_3 \end{vmatrix} , \quad y = \frac{1}{|M|} \begin{vmatrix} a_1 & k_1 & c_1 \\ a_2 & k_2 & c_2 \\ a_3 & k_3 & c_3 \end{vmatrix} , \quad z = \frac{1}{|M|} \begin{vmatrix} a_1 & b_1 & k_1 \\ a_2 & b_2 & k_2 \\ a_3 & b_3 & k_3 \end{vmatrix} .$$

EXERCISES

1. By using Cramer's rule, solve the following systems of equations.
   (a) $2x - 3y = 7$
        $3x + 5y = 1$.
   (b) $5x + 2y = \;\; 5$
        $x - 3y = 18$.

2. Use Cramer's rule to solve

$$x + 2y + 3z = 4$$
$$7x - y + 6z = 0$$
$$2x + y + z = 0 .$$

3. Use Cramer's rule to solve

$$4x + 5y + = 2$$
$$11x + y + 2z = 3$$
$$x + 5y + 2z = 1 .$$

4. Consider the following system of equations

$$x + y + z = 6$$
$$3x - y + 2z = 7$$
$$5x + 2y + 2z = 15 .$$

(a) Solve the system by Cramer's rule.

(b) Solve by performing elementary row operations.

Which method is more convenient to use?

5. Can Cramer's rule be used to solve the following system of linear equations?

$$2x + 3y + 4z = 8$$
$$x + 2y - z = 5$$
$$x + y + 5z = 3 .$$

Is the system consistent?

6. Use Cramer's rule to solve

$$x + y + z = a$$
$$x + (1 + a)y + z = 2a$$
$$x + y + (1 + a)z = 0 .$$

for any non-zero real number a.

7. For any real numbers a, b and c, solve the following system of linear equations by using Cramer's rule.

$$x - y + z = a$$
$$x + y - z = b$$
$$-x + y + z = c .$$

8. Consider triangle ABC with sides a, b, c.

(a) By constructing suitable right-angled triangles, show that

$$c \cos A + a \cos C = b$$
$$a \cos B + b \cos A = c$$
$$b \cos C + c \cos B = a .$$

(b) If the above system is thought of as one with unknowns $\cos A$, $\cos B$ and $\cos C$, show that the determinant of the coefficient matrix is non-zero.

(c) Use Cramer's rule to obtain the cosine laws.

NUMERICAL ANSWERS TO EXERCISES

Exercise 1.2 (p.9)

1. (a) $[4, -5]$
   (b) $[-3, 42]$
   (c) $[-6, -33]$
   (d) $[-3, -114]$
   (e) $[-4, -1]$

2. $[-1, 1]$

3. $c_1 = -3$, $c_2 = 2$

4. No, it is not true.

6. $\mathbf{a} = \frac{1}{2}(4\mathbf{u} + 5\mathbf{v})$
   $\mathbf{b} = \frac{1}{2}(2\mathbf{u} + 3\mathbf{v})$

Exercise 1.3 (p.14)

1. (a) $[3, 2] = 1 \cdot [1, 0] + 2 \cdot [1, 1]$
   (b) $[0, 0] = 0 \cdot [1, 0] + 0 \cdot [1, 1]$
   (c) $[-1, 1] = -2 \cdot [1, 0] + 1 \cdot [1, 1]$
   (d) $[0, 1] = -1 \cdot [1, 0] + 1 \cdot [1, 1]$

4. (a) $x = 2$
   (b) $x = 0$ or $x = -2$
   (c) $x = 2\sqrt{2}$ or $-2\sqrt{2}$
   (d) x can take any real number.

5. $[x_1, x_2] = \frac{5x_1 + x_2}{17}[3, 2] + \frac{3x_2 - 2x_1}{17}[-1, 5]$

7. Yes, $\mathbf{c}$ and $\mathbf{d}$ are linearly independent.

Exercise 1.4 (p.21)

1. $k = \frac{7}{3}$
2. $k = -8$
7. (a) $\mathbf{m} = \frac{1}{4}(\lambda\mathbf{p} + \mu\mathbf{q})$
10. (a) $\mathbf{p} = \frac{1}{2}(\mathbf{a}_1 + \mathbf{a}_2)$
    $\mathbf{q} = \frac{1}{2}(\mathbf{b}_1 + \mathbf{b}_2)$
    $\mathbf{r} = \frac{1}{2}(\mathbf{c}_1 + \mathbf{c}_2)$

Exercise 1.5 (p.29)

1. (a) -1
   (b) 0
   (c) 0
3. (a) 30
   (b) 19
   (c) 12
   (d) -13
4. $k = -3$ or $k = 5$
6. $k = \frac{6}{19}$
7. $\mathbf{a} \cdot \mathbf{b} = \cos\alpha\cos\beta + \sin\alpha\sin\beta$
9. $\cos A = \frac{8\sqrt{17}}{85}$, $\cos B = \frac{9\sqrt{61}}{305}$, $\cos C = \frac{26\sqrt{1037}}{1037}$
10. (a) $k = \frac{8}{3}$
    (b) $k = -\frac{3}{2}$
    (c) $k = \frac{96+50\sqrt{3}}{11}$ or $\frac{96-50\sqrt{3}}{11}$
11. (a) $[3, 0]$
    (b) $[0, -4]$
    (c) $[-\frac{1}{2}, -\frac{1}{2}]$
12. (a) $[0, 0]$
    (b) $\frac{-1}{\sqrt{10001}}[100, -1]$
    (c) $[-1, -2]$
13. 5

14. $[3, 15]$ or $[-5, -25]$

17. $|\mathbf{a} + \mathbf{b}| = 17$, $|\mathbf{a} - \mathbf{b}| = 17$

18. $|\mathbf{a} + \mathbf{b}| = 20$

22. (a) $t = -\frac{\mathbf{a} \cdot \mathbf{b}}{|\mathbf{b}|^2}$

23. $\beta - \alpha = \frac{\pi}{2}$

25. (c) 13 sq. units

Exercise 1.6   (p.37)

1. (a) $x - 2y + 5 = 0$
   (b) $x - 3 = 0$
   (c) $5x + 2y + 18 = 0$
   (d) $2x + 3y + 12 = 0$

2. (a) $[-4, -1] \cdot [x, y - 1] = 0$
   (b) $[1, 1] \cdot [x + 1, y - 1] = 0$
   (c) $[1, -1] \cdot [x - 3, y + 2] = 0$
   (d) $[3, -1] \cdot [x - 5, y - 1] = 0$

3. (a) $\frac{12}{13}x + \frac{5}{13}y - 1 = 0$
   (b) $\frac{8}{17}x + \frac{15}{17}y + \frac{2}{17} = 0$
   (c) $\frac{7}{\sqrt{53}}x - \frac{2}{\sqrt{53}}y + \frac{4}{\sqrt{53}} = 0$

4. (a) $\frac{46}{13}$
   (b) $\frac{15}{17}$
   (c) $\frac{30}{\sqrt{53}}$

5. $45°$

6. $k = \frac{1}{3}$ or $k = 2$

7. $3\sqrt{10}$

8. $(3 - \sqrt{5})x + (\sqrt{5} - 1)y = 0$, or
   $(3 + \sqrt{5})x - (1 + \sqrt{5})y = 0$

9. $27x - 99y + 23 = 0$ or $77x + 21y + 3 = 0$

Exercise 2.1   (p.42)

1. $d_x = \sqrt{10}$, $d_y = \sqrt{17}$, $d_z = 5$.
2. (a) $x = -5$ or   7
   (b) $y = 2$
   (c) no real solution for z.
3. $\left(0, 0, \frac{179}{40}\right)$
4. $(0, 1, -2)$
5. $\left(-2, 3, -\frac{5}{2}\right)$, radius $= \frac{3}{2}$

Exercise 2.2   (p.45)

1. (a) $[8, 2, 6]$
   (b) $[6, 10, -6]$
   (c) $[7, 11, -12]$
2. (a) $[1, 1, \frac{4}{3}]$
   (b) $[\frac{1}{5}, -\frac{9}{5}, \frac{18}{5}]$
   (c) $[\frac{4}{3}, \frac{1}{3}, \frac{13}{6}]$

Exercise 2.3   (p.52)

1. $a = 12 + 5\mu$, $b = -4 - 2\mu$ for real μ
8. $\mathbf{d} = -\frac{1}{3}\mathbf{a} + \frac{2}{3}\mathbf{b} + \frac{5}{3}\mathbf{c}$

Exercise 2.5   (p.62)

1. (a) -2
   (b) 2
   (c) 10
2. (a) $\frac{3\sqrt{87}}{29}$
   (b) $\frac{\sqrt{7}}{7}$

3. $t = -\frac{2}{3}$

4. **44**

5. (a) $|\mathbf{a}|^2 - |\mathbf{b}|^2$

   (b) $2\mathbf{b} \cdot \mathbf{c}$

   (c) $ms|\mathbf{a}|^2 + (mt + ns)\mathbf{a} \cdot \mathbf{b} + nt|\mathbf{b}|^2$

6. **240**

7. **36**

8. (a) $[\frac{4}{3}, \frac{4}{3}, \frac{2}{3}]$

   (b) $[\frac{45}{26}, \frac{30}{13}, -\frac{15}{26}]$

9. $-\frac{3}{2}$

10. $\sqrt{20 + 6\sqrt{3}}$

12. **76.8°**

Exercise 2.6   (p.70)

1. (a) $x - 3y + 2z + 12 = 0$

   (b) $y = 7$

   (c) $x + y - 2z + 2 = 0$

2. (a) $\frac{2}{13}$

   (b) $\frac{\sqrt{42}}{14}$

   (c) $\frac{3\sqrt{14}}{7}$

3. (a) **3**

   (b) **1**

4. (a) $-\frac{1}{21}$

   (b) $\frac{\sqrt{21}}{6}$

   (c) $\frac{2\sqrt{26}}{13}$

5. $3x + 2y + 6z + 7 = 0,$ or

   $3x + 2y + 6z - 7 = 0$

6. (a) $a = 4$ and $b = -6$

   (b) $a = -\frac{1}{7}$

   (c) $c = \frac{-2 \pm \sqrt{327}}{5}$

8. $4x - y + z - 4 = 0$

Exercise 2.7 (p.76)

1. $\mathbf{a} \times \mathbf{b} = -2[1, 1, 1]$
   $\mathbf{b} \times \mathbf{c} = [1, -3, 1]$
   $\mathbf{c} \times \mathbf{a} = [-3, 9, 5]$

2. $2\sqrt{2}$

3. $\pm[\frac{5}{\sqrt{30}}, \frac{2}{\sqrt{30}}, -\frac{1}{\sqrt{30}}]$

4. $x + 7y - 4z - 15 = 0$

5. $4x - 6y + 3z + 11 = 0$

6. $\frac{1}{2}\sqrt{166}$

7. $\sqrt{17}$

8. (a) -33
   (b) -2

9. 6

10. $m = 3$, $n = 2$, $\ell = -3$

12. 8

15. (a) $3\sqrt{3}$
    (b) 24

19. $|\mathbf{a}|^4 \cdot \mathbf{b}$

24. (a) $x + ay + z = 0$
    (b) $a = \sqrt{2}$
    (c) $a = \frac{\sqrt{17}}{4}$

Exercise 2.8 (p.83)

1. (a) $\mathbf{x} = [3, -1, 2] + t[-1, 2, -3]$, for real t.
   (b) $\mathbf{x} = [2, 3, 4] + t[-2, 0, 5]$, for real t.

2. $\mathbf{x} = [2, 3, -1] + t[-1, -8, 2]$
   $\frac{x-2}{-1} = \frac{y-3}{-8} = \frac{z+1}{2}$
   Points of intersection are $(\frac{3}{2}, -1, 0)$, $(0, -13, 3)$, and $(\frac{13}{8}, 0, -\frac{1}{4})$.

3. $\mathbf{x} = [2, 1, 0] + t[1, 3, 1]$

4. $\frac{x-3}{-8} = \frac{y+1}{6} = \frac{z+6}{2}$

5. $a = -3$

6. $(0, -4, 1)$

7. (a) $3x - 5y - 4z = 0$
   (b) $\mathbf{x} = t[1, 15, -18]$

12. (a) $\frac{\sqrt{85}}{3}$
    (b) $\frac{\sqrt{17}}{17}$

13. (a) $(49, 49, 49)$ and $(50, 46, 51)$
    (b) $\mathbf{x} = [49, 49, 49] + t[-1, 3, -2]$

14. (a) $(1, 2, 0)$
    (b) $5x + 4y + z = 0$

15. (a) $\mathbf{x} = t[1, 0, -1]$
    (b) $\mathbf{x} = t[0, 1, -1]$ or $\mathbf{x} = t[1, -1, 0]$

16. (a) $b = 3$, $(-2, 1, 2)$
    (b) $2x + 3y - z - 11 = 0$
    (c) $(-30, 22, -5)$

Exercise 2.9   (p.86)

1. (a) $(\frac{3\sqrt{3}}{2}, \frac{3}{2}, 9)$
   (b) $(2\sqrt{2}, 2\sqrt{2}, 8)$
   (c) $(\sqrt{3}, -1, 3)$

2. (a) $(0, 2, 2\sqrt{3})$
   (b) $(0, 3, 0)$
   (c) $(-\frac{\sqrt{2}}{2}, -\frac{\sqrt{6}}{2}, \sqrt{2})$

3. (a) $(4, \frac{\pi}{6}, 4)$, $(4\sqrt{2}, \frac{\pi}{6}, \frac{\pi}{4})$
   (b) $(\frac{\sqrt{2}}{2}, \frac{\pi}{6}, -\frac{\sqrt{2}}{2})$, $(1, \frac{\pi}{6}, \frac{3\pi}{4})$

4. Cylindrical coordinates $\rightarrow$ Spherical coordinates

$$r = \sqrt{r_1^2 + z^2}$$
$$\theta = \theta$$
$$\varphi = \cos^{-1} \frac{z}{\sqrt{r_1^2 + z^2}}$$

Spherical coordinates → Cylindrical coordinates

$$r_1 = r \sin \varphi$$
$$\theta = \theta$$
$$z = r \cos \varphi$$

5. (a) $x^2 + y^2 + z^2 = 16$
   (b) $x^2 + y^2 = 36$
   (c) $x^2 + y^2 - z^2 + 12z - 36 = 0$
6. (a) $\theta = \frac{\pi}{4}$ and $\theta = \frac{5\pi}{4}$
   (b) $r = 3$
   (c) $r^2 = 3z$
7. (a) $x^2 + y^2 + z^2 - ax = 0$
   (b) $(x^2 + y^2 + z^2)^2 - a^2x^2 - a^2y^2 = 0$
   (c) $z = a$

Exercise 3.2 (p.93)

1. (a) $(-2, 0)$, $x = 2$
   (b) $(\frac{1}{8}, 0)$, $x = -\frac{1}{8}$
   (c) $(0, 2)$, $y = -2$
2. (a) $x^2 = 20y$
   (b) $y^2 = 8x$
   (c) $y^2 = -4x$
3. $x^2 = 16y$ or $x^2 = -16y$
4. $y^2 = 4x$
6. (c) $\frac{4}{m^2}\sqrt{a(m^2 + 1)(a - mc)}$
7. (a) $P = (\frac{4a}{m^2}, \frac{4a}{m})$
   $Q = (4am^2, -4am)$
   (b) Equations of chord PQ is

   $$(1 - m^2)y - mx + 4am = 0 .$$

   The fixed point is $(4a, 0)$.

   (c) $y^2 = 2a(x - 4a)$, a parabola.

10. (b) $(x - \frac{m}{2})^2 + (y - \frac{m^2}{2} - 2m)^2 = \frac{1}{4}(m^2 + 1)(m^2 + 8m)$

    (c) $m = \frac{1}{2}$

Exercise 3.3 (p.105)

1. (a) $\frac{x^2}{100} + \frac{y^2}{36} = 1$

    (b) $\frac{x^2}{9} + \frac{y^2}{64} = 1$

    (c) $\frac{x^2}{9} + \frac{y^2}{45} = 1$

2. (a) $(-\sqrt{12}, 0)$ and $(\sqrt{12}, 0)$

    (b) $\frac{x^2}{20} + \frac{y^2}{8} = 1$

4. $\frac{x^2}{2a^2 - b^2} + \frac{y^2}{a^2} = 1$

7. $\frac{x^2}{a^2} - \frac{y^2}{b^2} = 1$

9. (b) $\frac{x^2}{(\frac{2a+b}{2})^2} + \frac{y^2}{(\frac{b}{2})^2} = 1$ or $\frac{x^2}{(\frac{2a-b}{2})^2} + \frac{y^2}{(\frac{b}{2})^2} = 1$

Exercise 3.4 (p.113)

1. (a) $\frac{x^2}{8} - y^2 = -1$

    (b) $\frac{x^2}{25} - \frac{y^2}{119} = 1$

    (c) $\frac{x^2}{9} - \frac{y^2}{27} = 1$

2. (a) $\frac{4x^2}{9} - \frac{y^2}{4} = 1$

    (b) $\frac{25x^2}{256} - \frac{25y^2}{144} = 1$

4. (a) $\frac{|bx_1 + ay_1|}{\sqrt{a^2 + b^2}}, \frac{|bx_1 - ay_1|}{\sqrt{a^2 + b^2}}$

5. $x = \frac{a(a^2 - b^2)}{a^2 + b^2}$

Exercise 3.7 (p.136)

1. $(3, 2), (0, 0), (2, 5)$

2. $a = 1$

   $b = 2$

3. $a = 1$

   $b = -5$

4. (a) $y^2 = 4x$

   (b) $x^2 + 4y^2 = 36$

   (c) $x^2 + y^2 = 25$

5. (a) $(x - 3)^2 - 24y - 48 = 0$

   (b) $(x + 2)^2 - 14y - 7 = 0$

   (c) $2(y + 3)^2 + 5x - 5 = 0$

6. (a) $\frac{(y-5)^2}{36} + \frac{(x+3)^2}{27} = 1$

   (b) $\frac{x^2}{36} + \frac{(y-2)^2}{20} = 1$

   (c) $\frac{(x-3)^2}{16} + \frac{(y-1)^2}{9} = 1$

7. $x^2 + \frac{y^2}{4} = 1$, an ellipse.

8. $y^2 = -\frac{5}{2}x$, a parabola.

9. $x^2 - y^2 = 0$, a pair of straight lines.

10. $\frac{x^2}{2} - \frac{y^2}{3} = 1$, a hyperbola.

11. Asymptotes:

$$5y - 3x - 1 = 0 , \quad 5y + 3x + 11 = 0 .$$

   Foci:

$$(-2, -1 + \sqrt{34}) , \quad (-2, -1 - \sqrt{34}) .$$

   Directrices:

$$y = \frac{9}{\sqrt{34}} - 1 , \quad y = -\frac{9}{\sqrt{34}} - 1 .$$

12. $\left(-\frac{b}{a}, c - \frac{b^2}{a}\right)$

13. $xy = \frac{bc - ad}{c^2}$, a rectangular hyperbola with centre $\left(-\frac{d}{c}, \frac{a}{c}\right)$.

Exercise 3.8   (p.146)

1. (a) $(9\sqrt{2}, 3\sqrt{2})$

   (b) $(3\sqrt{3} - 6, 3 + 6\sqrt{3})$

2. (a) $\theta = \sin^{-1}\frac{3}{5}$

   (b) $\theta = -\pi 4$

4. $\frac{x^2}{9} + \frac{y^2}{4} = 1$, an ellipse.

5. $5y^2 - 3x^2 = 6$, a hyperbola.

6. $xy = 4$, a rectangular hyperbola.

7. $\tan 2\theta = \frac{4}{3}$

   $\frac{x^2}{16} + \frac{y^2}{36} = 1$, an ellipse.

8. $\theta = \frac{\pi}{4}$

   $\frac{y^2}{12} - \frac{x^2}{36} = 1$, a hyperbola.

9. $\theta = \frac{\pi}{6}$

   $\frac{x^2}{3} + \frac{y^2}{2} = 1$, an ellipse.

10. $x^2 + 6y^2 = 22$, an ellipse.

11. $y^2 = \sqrt{2}x$, a parabola.

12. $\theta = \tan^{-1}\frac{B}{A}$, $h = -\dfrac{C}{\sqrt{A^2+B^2}}$

13. $y = \frac{m_1 - m}{1 + mm_1}x$

Exercise 3.9   (p.153)

1. (a) Elliptic.
   (b) Elliptic.
   (c) Parabolic.
   (d) Elliptic.

2. $2x - y + 2 = 0$ and $x + 2y - 1 = 0$

3. A parabola; $x^2 = y$.

Exercise 3.10   (p.164)

5. $(\frac{1}{2}, -2)$, $(\frac{9}{2}, 6)$

   Equation of tangent at $(\frac{1}{2}, -2)$ is

   $$y + 2x + 1 = 0 \ .$$

   Equation of tangent at $(\frac{9}{2}, 6)$ is

   $$3y - 2x - 9 = 0 \ .$$

   Point of intersection $= (-\frac{3}{2}, 2)$.

6. $m = -1, c = -2$

8. (a) $y + px = 2ap + ap^3$

14. (b) $x - py + ap^2 = 0$

19. (a) Tangent: $\sqrt{2}\cos\theta x + 2\sin\theta y = 4$.
    Normal: $2\sin\theta x - \sqrt{2}\cos\theta y = \sqrt{2}\sin 2\theta$.
    (b) Centre: $(0, \frac{\cos^2\theta}{\sin\theta})$.

20. (a) $y = \lambda x \pm \sqrt{a^2 m^2 + b^2}$

Exercise 4.1   (p.173)

1. (a) An ellipsoid of revolution.
   (b) An paraboloid of revolution.
   (c) A hyperboloid of revolution of two sheets.
   (d) A hyperboloid of revolution of one sheet.

4. $y = x^2 + z^2$

5. $x^2 + 2y^2 + 2z^2 = 1$

6. $4x^2 - y^2 - z^2 = 4$

7. $4x^2 + 4y^2 - z^2 = 4$

8. $x^2 + y^2 + z^2 - 4x = 21$

9. $x^2 + y^2 + z^2 + 2y = 3$

Exercise 4.2   (p.177)

1. (a) $x^2 + z^2 = 25$
   (b) $\frac{y^2}{4} + \frac{z^2}{25} = 1$
   (c) $\frac{x^2}{4} - \frac{y^2}{9} = 1$

2. $x^2 + y^2 = 4$

3. $y^2 - x^2 = k$

4. (a) xy-plane : $(0, 0, 0)$.
   yz-plane : $\frac{y^2}{4} - \frac{z^2}{9} = 0$.
   xz-plane : $\frac{x^2}{4} - \frac{z^2}{9} = 0$.
   (b) $x^2 + y^2 = \frac{4h^2}{9}$

(c) $\frac{y^2}{k^2} - \frac{z^2}{(\frac{3k}{2})^2} = -1$

$\frac{x^2}{\ell^2} - \frac{z^2}{(\frac{3\ell}{2})^2} = -1$

5. (a) $4y^2 + 9z^2 - 4x^2 = 0$

   (b) $x^2 + y^2 - z^2 = 0$

6. A hyperbolic cylinder.

Exercise 4.3   (p.182)

1. $\frac{x^2}{9} + \frac{y^2}{25} + \frac{z^2}{49} = 1$

2. $\frac{x^2}{25} + \frac{y^2}{9} + \frac{z^2}{16} = 1$

3. Foci : $(3, -\frac{8\sqrt{3}}{5}, 0)$ and $(3, \frac{8\sqrt{3}}{5}, 0)$.

5. $\frac{\sqrt{462}}{6}$

8. $k = \pm\frac{b}{c}\sqrt{\frac{a^2-c^2}{b^2-a^2}}$

Exercise 4.4   (p.185)

1. $\frac{x^2}{36} + \frac{y^2}{4} - \frac{z^2}{16} = -1$

2. (a) $\left(-\frac{\sqrt{k^2-c^2}\cdot\sqrt{a^2-b^2}}{c}, 0, k\right)$

   $\left(\frac{\sqrt{k^2-c^2}\cdot\sqrt{a^2-b^2}}{c}, 0, k\right)$

   (b) $\left(k, 0, -\frac{\sqrt{b^2+c^2}\cdot\sqrt{a^2+k^2}}{a}\right)$

   $\left(k, 0, \frac{\sqrt{b^2+c^2}\cdot\sqrt{a^2+k^2}}{a}\right)$

Exercise 4.5   (p.188)

1. $\frac{x^2}{25} + \frac{y^2}{25} - \frac{z^2}{4} = 1$

2. $\begin{cases} x - y + z - 1 = 0 \\ x + y - z - 1 = 0 \end{cases}$ and $\begin{cases} x - z = 0 \\ y - 1 = 0 \end{cases}$

3. $\begin{cases} 3x - 6y + 2z - 6 = 0 \\ 3x + 6y - 2z - 6 = 0 \end{cases}$ and $\begin{cases} 1 - y = 0 \\ 3x - 2z = 0 \end{cases}$

Exercise 4.6   (p.190)

1. $\frac{x^2}{2} + \frac{y^2}{18} = 4z$

2. Centre : $(0, 1, 2)$, radius $= 1$.

3. $k = -1$, centre : $(-1, 0, 1)$, radius $= \sqrt{2}$.

Exercise 4.7   (p.192)

1. $\begin{cases} 2x + y - 8 = 0 \\ 2x - y - 2z = 0 \end{cases}$ and $\begin{cases} 2x - y - 4 = 0 \\ 2x + y - 4z = 0 \end{cases}$

2. (b) $\begin{cases} x + y = 0 \\ z = 0 \end{cases}$ and $\begin{cases} x - y + 1 = 0 \\ x + y + 2z = 0 \end{cases}$

Exercise 4.8   (p.200)

1. $(-3, 2, -1)$, $(2, -1, 6)$, and $(5, 9, -7)$

2. $x = x' + 3$, $y = y' - 7$, $z = z' - 3$

3. $x = x' + 3$, $y = y' - 1$, $z = z' - 3$

4. (a) $2x - y + 3z = 2$
   (b) $\frac{x-2}{4} = \frac{y+6}{3} = \frac{z}{2}$
   (c) $x^2 + 3z^2 - 6xy - 52x + 12y + 36z + 218 = 0$

5. $(-1, -2, -3)$

7. $x' = x + \frac{O}{A}$, $y' = y + \frac{E}{B}$, $z' = z + \frac{F}{C}$

8. $\left(\frac{a_2 - a_1}{2}, \frac{b_2 - b_1}{2}, \frac{c_2 - c_1}{2}\right)$,
   $\left(\frac{a_1 - a_2}{2}, \frac{b_1 - b_2}{2}, \frac{c_1 - c_2}{2}\right)$

9. The equations of transformation are

$$x = \frac{1}{3}(-x' + 2y' + 2z')$$

$$y = \frac{1}{3}(2x' - y' + 2z')$$

$$z = \frac{1}{3}(2x' + 2y' - 2)$$

and

$$x' = \frac{1}{3}(-x + 2y + 2z)$$

$$y' = \frac{1}{3}(2x - y + 2z)$$

$$z' = \frac{1}{3}(2x + 2y - z) \ .$$

Therefore

$$(0,0,0) \rightarrow (0,0,0)$$
$$(5,-3,7) \rightarrow (1,9,-1)$$
$$(8,7,-6) \rightarrow (-2,-1,12) \ .$$

10. $e'_3 = [\frac{2}{3}, -\frac{1}{3}, \frac{2}{3}]$
    $x = \frac{1}{3}(2x' + 2y' - z')$
    $y = \frac{1}{3}(-x' + 2y' - 2z')$
    $z = \frac{1}{3}(2x' - y' + 2z')$

11. $3x^2 + y^2 - 2z = 0$, a hyperbolic paraboloid.

12. $\frac{x^2}{(\frac{4}{3})^2} + \frac{y^2}{(\frac{4}{3})^2} + \frac{z^2}{(\frac{\sqrt{2}}{3})^2} = 1$

14. (a) $x = x' - 1$
        $y = y' + 1$
        $z = z' - 2$
    (b) $x'' = \frac{1}{3}(2x' - y' + 2z')$
        $y'' = \frac{1}{3}(-x' + 2y' + 2z')$
        $z'' = \frac{1}{3}(2x' + 2y' - z')$

Exercise 5.2   (p.216)

1. (a) $[2,0,3,-4]$
   (b) $[11,2,1,8]$

2. (a) and (b).

3. $m = -4\frac{3}{5}, n = 0$

9. No.

10. $c = 0$

21. $\ell_1 = \frac{d}{ad-bc}, m_1 = \frac{-b}{ad-bc}$
    $\ell_2 = \frac{-c}{ad-bc}, m_2 = \frac{a}{ad-bc}$

Exercise 5.3   (p.220)

1. (a) Linearly dependent.
   (b) Linearly independent.

Exercise 5.4   (p.228)

1. (a) $\begin{pmatrix} 5 & 14 & 23 \\ 0 & -6 & -12 \\ 4 & 10 & 16 \end{pmatrix}$

   (b) $\begin{pmatrix} -10 & -3 & 6 \\ -14 & 40 & -5 \\ -18 & 8 & 6 \end{pmatrix}$

   (c) $\begin{pmatrix} -13 & 7 & 12 \\ 12 & -33 & 17 \\ -12 & -2 & 18 \end{pmatrix}$

2. (a) 2
   (b) 2
   (c) 3

3. (a) 2
   (b) 3
   (c) 3

Exercise 5.5   (p.235)

1. (a) $\mathbf{a}_1 \rightarrow \mathbf{a}_1 + 3\mathbf{a}_2$
      $\mathbf{a}_2 \rightarrow \mathbf{a}_2 \times (-2)$
      $\mathbf{a}_2 \rightarrow \mathbf{a}_2 + 2\mathbf{a}_1$
   (b) $\mathbf{a}_2 \leftrightarrow \mathbf{a}_1$, $\mathbf{a}_1 \rightarrow \mathbf{a}_1 + 2\mathbf{a}_2$, $\mathbf{a}_1 \rightarrow 2\mathbf{a}_1$,
      $\mathbf{a}_2 \rightarrow 4\mathbf{a}_2$, $\mathbf{a}_2 \rightarrow \mathbf{a}_2 + \mathbf{a}_3$, $\mathbf{a}_1 \rightarrow \mathbf{a}_1 - \mathbf{a}_2$
   (c) $\mathbf{a}_1 \rightarrow \mathbf{a}_1 + \frac{2}{3}\mathbf{a}_2$, $\mathbf{a}_2 \rightarrow \frac{1}{12}\mathbf{a}_2$, $\mathbf{a}_3 \rightarrow \mathbf{a}_3 - 2\mathbf{a}_2$, $\mathbf{a}_3 \rightarrow \mathbf{a}_3 + \mathbf{a}_1$,
      $\mathbf{a}_2 \rightarrow \mathbf{a}_2 + \mathbf{a}_1$, $\mathbf{a}_3 = 2\mathbf{a}_3$, $\mathbf{a}_2 \rightarrow 4\mathbf{a}_2$, $\mathbf{a}_1 \rightarrow 3\mathbf{a}_1$

2. (a) $\begin{pmatrix} 5 & 3 \\ -1 & 4 \end{pmatrix}$

(b) $\begin{pmatrix} 3 & -2 \\ 4 & 5 \\ 5 & -8 \\ 6 & 13 \\ 7 & 1 \end{pmatrix}$

(c) $\begin{pmatrix} 0 & 0 & 1 & 0 \\ 1 & 0 & 2 & 0 \\ 0 & 2 & 0 & 3 \\ 0 & 0 & 3 & 4 \end{pmatrix}$

3. (a) No. Yes.
   (b) No. No.
   (c) No.

Exercise 5.6 (p.240)

3. (a) Yes.
   (b) No.

4. (a) Yes.
   (b) No.

5. $\{[2,6,0,1,3],[1,1,0,-1,4],[0,0,0,1,1]\}$

8. (b) One possible answer : $\mathbf{u} = [1,0,0]$, $\mathbf{v} = [0,1,0]$.

Exercise 6.1 (p.245)

1. $\begin{pmatrix} 3 & 0 & 0 \\ 0 & 6 & 0 \\ 0 & 0 & 9 \end{pmatrix}$

2. $\begin{pmatrix} 1 & 2 & 3 & 4 \\ 2 & 2 & 6 & 4 \\ 3 & 6 & 3 & 12 \\ 4 & 4 & 12 & 4 \end{pmatrix}$

3. $\begin{pmatrix} 1 & 0 & 1 & 0 & 1 \\ 0 & 1 & 0 & 1 & 0 \\ 1 & 0 & 1 & 0 & 1 \\ 0 & 1 & 0 & 1 & 0 \\ 1 & 0 & 1 & 0 & 1 \end{pmatrix}$

4. $x = 1, y = 2, z = 3$

5. $a = -2b, d = -b, c = -5b$ for **any real** b

Exercise 6.2   (p.247)

1. (a) $\begin{pmatrix} 4 & -2 & 14 \\ 6 & 0 & 8 \end{pmatrix}$

   (b) $\begin{pmatrix} 5 & -5 & 9 \\ 4 & 2 & -1 \end{pmatrix}$

   (c) $\begin{pmatrix} -9 & 17 & 11 \\ 4 & -10 & 37 \end{pmatrix}$

2. (a) $\begin{pmatrix} 9 & 15 & 6 \\ 18 & 21 & 6 \\ 27 & 24 & 3 \end{pmatrix}$

   (b) $\begin{pmatrix} 4 & 7 & 5 \\ 13 & 1 & 2 \\ 11 & 13 & 9 \end{pmatrix}$

   (c) $\begin{pmatrix} 4 & 7 & 5 \\ 13 & 1 & 2 \\ 11 & 13 & 9 \end{pmatrix}$

4. (a) $A = \begin{pmatrix} 2 & 1 \\ 1 & 2 \end{pmatrix}$

   (b) $A = (2 \quad 15)$

5. $x = 1, y = \frac{1}{2}, w = 4, z = \frac{1}{2}$

6. $A = \begin{pmatrix} 3 & 2 & 2 \\ 1 & -1 & 2 \\ 3 & 5 & 4 \end{pmatrix}$ $\begin{pmatrix} -1 & -1 & 1 \\ -1 & -6 & 2 \\ 2 & 3 & 2 \end{pmatrix}$

Exercise 6.3   (p.251)

1. (a) $(2 \quad 3)$

   (b) $\begin{pmatrix} 18 & 2 \\ 8 & -1 \end{pmatrix}$

(c) $\begin{pmatrix} 2 & 8 \\ 8 & 15 \end{pmatrix}$

2. (a) $A^2 = \begin{pmatrix} 4 & 0 \\ 0 & 1 \end{pmatrix}$, $A^3 = \begin{pmatrix} 8 & 0 \\ 0 & 1 \end{pmatrix}$

(b) $A^2 = \begin{pmatrix} 2 & 2 \\ 2 & 2 \end{pmatrix}$, $A^3 = \begin{pmatrix} 4 & 4 \\ 4 & 4 \end{pmatrix}$

(c) $A^2 = \begin{pmatrix} a^2 + bc & ac + cd \\ ab + bd & bc + d^2 \end{pmatrix}$

$A^3 = \begin{pmatrix} a^3 + 2abc + bcd & a^2c + bc^2 + abc + cd^2 \\ a^2b + abd + b^2c + bd^2 & abc + 2bcd + d^3 \end{pmatrix}$

3. (a) $\begin{pmatrix} 4 & 7 \\ 6 & 1 \end{pmatrix}$

(b) $\begin{pmatrix} 18 & 21 \\ 21 & 24 \\ 24 & 27 \end{pmatrix}$

(c) $\begin{pmatrix} 3 & 1 & 0 \\ 5 & 5 & 1 \\ 4 & 4 & 1 \end{pmatrix}$

4. (a) $A^2 = \begin{pmatrix} 0 & 0 & ac \\ 0 & 0 & 0 \\ 0 & 0 & 0 \end{pmatrix}$, $A^3 = \begin{pmatrix} 0 & 0 & 0 \\ 0 & 0 & 0 \\ 0 & 0 & 0 \end{pmatrix}$

(b) $A^2 = \begin{pmatrix} 1 & 0 & 2a \\ 0 & 1 & 0 \\ 0 & 0 & 1 \end{pmatrix}$, $A^3 = \begin{pmatrix} 1 & 0 & 3a \\ 0 & 1 & 0 \\ 0 & 0 & 1 \end{pmatrix}$

7. $A = B = \begin{pmatrix} 1 & 0 \\ 0 & 0 \end{pmatrix}$, $C = \begin{pmatrix} 1 & 0 \\ 1 & 0 \end{pmatrix}$

8. $A = \begin{pmatrix} 1 & 0 \\ 0 & 0 \end{pmatrix}$, $B = \begin{pmatrix} 0 & 0 \\ 1 & 0 \end{pmatrix}$

9. $A = \begin{pmatrix} 0 & 1 \\ 0 & 0 \end{pmatrix}$, $B = \begin{pmatrix} 0 & 0 \\ 1 & 0 \end{pmatrix}$

10. (a) $A = \begin{pmatrix} 0 & 1 \\ 0 & 0 \end{pmatrix}$, $B = \begin{pmatrix} 0 & 0 \\ 1 & 0 \end{pmatrix}$

14. (a) $A^4 = \begin{pmatrix} x^4 & 0 \\ y(x+z)(x^2+z^2) & z^4 \end{pmatrix}$

(b) $a = 3, b = 2, c = 8,$ or
    $a = -3, b = -2, c = -8$

Exercise 6.4   (p.257)

1. (a) $E = \begin{pmatrix} 0 & 1 \\ 1 & 0 \end{pmatrix}$

   (b) $E = \begin{pmatrix} 1 & 3 \\ 0 & 1 \end{pmatrix}$

   (c) $E = \begin{pmatrix} 1 & 0 & 0 \\ 0 & 1 & 0 \\ 0 & 0 & \frac{1}{2} \end{pmatrix}$

2. (a) $E = \begin{pmatrix} 0 & 1 \\ 1 & 0 \end{pmatrix}$

   (b) $E = \begin{pmatrix} 2 & 0 \\ 0 & 1 \end{pmatrix}$

   (c) $E = \begin{pmatrix} 1 & 0 & 0 \\ 0 & 1 & 0 \\ 0 & 1 & 1 \end{pmatrix}$

3. (a) $\begin{pmatrix} 1 & 2 \\ 0 & 1 \end{pmatrix} \begin{pmatrix} 0 & 1 \\ 1 & 0 \end{pmatrix} = \begin{pmatrix} 2 & 1 \\ 1 & 0 \end{pmatrix}$

   (b) $\begin{pmatrix} 1 & -3 \\ 0 & 1 \end{pmatrix} \begin{pmatrix} 1 & 0 \\ 2 & 1 \end{pmatrix} \begin{pmatrix} 1 & 0 \\ -1 & 1 \end{pmatrix} = \begin{pmatrix} -2 & -3 \\ 1 & 1 \end{pmatrix}$

5. (b) Take $A = B = \begin{pmatrix} 1 & 0 \\ 0 & 1 \end{pmatrix}.$

Exercise 6.5   (p.264)

1. (a) $\begin{pmatrix} 3 & -2 \\ -1 & 1 \end{pmatrix}$

   (b) $\begin{pmatrix} -5 & -3 \\ -3 & -2 \end{pmatrix}$

   (c) $\begin{pmatrix} -\frac{4}{7} & \frac{3}{7} \\ \frac{5}{7} & -\frac{2}{7} \end{pmatrix}$

2. (a) $\begin{pmatrix} -\frac{1}{3} & \frac{9}{2} & \frac{17}{3} \\ -\frac{1}{3} & -\frac{3}{2} & -\frac{5}{3} \\ \frac{2}{3} & -1 & -\frac{4}{3} \end{pmatrix}$

(b) Not invertible.

(c) $\begin{pmatrix} 1 & -4 & -3 \\ 1 & -5 & -3 \\ -1 & 6 & 4 \end{pmatrix}$

3. (a) $\begin{pmatrix} \frac{1}{2} & 0 \\ 0 & 1 \end{pmatrix} \begin{pmatrix} 1 & -5 \\ 0 & 1 \end{pmatrix} A = I$

(b) $A^{-1} = \begin{pmatrix} \frac{1}{2} & 0 \\ 0 & 1 \end{pmatrix} \begin{pmatrix} 1 & -5 \\ 0 & 1 \end{pmatrix}$

8. (a) $\begin{pmatrix} \frac{1}{a} & 0 & 0 & 0 \\ 0 & \frac{1}{b} & 0 & 0 \\ 0 & 0 & \frac{1}{c} & 0 \\ 0 & 0 & 0 & \frac{1}{d} \end{pmatrix}$

(b) $\begin{pmatrix} 0 & 0 & 0 & \frac{1}{d} \\ 0 & 0 & \frac{1}{c} & 0 \\ 0 & \frac{1}{b} & 0 & 0 \\ \frac{1}{a} & 0 & 0 & 0 \end{pmatrix}$

(c) $\begin{pmatrix} \frac{1}{e} & 0 & 0 & 0 \\ -\frac{1}{e^2} & \frac{1}{e} & 0 & 0 \\ \frac{1}{e^3} & -\frac{1}{e^2} & \frac{1}{e} & 0 \\ -\frac{1}{e^4} & \frac{1}{e^3} & -\frac{1}{e^2} & \frac{1}{e} \end{pmatrix}$

10. $A^{-1} = \frac{3}{4}I - \frac{1}{2}A - \frac{1}{4}A^2$

12. (a) $A = \begin{pmatrix} 1 & 0 \\ 0 & 1 \end{pmatrix}, \quad B = \begin{pmatrix} -1 & 0 \\ 0 & -1 \end{pmatrix}$

Exercise 6.6   (p.266)

1. (a) 1
   (b) 6
   (c) 0

3. (b) Area $= \begin{vmatrix} 1 & 6 \\ -2 & 3 \end{vmatrix} = 3 - (-12) = 15$

   (b) Area $= \begin{vmatrix} 1 & -15 \\ 4 & 11 \end{vmatrix} = 11 - (-60) = 71$

5. (a) 21
   (b) −7
   (c) 7

Exercise 6.7   (p.277)

1. (a) 54
   (b) −64
   (c) −142

2. (a) $a = 2$ or $a = 3$
   (b) $a = 2$ or $a = 6$

4. (a) −21
   (b) 18

5. (a) 4
   (b) −60
   (c) −4

7. (a) Invertible.
   (b) Invertible.
   (c) Note invertible.

8. (a) $\begin{pmatrix} \frac{9}{20} & \frac{1}{10} & -\frac{3}{20} \\ \frac{1}{10} & -\frac{1}{5} & \frac{1}{5} \\ -\frac{3}{20} & \frac{3}{10} & \frac{1}{20} \end{pmatrix}$

   (b) $\begin{pmatrix} -\frac{1}{2} & -3 & \frac{5}{2} \\ \frac{1}{4} & \frac{5}{2} & \frac{7}{4} \\ \frac{1}{4} & \frac{1}{2} & \frac{1}{4} \end{pmatrix}$

10. $x = 1$ or $x = -\frac{1}{2}$

11. $q - r$

13. (a) $4a^2b^2c^2$

18. (a) $\det A = 1$ or $\det A = 0$.
    (b) $\det A = 0$.
    (c) $\det A = k^3$ or $\det A = 0$.
    (d) No such A exists.
    (e) $\det A = 0$ or $\det B = 0$.

Exercise 7.1 (p.283)

1. A system of linear equations may have more than one solution.

Exercise 7.2 (p.288)

2. (a) $\begin{pmatrix} 5 & -2 & 1 \\ 1 & 4 & 6 \end{pmatrix}$

   (b) $\begin{pmatrix} 1 & -3 & -4 \\ 0 & 1 & 1 \end{pmatrix}$

   (c) $\begin{pmatrix} 1 & -1 & 1 & 4 \\ 1 & 0 & -1 & -1 \\ 0 & 1 & 2 & 1 \end{pmatrix}$

3. (a)
$$\begin{aligned} x_1 + 2x_2 - 4x_3 &= 2 \\ x_2 - 2x_3 &= -1 \\ 3x_1 + x_2 \quad\quad &= 7 \end{aligned}$$

   (b)
$$\begin{aligned} x_1 \quad\quad + 4x_3 + 7x_4 &= 10 \\ x_2 - 3x_3 - 4x_4 &= -2 \\ x_3 + x_4 &= 2 \end{aligned}$$

4. (a) Solvable.
   (b) Not solvable.
   (c) Solvable.

5. $k = 4$

6. Either (a) $ad - bc \neq 0$, or (b) if $ad - bc = 0$, $d = 0$, $b = 0$, at least one of a, $c \neq 0$.

Exercise 7.3 (p.298)

1. (a) and (d) are. Yes.

2. (a) $\begin{pmatrix} 1 & 0 \\ 0 & 1 \end{pmatrix}$

   (b) $\begin{pmatrix} 1 & 0 & \frac{1}{4} & 0 \\ 0 & 1 & \frac{1}{4} & 1 \end{pmatrix}$

(c) $\begin{pmatrix} 1 & 0 & 0 & 1 \\ 0 & 1 & 0 & 1 \\ 0 & 0 & 1 & 0 \\ 0 & 0 & 0 & 0 \end{pmatrix}$

3. (a) $[-\frac{17}{9}t, -\frac{4}{9}t, t]$, for any real number t.
   (b) $[-\frac{1}{4}s, -\frac{1}{4}s - t, s, t]$, for any real numbers s, t.

4. $x_1 = 0$, $x_2 = 0$, $x_3 = 0$

5. $[-t, 0, t]$, for real number t.

6. $[-t, -t, 0, t]$, for any real number t.

9. $a = -1$; solution is $[t, t]$, for any real number t.
   $a = 4$; solution is $[-\frac{2}{3}t, t]$, for any real number t.

10. $k = 1$; solution is $[t, -t, 0]$, for any real number t.
    $k = -1$; solution is $[t, 0, t]$, for any real number t.

11. Linearly independent.

12. Linearly independent.

13. Linearly independent.

14. $[\frac{2t-135}{5}, \frac{225-3t}{5}, s, t]$ for any real numbers s and t.

15. (a) $\{[3, -3, 1]\}$
    (b) $\{[-\frac{1}{2}, -1, 0, 1], [\frac{3}{2}, 0, 1, 0]\}$

Exercise 7.4   (p.306)

1. $[-16, \frac{17}{2}, 5]$

2. $[2, 1, 4]$

3. $[\frac{20-4t}{13}, \frac{3t-28}{13}, \frac{9t-45}{13}, t]$ for any real number t

4. $[-\frac{17}{7}, \frac{11}{7}, \frac{6}{5}, \frac{117}{35}]$

5. 3.75

6. $k = 16$; solution is $[\frac{37-6t}{8}, \frac{2t-19}{8}, t]$ for any real number t.

7. $x_1 = a_1 + a_2 + a_3 + a_4 + t$
   $x_2 = a_2 + a_3 + a_4 + t$
   $x_3 = a_3 + a_4 + t$
   $x_4 = a_4 + t$
   $x_5 = t$, for any real number t.

8. $[\frac{13}{5}, \frac{4}{5}, 0] + t[\frac{4}{5}, -\frac{3}{5}, 1]$, for any real number t, so that it is a straight line in space.

9. $-11x + 17y + 13z - 3 = 0$

10. (b) $[\frac{3+2t}{5}, \frac{19t-4}{5}, t]$, for any real number t.

11. (i) $x_1 = -\frac{k+1}{k+2}$, $x_2 = \frac{1}{k+2}$, $x_3 = \frac{(k+1)^2}{k+2}$
    (ii) No solution.
    (iii) $x_1 = 1 - s - t$, $x_2 = s$, $x_3 = t$, for any real numbers s and t.

12. (i) $x_1 = -\frac{1}{k}$, $x_2 = \frac{1}{k}$, $x_3 = 0$
    (ii) No solution.
    (iii) $x_1 = \frac{-21t-4}{8}$, $x_2 = \frac{t+4}{8}$, $x_3 = t$, for any real number t.

13. b can be any real number if $a = 2$, and $b = 0$ for $a = -2$.

14. $x_1 = \frac{5a-b+19}{13}$

    $x_2 = \frac{4a-15b+36c}{26}$

    $x_3 = \frac{-3a+8b-14c}{13}$

    $A^{-1} = \begin{pmatrix} \frac{5}{13} & -\frac{9}{13} & \frac{19}{13} \\ \frac{2}{13} & -\frac{15}{26} & \frac{18}{13} \\ -\frac{3}{13} & \frac{8}{13} & -\frac{14}{13} \end{pmatrix}$

15. $a = 3$, $b = 2$, $c = 1$

16. (a) $x_1 = 1$, $x_2 = -3$, $x_3 = 2$, $x_4 = -1$
    (b) $[3, -6, -7, 1]$
    $= [1, 2, -3, 5] - 3[1, 6, 0, 1] + 2[2, 3, -1, 0] - [-1, -4, 2, 1]$

Exercise 7.5   (p.311)

1. (a) $x = 2$, $y = -1$
   (b) $x = 3$, $y = -5$

2. $x = -\frac{14}{15}$, $y = \frac{2}{3}$, $z = \frac{6}{5}$

3. $x = \frac{3}{11}$, $y = \frac{2}{11}$, $z = -\frac{1}{11}$

4. $x = 1$, $y = 2$, $z = 3$

5. We cannot use Cramer's rule.
   $(1 - 11t, 2 + 6t, t)$, for all real numbers t.

6. $x = a$, $y = 1$, $z = -1$

7. $x = \frac{a+b}{2}$, $y = \frac{b+c}{2}$, $z = \frac{c+a}{2}$

INDEX

341